Chemistry of the Solar System

Chemistry of the Solar System

Katharina Lodders and Bruce Fegley, Jr.
Department of Earth & Planetary Sciences, Washington University, St. Louis, MO, USA

RSCPublishing

ISBN: 978-0-85404-128-2

A catalogue record for this book is available from the British Library

Published by The Royal Society of Chemistry,
Thomas Graham House, Science Park, Milton Road,
Cambridge CB4 0WF, UK

Registered Charity Number 207890

For further information see our web site at www.rsc.org

Preface

In a small book like this it is impossible to cover all the possible topics on chemistry within and of the solar system in much detail – the solar system is just too diverse for that. The Sun and the planets with their moons each have their own special chemical histories and unique states as well as ongoing chemical processes. The meteorites as fragments from some of the thousands of smaller rocky objects in the asteroid belt bear witness to the intricate chemical processes in the early solar system. The icy comets from the regions beyond Pluto with the plethora of Kuiper belt objects may even carry chemical signatures from the boundary of the solar system with the interstellar medium. Then there is the chemistry of the interplanetary dust, and that of the interstellar dust. Since the nineteenth century, the knowledge gathered by dedicated researchers has increased exponentially as a stroll through miles of library isles framed with bookshelves of monographs, conference proceedings and scientific journals easily reveals. To this, we can add the terabytes of information accumulating on the internet.

Therefore, we had to make choices on what topics to cover, and on the level of detail. We aimed to provide general introductions and to describe processes that are common to various planetary objects despite the differences that they may display now. We hope that our selection of topics generates more curiosity about planetary science and meteoritics from readers not yet very familiar with the diversity of the applications of chemistry to astronomical objects, including the Earth.

The learned experts may find that the latest results from their areas of expertise are not covered in enough depth, or that their favorite theories are only briefly mentioned or not mentioned at all. However, this would

Chemistry of the Solar System
By Katharina Lodders and Bruce Fegley, Jr.
© K. Lodders and B. Fegley, Jr. 2011
Published by the Royal Society of Chemistry, www.rsc.org

have expanded the book beyond the introductory level. Moreover, is it wise to debate the obscure details of a single nuclide in a minor mineral phase before having a clue about the rock that contains it? We refer interested readers to monographs, reviews and internet resources for more in-depth studies, technical details and for finding individual research papers. We included several tables and illustrations that the reader may find practical, and we thank our colleagues who kindly provided some of the photographs.

We include a few historical accounts on how ideas developed, and, together with new discoveries, shaped our current understandings. Aside from the entertaining aspects that history can occasionally provide, "the best way to obtain new ideas is to find weaknesses in old ones" as D. ter Haar and A. G. W. Cameron once put it. Some may say that there is too much of the old stuff here, and not enough about the latest theories and results. However, theories come and go, and are as cyclic as the seasons. A theory that seems to be a recent new insight may well have been described in its principles by some researcher who is long since gone and forgotten. This does not mean that all old theories remain valid, especially if clear scientific evidence to the contrary assigns these theories to the realms of fiction. Currently accepted theories, whether old, recycled or new, will have to stand the test of time. The ideas advocated by some strong voices today may also not be the ones that will be remembered in the future, and the half-life of an idea may only be proportional to the strength and lifetime of the voice sounding them. Speaking of old insights and scientific theories, the reader may remember the words of Samuel Pierpont Langley, the aviation pioneer and president of the American Association for the Advancement of Science, from 1889:

> "Science is an onrushing pack of hounds which in the long run, perhaps, catches its game, but where, nevertheless, when at fault, each individual goes his own way, by scent, not by sight, some running back and some forward; where the louder-voiced bring many to follow them, nearly as often in the wrong path as in the right one; where the entire pack even has known to move off bodily on a false scent."

With these precautions on scientific results and theories, we invite the reader to enjoy our chemical exploration of the solar system.

Katharina Lodders and Bruce Fegley, Jr.

Contents

Chemistry of the Solar System
By Katharina Lodders and Bruce Fegley, Jr.
© K. Lodders and B. Fegley, Jr. 2011
Published by the Royal Society of Chemistry, www.rsc.org

Dedicated to the memory of our friends and colleagues Al Cameron and Gero Kurat.

Some Milestones Up to the Beginning of the Space Age

Year	Discoveries
	C, S, Fe, Cu, Ag, Sn, Sb, Au, Hg, Pb are known since antiquity
1664	Robert Hooke (1635–1703) discovers the big red spot on Jupiter
1669	Henning Brand discovers phosphorus
1672	Giovanni Domenico Cassini (1625–1712) calculates the distances of the known planets from the Sun
1735	Henning Brand discovers cobalt
	Antonio de Ulloa (1716–1795) discovers platinum
1746	Andreas Marggraf (1709–1782) discovers zinc
1751	Axel Cronstedt discovers nickel
1753	Claude Geoffroy (the younger) discovers bismuth
1755	Joseph Black discovers magnesium
1755	Emanuel Kant proposes the formation of the solar system out of a gaseous and dusty nebula
1757	Alexis Claude Clairaut (1713–1765) determines the mass of the Moon and of Venus
1761	Michail Vassilevic Lomonossov (1711–1765) and Joseph Nicolas Delisle (1688–1768) recognize the presence of a cloudy atmosphere on Venus.
1766	Henry Cavendish (1731–1810) discovers hydrogen
1766	Johann Daniel Tietz ("Titius," 1729–1796) and, in 1772, Johann Elert Bode (1747–1826) formulate an empirical rule of planetary distances from the Sun
1771–1774	Joseph Priestley and Carl Wilhelm Scheele discover oxygen
1772	Daniel Rutherford discovers nitrogen

Chemistry of the Solar System
By Katharina Lodders and Bruce Fegley, Jr.
© K. Lodders and B. Fegley, Jr. 2011
Published by the Royal Society of Chemistry, www.rsc.org

(*Continued*).

Year	Discoveries
1774	Johan Gottlieb Gahn (1745–1818) discovers manganese
	Karl Wilhelm Scheele (1742–1786) discovers chlorine
1778	Scheele discovers molybdenum
1781	William Herschel (1738–1822), discovers Uranus on March 31st. The known size of the solar system doubles.
1781	Scheele discovers tungsten
1782	Franz-Joseph Müller von Reichenstein discovers tellurium
1783	Cavendish gives the first analysis of air as 79.2% N_2 and 20.8% O_2 by volume
1784	Herschel recognizes the polar ice caps on Mars
1787	Herschel discovers Uranus' moons Titania and Oberon
1789	Martin Klaproth discovers Uranium
	Klaproth discovers zirconia (ZrO_2), which was first isolated as the element by Jöns Jacob Berzelius (1779–1848) in 1824
1791	William Gregor discovers titanium
1794	Ernst Florens Friedrich Chladni (1756–1827) recognizes the extraterrestrial origin of meteorites
1796	Pierre Simon Marquis de Laplace (1749–1827) postulates his nebula hypothesis for the origin of the solar system
1797	Louis Vauquelin discovers chromium
1798	Vauquelin discovers beryllium
1798	Henry Cavendish determines the Earth's mass (6.6×10^{24} kg) and density ($5.5\,\mathrm{g\,cm^{-3}}$)
1800	Herschel discovers infrared radiation in the solar spectrum
1801	Charles Hatchett (1765–1847) discovers niobium (columbium)
1801	Guiseppe Piazzi (1746–1826) discovers Ceres, the largest asteroid (or dwarf planet)
1801	Johann Wilhelm Ritter (1776–1810) discovers ultraviolet radiation in the solar spectrum
1802	Anders Gustaf Ekeberg (1767–1813) discovers tantalum
1803	John Dalton presents his atomic hypothesis
1803	Martin Klaproth, Jöns Jacob Berzelius and Wilhelm von Hisinger discover cerium
	Smithson Tennant discovers osmium and iridium
	William Wollaston discovers palladium and rhodium
1805	Dalton gives the first table of atomic weights
1805	Andrés Del Rio discovers vanadium
1807	Humphrey Davy discovers sodium and potassium
1808	Davy discovers calcium, strontium, and barium
	Humphrey Davy, Joseph-Louis Gay-Lussac and Louis-Jacques Thenard discover boron
1811	Bernard Courtois discovers iodine in dried brown algae
	Gay-Lussac and Thenard discover silicon

(*Continued*).

Year	Discoveries
1814	Joseph von Fraunhofer discovers the "Fraunhofer" absorption lines in the solar spectrum
1816	Based on the available data on atomic weights, William Prout suggests that all elements were formed from H
1817	Johan A. Arfvedson discovers lithium
	Jöns Jacob Berzelius discovers selenium
	Friedrich Strohmeyer discovers cadmium
1826	Antoine J. Balard discovers bromine
1827	Friedrich Wöhler discovers aluminum
1828	Berzelius discovers thorium
1838	Christian Friedrich Schönbein (1799–1869) coins the term "geochemistry"
1839	Carl Mosander discovers lanthanum
1840	Schönbein discovers ozone during electrolysis of water
1841	Foundation of the "Chemical Society London"
1843	Mosander discovers ytterbium, terbium and erbium
1844	Karl Ernst Claus discovers ruthenium
1846	Johan Gottfried Galle discovers Neptune using the calculations by John C. Adam and Urbain J. J. Leverrier
1847	Elie de Beaumont (1798–1874) summarizes the occurrence of the known elements in rocks, meteorites and organic matter
1851	William Lassell discovers Uranus' moons Oberon and Ariel
1851–1858	Optical microscopy and use of thin-sections of rocks begins
1860	Gustav Kirchhoff and Robert Wilhelm Bunsen (1811–1899) develop spectral analysis and recognize the importance of the Fraunhofer lines
1860	Bunsen and Kirchhoff discover caesium
1861	Sir William Crookes discovers thallium
	Bunsen and Kirchhoff discover rubidium
1863	Ferdinand Reich and Hieronymus T. Richter discover indium
1863–1866	John Alexander Reina Newlands publishes a series of papers on the periodic relations of the elements with atomic weights, "law of octaves"
1868	Pierre Janssen and Joseph Norman Lockyer discover helium in the Sun
1869	Dmitri Mendeleev (1834–1907) publishes his paper on "The correlation between properties of elements and their atomic weights" and predicts elements to be discovered
1869/1870	Lothar Meyer (1830–1895) publishes his observations on the periodic system of the elements
1875	Paul Emile Lecoq de Boisbaudran discovers gallium
1876	Foundation of the American Chemical Society

(*Continued*).

Year	Discoveries
1878	Marc Delafontaine and Jacques-Lois Soret as well as Per Cleve discover holmium
1879	Cleve discovers thulium
	Lecoq de Boisbaudran discovers samarium
	Lars Nilson discovers scandium
1880	Jean Marignac discovers gadolinium
1885	Carl Auer von Welsbach discovers praseodymium and neodymium
1886	Lecoq de Boisbaudran discovers dysprosium
	Henri Moissan (1852–1907) discovers fluorine (Nobel Prize in Chemistry 1906)
	Clemens Winkler discovers germanium
1889	Frank Wigglesworth Clarke (1847–1931) determines the elemental composition of the Earth's crust. In 1908, he re-introduces the term "geochemistry" in his "Data on Geochemistry"
1892	Marie Curie (1867–1934) discovers polonium (Nobel Prize in Chemistry 1911)
1894	Lord John William Rayleigh (1842–1919) and Sir William Ramsay (1852–1916) discover argon (Nobel Prize in Chemistry 1904)
1895	Wilhelm Conrad Röntgen (1847–1923) discovers X-ray radiation (Nobel Prize in Physics 1901)
1896	Henri Antoine Becquerel (1852–1908) discovers natural radioactivity (Nobel Prize in Physics 1903)
1896	Eugéne Demarçay discovers europium
1897	Sir Joseph John Thomson (1856–1940, Nobel Prize in Physics 1906) discovers the electron
1898	Pierre & Marie Curie discover radium
	William Ramsay and Morris Travers discover Ne, Kr, and Xe
1900	Max K. E. L. Planck (1858–1947, Nobel Prize in Physics 1918) radiation formula
1900	Friedrich Ernst Dorn discovers radium
1902	Sir Ernest Rutherford (1871–1937, Nobel Prize in Chemistry 1908) and Frederick Soddy (1877–1956, Nobel Prize in Chemistry 1921) propose atom-disintegration as explanation for radioactive decay
1905	Albert Einstein (1879–1955, Nobel Prize in Physics 1921) special relativity
1907	Georges Urbain discovers ytterbium and lutetium
1909	Rutherford finds identity of α particle and He^{2+}
1911	Rutherford's nuclear atom structure
1912	Victor Franz Hess (1883–1964) discovers cosmic ray radiation (Nobel Prize in Physics 1936)

(*Continued*).

Year	Discoveries
1912	Max von Laue (1879–1960) discovers X-ray diffraction on crystal lattices (Nobel Prize in Physics 1914)
1913	Rutherford identifies proton
1913	Niels Henrik David Bohr (1885–1962) presents atomic model (Nobel Prize in Physics 1922)
1913	Henry Moseley finds that atomic numbers equal nuclear charges
1913	Soddy introduces the term "isotope" for atoms of the same element with different atomic weights. From *isos-topos* for equal place, or located at the same place (in the periodic table). The term isotope is reserved for nuclides with the same proton number (charge) and different numbers of neutrons
1915/16	Einstein's general relativity
1917	William D. Harkins recognizes the differences in elemental abundance distributions between the elements with even and odd atomic numbers
1918	Otto Hahn (1879–1968, Nobel Prize in Chemistry 1944) and Lise Meitner discover protactinium
1919	Francis William Aston (1877–1945, Nobel Prize in Chemistry 1922) develops the first mass spectrometer
1920	Rutherford postulates neutron
1923	Dirk Coster and Georg von Hevesy discover hafnium
1925	Walter Noddack, Ida Tacke and Otto Berg discover rhenium
1929	William Francs Giauque (1895–1982, Nobel Prize in Chemistry 1949) discovers the oxygen isotopes
1929	Henry Norris Russell publishes the first comprehensive elemental abundance determinations for the Sun
1931	Ernst Ruska and Max Knoll develop the electron-microscope
1931	Harold Clayton Urey (1893–1981, Nobel Prize in Chemistry 1934) discovers deuterium
1932	Sir James Chadwick (1891–1974) discovers the neutron (Nobel Prize in Physics 1935) Carl David Anderson (1905–1991) discovers the positron (Nobel Prize in Physics 1936)
1933	Werner Heisenberg (1901–1976, Nobel Prize in Physics 1932) describes the make-up of atomic nuclei from protons and neutrons
1935	Mass-spectrometric discovery of ^{235}U by Dempster
1937	Victor M. Goldschmidt (1888–1947) publishes his papers on abundances and geochemical properties of the elements
1937	Carlo Perrier and Emilio Segrè discover artificially produced technetium
1937/38	Hans Albrecht Bethe (1906–2005, Nobel Prize in Physics 1967) and Carl Friedrich von Weizsäcker describe H fusion through the p-p chain and the CNO-catalyzed cycle

(*Continued*).

Year	Discoveries
1945	J. A. Marinsky, L. E. Glendenin and C. D. Coryell discover artificially produced promethium
1946	Willard Frank Libby (1908–1980) introduces radiocarbon (^{14}C) dating (Nobel Prize in Chemistry 1960)
1948	Gerard Kuiper discovers Uranus' moon Miranda
1949	W. Gentner develops methods for Rb/Sr and K/Ar age determinations
1952	Paul Merrill discovers Tc in certain red giant stars
1957	Rudolf Ludwig Mössbauer discovers recoil-free nuclear resonance radiation (Mössbauer effect, Nobel Prize in Physics 1961)
1957	E. M. Burbidge (1919), G. R. Burbidge (1925), William A. Fowler (1911–1995, Nobel Prize in Physics 1983) and Fred Hoyle (1915–2001) and, independently, Alastair G. W. Cameron (1925–2005) publish their work on nucleosynthesis
1957	Sputnik-1, the first artificial satellite was launched on 4th October

CHAPTER 1

The Elements in the Solar System

"Of course I was not there when the solar system originated and
I do not know how it originated. I am only a student of the subject and I modify my ideas as new evidence appears or as new
ideas occur to me."

Harold Urey, 1963

A founding father of cosmochemistry

1.1 INTRODUCTION

Modern evidence leads to the insight that stars and their planetary
systems form by gravitational collapse of interstellar molecular clouds.
Much of the chemistry in our solar system is governed by the original
element inventory that the solar system inherited from its presolar
molecular cloud about 4.6 billion years ago.

Molecular clouds are cosmic recycling bins for the elements produced
in many stars from different generations. The big-bang endowed the
universe with H, D, He (both isotopes ^{3}He and ^{4}He) and a dash of
Li (mainly ^{7}Li). Hydrogen and He serve as major nuclear fuel in stars.
The light elements Li, Be and B have particular histories of their own
(see below) but all other elements came into being through stellar
nucleosynthesis over time.

Stars like our Sun and ones that are more massive do not exist forever.
In its final evolutionary stage, a star returns most of its mass – including
mass in the form of freshly synthesized elements – back to the interstellar

Chemistry of the Solar System
By Katharina Lodders and Bruce Fegley, Jr.
© K. Lodders and B. Fegley, Jr. 2011
Published by the Royal Society of Chemistry, www.rsc.org

medium. The efficiency and yields for element production depends on the initial mass of the star, which also determines whether the newly produced elements are released through a stellar explosion as a super-nova, or by less violent stellar winds. The result is that the stellar "ashes" can become part of the molecular clouds in the interstellar medium from which new generations of stars can rise.

Element production in generations of stars has been ongoing since the time the universe formed some 14 billion years ago. It is still ongoing in the stars that we can see in our galaxy and in stars of other galaxies dispersed in the universe. There is no need to worry that stars will dis-appear soon because there is no H and He left as initial stellar fuel. On a universal scale, H and He remain the most abundant elements. At the Sun's birthplace in our galaxy, the mass fraction of all elements heavier than helium had increased only to about 1.5% between the time when element production in stars of our Milky Way Galaxy started and the solar system formed. The rather low abundance of all heavy elements compared to H and He is one reason why astronomers collectively call all elements heavier than helium "metals." This definition of "metals" is not the traditional characterization that one normally associates with metals. On the other hand, there is the natural bias from daily life on a planet that essentially only consists of these "metals." Typically, there are no daily concerns about the fact that most matter in the Universe was, is and will be so for quite some time H and He. We refrain here from using the astronomer's definition of metals too much to avoid confusion.

The investigations of what elements exist and what their abundances are went hand in hand. Already by 1847, the French geologist Élie de Beaumont (1798–1874) assembled a list (Figure 1.1) for the occurrences of the elements known at the time.[1] He listed the elements in order of increasing electro-negativity as suggested by Berzelius. Among the elements in his list, three entries seem unusual from the current point of view: glucinium is the old designation for Be, didymium turned to be a mixture of Pr and Nd, and pelopium, thought to be a new element found in the mineral columbite, was later shown to be impure columbium ($=$ Nb). In his first entry column, De Beaumont used stars to mark the 16 elements that Henry de la Bèche found to be the most widely dis-tributed over the surface of the Earth. In subsequent columns, he indi-cated which elements have been found in modern and ancient volcanoes, basic rock, granites, stanniferous veins, normal ore veins and geodes, mineral waters, volcanic emanations, native metals, meteorites, and organic matter. De Beaumont emphasizes that out of the 16 most abundant elements on the Earth's surface, 15 are also those that

Tableau de la distribution des corps simples dans la nature.

	1 Corps les plus répandus sur la surface du globe.	2 Roches volcaniques actuelles.	3 Roches volcaniques anciennes.	4 Roches basiques.	5 Granites.	6 Filons stannifères.	7 Filons ordinaries et geodes.	8 Sources minerales.	9 Émanations volcaniques.	10 Radicaux natifs	11 Aérolithes.	12 Corps organizés.
1 Potassium.	*	*	*	*	*	*	*	*	*	…	*	*
2 Sodium.	*	*	*	*	*	*	*	*	*	…	*	*
3 Lithium.	…	…	…	…	*	*	…	*				
4 Barium.	…	…	…	…	…	*	*	*				
5 Strontium.	…	…	…	…	…	…	*	*				
6 Calcium.	*	*	*	*	*	*	*	*	*	…	*	*
7 Magnesium.	*	*	*	*	*	*	*	*	*	…	*	*
8 Yttrium.	…	…	…	…	*	*						
9 Glucinium.	…	…	…	…	*	*	*					
10 Aluminium.	*	*	*	*	*	*	*	*	*	…	*	*
11 Zirconium.	…	…	…	…	*	*						
12 Thorium.	…	…	…	…	*							
13 Cerium.	…	…	…	…	*	*						
14 Lanthane.	…	…	…	…	*	*						
15 Didymium.	…	…	…	…	*	*						
16 Urane.	…	…	…	…	*	*	*					
17 Manganèse.	*	*	*	*	*	*	*	*	*	…	*	*
18 Fer.	*	*	*	*	*	*	*	*	*	…	*	*
19 Nickel.	…	…	…	…	…	*	*	…	…	…	*	
20 Cobalt.	…	…	…	*	*	*	*	…	*	…	*	
21 Zinc.	…	…	…	*	*	*	*					
22 Cadmium.	…	…	…	…	…	*	*					
23 Etain.	…	…	…	…	*	*	*					
24 Plomb.	…	…	…	*	*	*	*	…		*		
25 Bismuth.	…	…	…	*	*	*	*	…	…	*		
26 Cuivre.	…	…	…	*	*	*	*	*?	*	*	*	
27 Mercure.	…	…	…	…	…	…	*	…	…	*		
28 Argent.	…	…	…	*	*	*	*	…	…	*		
29 Palladium.	…	…	…	*	*?	*	*	…	…	*		
30 Rhodium.	…	…	…	*	…	…	…	…	…	*		
31 Ruthenium.	…	…	…	*	…	…	…	…	…	*		
32 Iridium.	…	…	…	*	…	…	…	…	…	*		
33 Platine.	…	…	…	*	…	…	*	…	…	*		
34 Osmium.	…	…	…	*	…	…	…	…	…	*		
35 Or.	…	…	…	*	*	*	*	…	…	*		
36 Hydrogène.	*	*	*	*	*	*	*	*	*	…	*	*
37 Silicium.	*	*	*	*	*	*	*	*	*	…	*	*
38 Carbone.	*	…	…	…	*	*	*	*	*	*	*	*
39 Bore.	…	…	…	…	*	*	*	*	*			
40 Titane.	…	*	*	*	*	*	*					
41 Tantale.	…	…	…	…	*	*						
42 Nobium.	…	…	…	…	*	*						
43 Pelopium.	…	…	…	…	*	*						
44 Tungstène.	…	…	…	…	*	*	*					
45 Molybdène.	…	…	…	…	*	*	*	*				
46 Vanadium.	…	…	…	…	*	*	*					
47 Chrome.	…	…	…	*	*	*	*	…	…	…	*	
48 Tellure.	…	…	…	…	…	*	*	*	…	*		
49 Antimoine.	…	…	…	…	…	*	*	*	…	*	*	
50 Arsenic.	…	…	…	*	*	*	*	*	*	*		
51 Phosphore.	*	…	*	*	*	*	*	*	…	…	*	*
52 Azote.	*	…	…	…	…	…	…	*	*	*	*	*
53 Sélénium.	…	…	…	…	…	*	*	…	*	*	*	
54 Soufre.	*	*	*	*	*	*	*	*	*	*	*	*
55 Oxygène.	*	*	*	*	*	*	*	*	*	*	*	*
56 Iode.	…	…	…	…	…	…	*	*				
57 Brome.	…	…	…	…	…	…	*	*				
58 Chlore.	*	*	*	*	*	*	*	*	*	…	*	*
59 Fluor.	*	*	*	*	*	*	*	*	…			
	16	14	15	30	42	48	43	24	19	20	21	16

Figure 1.1 Élie de Beaumont's 1847 summary of the natural occurrence of the elements.

Angelot, whose data de Beaumont included in his table, had found in meteorites. In the last column of his table, de Beaumont indicates the general occurrence of the elements in "organized bodies," or biotic matter. He observes:[1]

> "These elements are 16 by number and they are precisely the same as the 16 elements which, after De La Bèche in the first column of the table, are the most distributed ones over the Earth's surface. This identity shows that the surface of the Earth encloses in all its parts everything that is essential for the existence of organized beings; it provides a new and striking example of the harmony that exists in all parts of nature. The 16 elements can be found in volcanic productions, in the mineral waters, and one sees that nature has provided not only a settlement but also the conservation of this indispensable harmony. The aging Earth will never cease to furnish all the elements to the organized beings necessary for their existence" (authors' translation).

De Beaumont put his table together some time before papers on the periodic table began to appear. Between 1863 and 1866, John Alexander Reina Newlands published a series of papers on the periodic relations of the elements with atomic weights. In 1869, Lothar Meyer and Dmitri Mendeleev published their notes on the periodic table. Only Mendeleev[2] boldly predicted the existence of several "missing" elements, which were subsequently discovered and bestowed with patriotic names (*e.g.*, Ga, Ge, Sc). The subject about the element occurrences and their relative abundances continued. Knowing the abundances and distribution of the elements would shed clues on the basic make-up and origins of matter. Important issues were to find the representative elemental compositions for the Earth, the Sun and the cosmos. The searches and discoveries of the missing elements in Mendeleev's periodic table involved mainly analyses of terrestrial materials. The development of optical spectral analysis by R. Bunsen and G. Kirchhoff in the early 1860s made it possible to access elements of low abundance and to quantitatively determine their concentrations in different minerals and rocks. After 1913, when H. G. J. Mosley provided the theoretical understanding for X-ray spectra, X-ray spectroscopy became another valuable analytical tool. Starting in 1919, Aston's developments of the first mass-spectrometers began to reveal the isotopic nature of the elements. The advances in micro-analytical instrumentation since the 1950s, such as the electron microprobe, neutron activation analysis and gamma-ray spectroscopy, led to a wealth of elemental and isotopic data from terrestrial, lunar and meteoritic rocks. More recently, ion-probe mass-spectrometric methods permit the analyses of the elemental and isotopic compositions on samples with micro- to nano-scale resolution that even include genuine mineral particles that formed around other stars and are hidden in meteorites.

Spectroscopy also advanced the discovery and quantitative assessment of the elements beyond the Earth. Starting in the 1860s, the analyses of the Fraunhofer absorption lines in solar photospheric spectrum, and spectral analysis of other stars, interstellar nebulae, and of comets soon revealed that the occurrence of the chemical elements is not restricted to the Earth. Analyses of meteorites, already recognized as extra-terrestrial materials by E. F. Chaldni in 1794, showed similar results. The same chemical elements as found on the Earth constitute the normal matter in other objects in the solar system and beyond in the stars. There were no stable elements in the Sun, other stars, in comets or meteorites that could not be found naturally occurring on Earth as well. Even the well-known case of He, discovered by Janssen and Lockyer in the solar spectrum in 1868, was no exception since its detection on Earth followed by Cleve and Langlet in 1895 and, independently, by Ramsey at the same time.

The story is a little different for the "missing" elements Tc (first called masurium) and Pm. Unlike the elements beyond atomic number 83 (Bi), Tc and Pm have their place between stable elements in the periodic table. All Tc and Pm isotopes are unstable and undergo radioactive decay; the half-lives of the longest-lived isotope are 4.2 million years for ^{98}Tc and 17.7 years for ^{145}Pm. On Earth, these elements were first known as products of artificial nuclear reactions, which succeeded in producing Tc in 1937 and Pm in 1947. These elements only occur naturally in stars that make them. In 1952, Merrill[3] found Tc in several red giant stars, and Aller and Cowley likely detected Pm in a chemically peculiar star in 1970, as seems to be confirmed recently.[4]

Merrill's discovery of Tc in stars showed that element synthesis does happen in stars, and it came at a time when models of element synthesis in stars had gotten their first foundations. The discovery of the neutron by J. Chadwick in 1932 facilitated the understanding of element synthesis, and nucleosynthesis models started to evolve in the late 1930s, when H. A. Bethe[5] and C. F. von Weizäcker[6] independently proposed models how thermonuclear fusion reactions convert H into He using C, N and O as catalysts in stars. The Bethe–Weizsäcker cycle, or CNO cycle, is one of the processes for H fusion in the Sun, the other is fusion through the proton–proton reaction chains where D (just discovered by H. C. Urey in 1932) occurs as an interim product.

Henry N. Russell carried out the first quantitative analysis of the elements in the Sun in 1929.[7] V. M. Goldschmidt's classical papers on elemental and isotopic distributions appeared in 1937[8] where he devised basic geochemical principles that govern the observed abundances in terrestrial and meteoritic rocks. The knowledge about the abundances as well as the isotopic composition of the elements had become fairly detailed by 1956, when H. E. Suess and H. C. Urey did their classical work on cosmic abundances.[9] In concert with abundance determinations, Alpher, Bethe and Gamov as well as ter Haar and Salpeter as well as many other researchers had provided more important groundwork on the theory of element synthesis through the 1940s and early 1950.[10] In 1957, comprehensive nucleosynthesis models by E. M. Burbidge, G. R. Burbidge, W. A. Fowler, and F. Hoyle[11] ("B²FH") and, independently, A. G. W. Cameron[12] appeared to explain the synthesis of the heavy elements. Today, the refined but still growing knowledge of the elemental and isotopic abundances in our solar system, in stars of our Galaxy and beyond continues to inspire nucleosynthesis and galactic chemical evolution models, and the abundances of the elements of the solar system remain a critical test for these models.

1.2   ABUNDANCES OF THE ELEMENTS

In 1885, when most known elements populated the periodic table, the Russian scientist I. A. Kleiber published a paper on the chemical composition of celestial bodies.[13] Figure 1.2 shows his qualitative synthesis for the general composition of celestial objects in the form of a plane periodic system following atomic numbers. This may be the first published "cosmochemical" periodic table of the elements. Note that the group for the noble gases was not yet included. Kleiber searched for periodic trends in elemental abundances in meteorites, the Sun, comets, fixed stars and meteors. He concluded that the composition of cosmic bodies is not the result of some random contribution of the elements.

Like Mendeleyev and others at the time, Kleiber noticed that elements with low atomic weights up to the iron group are more abundant than the heavier ones, with the notable exception of B. In particular, he noticed that the iron group elements are abundant in cosmic objects whereas the Pt-group elements are not, which suggested that the elements with similar chemistry are not necessarily similar in abundance.

The first comprehensive report on a quantitative determination of the relative abundances of the elements dates back to 1889 and was performed for the Earth's crust.[14] Frank Wigglesworth Clarke investigated the trend of elemental abundances in the Earth's crust as a function of

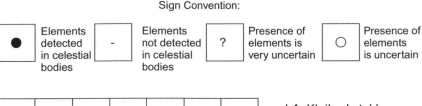

Sign Convention:

| ● | Elements detected in celestial bodies | - | Elements not detected in celestial bodies | ? | Presence of elements is very uncertain | ○ | Presence of elements is uncertain |

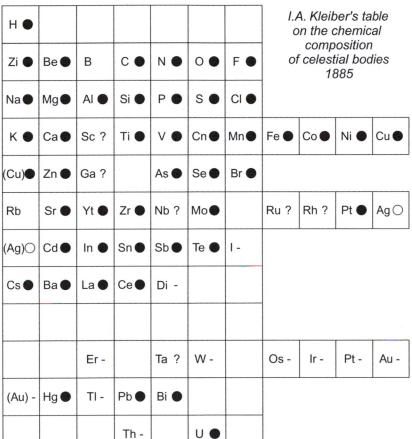

I.A. Kleiber's table on the chemical composition of celestial bodies 1885

Figure 1.2 Kleiber's qualitative synthesis for the general composition of celestial objects in the form of a plane periodic system following atomic numbers.[13] His diagram is probably one of the earliest versions of a "cosmochemical periodic table."

their atomic weights. With abundant accessible samples for chemical analyses, the rocks of the Earth's crust were a useful starting point to determine the relative elemental abundances, which might reveal something about the origins of the elements, aside of course from the economic interest of knowing the crust's elemental inventory. However,

Clarke was somewhat frustrated not to find discernable elemental abundance trends. He was able to refine the conclusion that the lighter elements up to Fe are much more abundant than the heavier elements, but not much more that could shed some light on the causes for the observed element abundances.

In any case, the composition of the $\sim$5–50 km thick crust of the Earth cannot be representative of the composition of the entire Earth, let alone the composition of the solar system. Most of the Earth' mass is in its silicate mantle and the core, and the crust is only about 0.4% of Earth's total mass. As a whole, the Earth mainly consists of the elements O, Mg, Si and Fe, and the other, less abundant elements in the Earth are distributed between the Fe-rich metallic core, the Mg- and Fe-silicate mantle rocks, the crust, oceans and atmosphere according to their geochemical affinities.

By 1917, several good meteorite analyses were available, and the relatively primitive nature of meteorites in contrast to the complex history of planetary rocks had already been established. Following Farrington's 1915[15] suggestion that the mean composition of meteorites may resemble that of the Earth as a whole, William D. Harkins (1873–1951) averaged the composition of different stony and iron meteorites to obtain such a representative composition for the Earth.[16] Then he plotted the abundance as a function of atomic number to make a fundamental discovery. The meteoritic abundances showed systematic trends with atomic number, quite different from those in the Earth's crust. Figure 1.3 is a redrawn version of his plot, which was the first illustration of the odd-even elemental abundance trend with atomic number. This trend is now known as "Harkins' rule."

Harkins[16] describes his observations with regard to the elemental abundances in meteorites: ... the elements with even atomic numbers "are in every case more abundant than their adjacent odd-numbered elements." Harkins found that the abundant elements Mg, Si, Ca, Fe and Ni "do not only have even atomic numbers, but in addition they make up 98.6% of the material in meteorites." He observed that odd-numbered elements with higher abundances such as Al and Co are in "between two extremely abundant even-numbered elements."

His diagram is limited to the abundant elements from C to the Fe group for which abundances were easier to determine than for the heavier, less abundant elements. In 1917, not all the rare earth elements had been discovered and their abundances in terrestrial rocks and meteorites were sketchy at best. However, from the data available, Harkins concluded that the abundance trend in the relative odd-even abundances should apply to the lanthanides as well, which was nicely confirmed by later analyses (Figure 1.4).

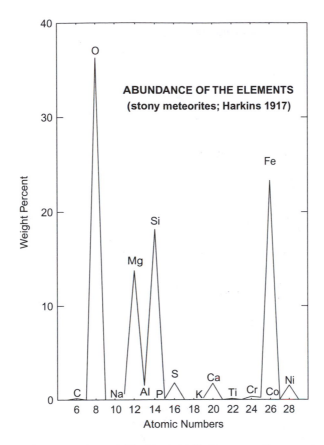

Figure 1.3 Redrawn version of Harkins' 1917 discovery diagram that shows the higher abundances of even-numbered elements compared to their neighboring odd-numbered elements.[16]

Looking at meteorites and terrestrial rocks Harkins found that the abundant elements are restricted to atomic numbers between 6(C) and 30(Zn), similar to Clarke's finding that these are the most abundant elements in the Earth's crust. The absence of volatile H and He in rocks is not too surprising; however, Harkins independently repeats Kleiber's observation of the relative paucity of the light elements Li, Be and B, as well as that of the heavy elements beyond the Fe group in meteorites. Explanations for these observations and Harkin's conclusion that "in the evolution of the elements, more material has gone into the even-numbered elements than those which are odd"[16] only became possible when more analytical data for solar system abundances became available and theories on nucleosynthesis evolved in the 1930–1950s.

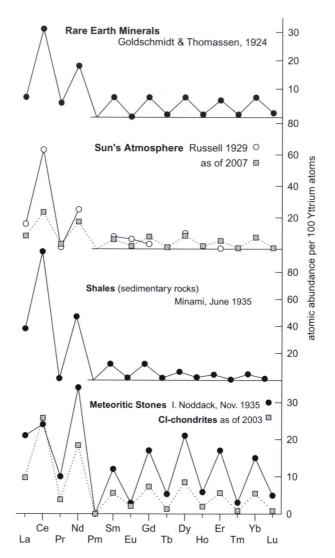

Figure 1.4 In 1937, V. M. Goldschmidt summarized data for the abundances of the rare earth elements (REE) in terrestrial rocks, the Sun and meteorites.[8] We have added recent data to his diagram for comparison. All concentrations are normalized to 100 Yttrium atoms (not shown). The REE abundances show Harkin's rule of the abundance variations of odd and even elements quite nicely.

1.2.1   Sources for the Solar System Composition

Most of the mass, 99.86%, in our solar system resides in the Sun. Thus, the composition of the Sun should provide a good average of the

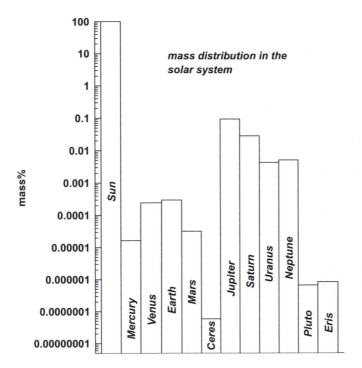

Figure 1.5 Mass distribution in the solar system.

element inventory that the solar system inherited from its parental molecular cloud. The gas-giant planets Jupiter, Saturn, Uranus and Neptune are the next most massive objects and make up $\sim 0.13\%$ of the total mass in the solar system (Figure 1.5). The terrestrial planets Mercury, Venus, Earth and Mars contribute $\sim 0.0009\%$ to the solar system's mass; the Earth alone makes up $\sim 0.0003\%$. The dwarf planets Ceres, Pluto and Eris make negligible fractions of the total solar system mass.

Ultimately, the cloud material also provided the elemental starting composition for the planets and the differences in the composition of the Sun and the planets as we see them today must reflect the chemical and physical element fractionation processes during planet formation and subsequent planetary evolution. These fractionations make it challenging to derive the overall chemical composition of the solar system from terrestrial crust and mantle rocks. Instead, it is more practical to compare planetary compositions to the solar system's composition to decipher how the element inventories of the planets were established and what differentiation processes occurred. In theory, it would be ideal to have good quantitative analyses of the elemental abundances in the Sun.

In practice, this is currently only possible for a subset of the elements in the Sun's photosphere (see below).

The composition of meteorites provides a second source of information for solar system abundances. Most meteorites do not show signs of large-scale differentiation of silicates and metal as planetary objects do and it was recognized early that meteorites are the closest samples of the original, "world making" material (*e.g.*, Merrill, 1909[17]). It is not possible to use the meteorite compositions to constrain the abundances of the noble gases or the abundances of elements like H, C, N and O, which form volatile compounds. However, for the "non-volatile" or "rocky" elements, meteorite analysis are usually more precise than photospheric measurements, and for several elements meteorites are the sole source of information on solar system elemental abundances.

As described in a separate chapter, there are compositionally different groups of meteorites. Consequently, one problem had been how the analyses of the different meteorites should be treated to derive representative abundances for the non-volatile elements in the solar system. A related issue is if there is a single, particular group of meteorites that can serve as an abundance standard.

Neither the solar photosphere nor meteorites can provide reasonable abundance estimates for some elements like the noble gases. In such cases, one can make use of the fact that most of the other normal dwarf stars in the H-burning stage have very similar relative abundances as the Sun. The younger, more massive and hotter B dwarf stars are particularly useful to derive "missing" solar elemental abundances and to check the solar abundance determinations for elements up to Fe. Yet another source for supplementing and checking the solar abundances is the analysis of interstellar diffuse nebulae such as the so-called HII regions (the notation HII indicates that ionized H dominates in contrast to HI regions with mainly neutral H). These regions are often associated with young stars, and the composition of these regions should be similar to the young B stars nearby. The elemental abundances from absorption spectroscopy of evolved giant stars or from emission spectroscopy of their descendants, the planetary nebulae, are useful for complementing the solar abundance data. (The name planetary nebulae has nothing to do with planetary systems. Through early telescopes, these dust and gas enshrouded stellar remnants appeared as "fuzzy" as the planets.) However, their usefulness to constrain solar abundances is limited because certain elements are among the giant stars' interior nucleosynthesis products that are eventually mixed to the stellar surfaces.

In the following, "meteoritic" or "CI-chondrite" abundances refer to elemental abundances from type CI-carbonaceous chondrites, "photospheric" abundances refer to abundance determinations of the present solar photosphere and "proto-solar" or "solar-system abundances" refer to elemental abundances of the proto-sun at the time of its formation.

For completeness, it needs to be mentioned that the solar abundances occasionally were called "cosmic" abundances in the older literature. While the solar system abundances derived from the solar photosphere and the meteorites only strictly apply to the solar system, these abundances are quite similar to those found in other stars like the Sun. However, there are systematic variations in the abundances of the heavy elements (the astronomers' "metals") in normal stars as a function of radial distance from the galactic center. These variations do not warrant making the solar system abundances representative for "cosmic" abundances.

1.2.2 Elemental Abundance Scales

There are two widely used relative atomic abundance scales for the elements. The astronomical abundances scale is fixed to the abundance of the most abundant element in the photosphere, hydrogen, at $\varepsilon(H) = 10^{12}$ atoms. Since abundances vary over twelve orders of magnitude, this relative scale avoids dealing with negative exponents in the abundances. It is often more convenient to use the decadic logarithm for the abundances so that $A(H) = \log \varepsilon H = 12$, or, for any element "El," we have $A(El) = \log(\varepsilon El/\varepsilon H) + 12$, which is the notation frequently encountered in the astronomical literature.

Geochemists and cosmochemists prefer to use a linear atomic abundance scale fixed to a silicon abundance of $N(Si) = 10^6$ atoms. The cosmochemical abundance scale is more convenient when dealing with planetary and meteoritic compositions because Si is a major element in rocks. The linear scale is still practical because the most abundant elements in the Sun H, He, Ne, Ar, C, N and, to some extent, O and their compounds are volatile and therefore typically less abundant in rocky planets and meteorites.

Uncertainties of photospheric and meteoritic abundance determinations are compared using the relationship $U(\%) = \pm 100(10^{\pm a} - 1)$ where "a" is the uncertainty in dex-units quoted for abundances on the logarithmic scale and "U" is the uncertainty on the linear scale in percent. The uncertainty in logarithmic units ("dex") is an uncertainty factor, hence the uncertainty in percent is smaller for $-a$ than for $+a$, or *vice versa*, a given percent uncertainty yields two different uncertainty factors.

1.2.3 Sun's Photospheric Composition

In principle, the best average for the solar system's composition would come from the analysis of the solar photosphere because the Sun contains most of the mass of the solar system. In practice, there were, and still are, various technical difficulties that prevent well-defined quantitative analyses for all elements.

The photospheric abundances are derived from the Fraunhofer lines of the solar absorption spectrum, which samples the wavelength range from the near-UV to the IR. The Fraunhofer lines mainly form in the photosphere and in the overlying photosphere-chromosphere transition region. Sixty-eight elements have been detected in the photosphere, and they are mainly present as monatomic or singly ionized ions. The quantitative determination of the photospheric abundances by spectroscopy is more elaborate than spectrochemical analysis in the laboratory, where samples and well-known standards can be measured under the same conditions. To derive quantitative abundances from the photospheric spectrum, it is necessary to know the atomic properties of the elements and the physical conditions in the solar photosphere. The information necessary and the challenges to derive quantitative abundances from stellar spectroscopy include, but are not limited to: identifications of element line positions in the spectra; presence or absence of suitable lines for certain elements in given spectral range or line blending in the measured spectra; knowledge of atomic transition probabilities and oscillator strengths (the f values); the sufficiency of the line broadening theories; and the adequacy of atmospheric structure models to describe the temperature and total pressure with depth in the atmosphere where the absorption lines originate. Finally, a still much-discussed issue is whether deviations of the excitation and ionization conditions as given by the Boltzmann and Saha equations for the local kinetic temperature (typically referred to as "local thermodynamic equilibrium" or LTE) must be considered in the abundance determinations.

The first comprehensive determination of the solar photospheric abundances by Russell in 1929[7] for 56 elements has seen many important revisions and updates over the years. Table 1.1 summarizes current analytical data. To date, out of the 83 elements that naturally occur in the solar system (all stable elements plus Th and U) the abundances for 68 elements have been measured in the Sun with varying

Table 1.1 Elemental abundances in CI-chondrites and in the solar photosphere (from ref. 18).[a]

Element	CI-chondrites $N(El)\ N(Si)$ $=10^6$	$A(El)\ log\ N(H)$ $=12$	Solar photosphere $N(El)^b$ $(N=10^{[A(El)-1.533]})$	$A(El)\ [log\ \varepsilon(H)$ $=12]$
H	5.13×10^6	8.24 ± 0.05	2.93×10^{10}	$\equiv12$
He	0.60	1.31	2.47×10^9	10.925 ± 0.02
Li	55.6	3.28 ± 0.05	0.369	1.10 ± 0.10
Be	0.612	1.32 ± 0.03	0.703	1.38 ± 0.09
B	18.8	2.81 ± 0.04	14.7	2.70 ± 0.17
C	7.60×10^5	7.41 ± 0.04	7.19×10^6	8.39 ± 0.04
N	5.53×10^4	6.28 ± 0.06	2.12×10^6	7.86 ± 0.12
O	7.63×10^6	8.42 ± 0.04	1.57×10^7	8.73 ± 0.07
F	804	4.44 ± 0.06	1060	4.56 ± 0.30
Ne	2.35×10^{-3}	-1.10	$[3.29\times10^6]$	$[8.05\pm0.10]$
Na	5.70×10^4	6.29 ± 0.02	5.85×10^4	6.30 ± 0.03
Mg	1.03×10^6	7.55 ± 0.01	1.02×10^6	7.54 ± 0.06
Al	8.27×10^4	6.45 ± 0.01	8.65×10^4	6.47 ± 0.07
Si	$\equiv1.00\times10^6$	$\equiv7.53\pm0.01$	0.970×10^6	7.52 ± 0.06
P	8195	5.45 ± 0.04	8410	5.46 ± 0.04
S	4.48×10^5	7.17 ± 0.02	4.04×10^5	7.14 ± 0.01
Cl	5168	5.25 ± 0.06	9270	5.50 ± 0.30
Ar	9.6×10^{-3}	-0.48	$[9.27\times10^4]$	$[6.50\pm0.10]$
K	3652	5.10 ± 0.02	3860	5.12 ± 0.03
Ca	6.04×10^4	6.31 ± 0.02	6.27×10^4	6.33 ± 0.07
Sc	34.4	3.07 ± 0.02	36.9	3.10 ± 0.10
Ti	2473	4.93 ± 0.03	2330	4.90 ± 0.06
V	280	3.98 ± 0.02	293	4.00 ± 0.02
Cr	1.33×10^4	5.66 ± 0.01	1.28×10^4	5.64 ± 0.01
Mn	9221	5.50 ± 0.01	6870	5.37 ± 0.05
Fe	8.70×10^5	7.47 ± 0.01	8.26×10^5	7.45 ± 0.08
Co	2254	4.89 ± 0.01	2440	4.92 ± 0.08
Ni	4.83×10^4	6.22 ± 0.01	4.98×10^4	6.23 ± 0.04
Cu	541	4.27 ± 0.04	475	4.21 ± 0.04
Zn	1296	4.65 ± 0.04	1220	4.62 ± 0.15
Ga	36.6	3.10 ± 0.02	22.2	2.88 ± 0.10
Ge	118	3.60 ± 0.04	110	3.58 ± 0.05
As	6.10	2.32 ± 0.04	–	–
Se	67.5	3.36 ± 0.03	–	–
Br	10.7	2.56 ± 0.06	–	–
Kr	1.64×10^{-4}	-2.25	$[55.8]$	$[3.28\pm0.08]$
Rb	7.10	2.38 ± 0.03	11.7	2.60 ± 0.10
Sr	23.4	2.90 ± 0.03	24.4	2.92 ± 0.05
Y	4.52	2.19 ± 0.04	4.75	2.21 ± 0.02
Zr	10.4	2.55 ± 0.04	11.1	2.58 ± 0.02
Nb	0.788	1.43 ± 0.04	0.771	1.42 ± 0.06
Mo	2.66	1.96 ± 0.04	2.44	1.92 ± 0.05
Ru	1.78	1.78 ± 0.03	2.03	1.84 ± 0.07
Rh	0.355	1.08 ± 0.04	0.386	1.12 ± 0.12
Pd	1.38	1.67 ± 0.02	1.34	1.66 ± 0.04
Ag	0.489	1.22 ± 0.02	(0.255)	(0.94 ± 0.30)
Cd	1.57	1.73 ± 0.03	1.73	1.77 ± 0.11

Table 1.1 (*Continued*).

Element	CI-chondrites		Solar photosphere	
	$N(El)\ N(Si)$ $=10^6$	$A(El)\ \log N(H)$ $=12$	$N(El)^b$ $(N=10^{[A(El)-1.533]})$	$A(El)\ [\log \varepsilon(H)$ $=12]$
In	0.178	0.78 ± 0.03	(0.927)	(<1.50)
Sn	3.60	2.09 ± 0.06	2.93	2.00 ± 0.30
Sb	0.313	1.03 ± 0.06	0.293	1.00 ± 0.30
Te	4.69	2.20 ± 0.03	–	–
I	1.10	1.57 ± 0.08	–	–
Xe	3.48×10^{-4}	-1.93	[5.46]	$[2.27 \pm 0.08]$
Cs	0.371	1.10 ± 0.02	–	–
Ba	4.61	2.20 ± 0.03	4.33	2.17 ± 0.07
La	0.457	1.19 ± 0.02	0.405	1.14 ± 0.03
Ce	1.17	1.60 ± 0.02	1.19	1.61 ± 0.06
Pr	0.176	0.78 ± 0.03	0.169	0.76 ± 0.04
Nd	0.857	1.47 ± 0.02	0.826	1.45 ± 0.05
Sm	0.265	0.96 ± 0.02	0.293	1.00 ± 0.05
Eu	0.0998	0.53 ± 0.02	0.0970	0.52 ± 0.04
Gd	0.342	1.07 ± 0.02	0.378	1.11 ± 0.05
Tb	0.0634	0.34 ± 0.03	0.0558	0.28 ± 0.10
Dy	0.412	1.15 ± 0.02	0.395	1.13 ± 0.06
Ho	0.0910	0.49 ± 0.03	0.0948	0.51 ± 0.10
Er	0.256	0.94 ± 0.02	0.267	0.96 ± 0.06
Tm	0.0406	0.14 ± 0.03	0.0405	0.14 ± 0.04
Yb	0.256	0.94 ± 0.02	0.212	0.86 ± 0.10
Lu	0.0380	0.11 ± 0.02	0.0386	0.12 ± 0.08
Hf	0.156	0.73 ± 0.02	0.222	0.88 ± 0.08
Ta	0.0210	-0.14 ± 0.04	–	–
W	0.137	0.67 ± 0.04	(0.378)	(1.11 ± 0.15)
Re	0.0554	0.28 ± 0.04	–	–
Os	0.680	1.37 ± 0.03	0.826	1.45 ± 0.11
Ir	0.640	1.34 ± 0.02	0.703	1.38 ± 0.05
Pt	1.27	1.64 ± 0.03	(1.61)	(1.74 ± 0.30)
Au	0.195	0.82 ± 0.04	(0.300)	(1.01 ± 0.18)
Hg	0.458	1.19 ± 0.08	–	–
Tl	0.182	0.79 ± 0.03	(0.261)	(0.95 ± 0.20)
Pb	3.33	2.06 ± 0.03	2.93	2.00 ± 0.06
Bi	0.138	0.67 ± 0.04	–	–
Th	0.0351	0.08 ± 0.03	(< 0.0352)	(< 0.08)
U	8.93×10^{-3}	-0.52 ± 0.0	$< 9.93 \times 10^{-3}$	<-0.47

[a]Note: Data in parenthesis are uncertain. Data in square brackets are not from solar measurements
and are determined indirectly.
[b]The conversion from the logarithmic astronomical scale to linear cosmochemical abundance scale
uses an average constant of 1.533 so that $N = 10^{[A(El)-1.533]}$.

degrees of accuracy. Most data are from photospheric analysis, and the abundances of F, Cl and Tl are actually derived from sun spot spectra.

The noble gas abundances cannot be directly measured in the solar photosphere because suitable absorption lines are lacking. Their "photospheric" abundances are determined indirectly, which is the reason why their elemental abundances in Table 1.1 are placed in square brackets. Although He is detected in the Sun, the He abundance is mainly based on helioseismic measurements, solar interior and evolution models, or on measurements of He/H abundances in other stars. The Ne and Ar abundances are estimated from the composition of the solar corpuscular radiation (solar wind and solar energetic particles), and from Ne and Ar measurements in hotter stars, where Ne and Ar lines are accessible. The Kr and Xe abundances are quite low and their abundances are usually derived from interpolations of neighboring element abundances and nucleosynthetic arguments.

The present-day elemental abundances in the outer solar atmosphere are not affected by the ongoing H-burning nucleosynthesis in the Sun's core because there is a non-convective layer between the core and the convective layer starting right below the photosphere. However, relative to H, the heavy elements can diffusively settle from the photosphere, which needs to be considered when current photospheric abundances are used to derive the proto-solar abundances (see below).

1.2.4 Elemental Abundances in Carbonaceous CI-Chondrites

The so-called CI-chondrites are a rare group of chondritic meteorites, which have become an important reference standard for solar system elemental abundances. More about meteorites is given in a subsequent chapter; however, because of their use as abundance standards, the data for CI chondrites are given in Table 1.1 for comparison.

1.2.5 Comparison of Meteoritic and Solar Abundances

Table 1.1 gives the photospheric and CI-chondrite abundances on the astronomical (normalized to log $\varepsilon H = 12$) and cosmochemical (normalized to $Si = 10^6$ atoms) scales.[18] One important issue is how these different *relative* scales are linked. The problem is that the relative H abundance in meteorites is much less than in the photosphere, but the astronomical scale is tied to element/H ratios in the photosphere. However, since H abundances in meteorites are low, their element/H ratios are much higher than those in the photosphere. On the other hand, it is not a problem to compare the photospheric abundances in their usual logarithmic notation in the astronomical scale to the meteorite data on the linear cosmochemical Si-based scale. For several elements, there is a more or less constant difference when the logarithmic meteoritic values are subtracted from the photospheric values.[18] The average difference (which corresponds to a factor if the scales were taken linear in the comparison) from about 35 well-determined elements is 1.533, so that the abundances on the astronomical scale (log εEl) are related to the abundances on the cosmochemical scale as: log N(El) = log εEl − 1.533.

With this relation, the *meteoritic* abundances are converted into the astronomical scale that is normalized to the *photospheric* H abundance of log εH = 12. Obviously, the meteoritic H abundances from the relation are less than this, because the element/H ratios are higher in the meteorites.

One could also simply re-normalize the photospheric data to Si, but this introduces uncertainty in linking the scales, because it becomes solely tied to the correctness and the quality of the Si abundances in the photosphere and in meteorites. A small change in the Si abundances from new measurements in either the photosphere or meteorites then would require re-computing the linked abundance scales. Except for historical sentiment to use Si for normalizing the cosmochemical abundance scale, there is no reason why honor is given solely to Si to link the scales. One may pick any other rocky, non-volatile element that should be fully retained in the meteorites and that is measured in the Sun. In the past, when photospheric abundance determinations were less certain, it was more reliable to calculate an average scaling factor from several elements that are well determined. The relative abundances of 35 elements agree within 10% in the photosphere and CI-chondrites, and their individual scaling factors give an average of 1.533 for linking the astronomical and cosmochemical abundance scales.

Figure 1.6 shows a plot of the photospheric *versus* the CI-chondritic abundances on the cosmochemical abundance scale. In most cases, the uncertainties in the abundances are smaller than the element symbols shown, but error bars are considerable for several element determinations in the Sun (Table 1.1). If the relative abundances were identical in the solar photosphere and CI-chondrites, the abundances would plot on the dotted 1 : 1 reference line. The agreement is within 10% (15%) for 31 (41) elements. This correspondence justifies the use of the CI-chondrites as an abundance standard, and to give preference to the meteoritic values of, for example, F, Cl, Ga, Rb, Ag, Cd, Au and Tl over the photospheric values because of their often-higher analytical precision. As a corollary, the CI-chondrite abundances are a useful abundance standard of elements that cannot be measured well at all in the solar photosphere (*e.g.*, As, Se, Br, Te, I, Cs, Ta, Re, Bi, Th, U).

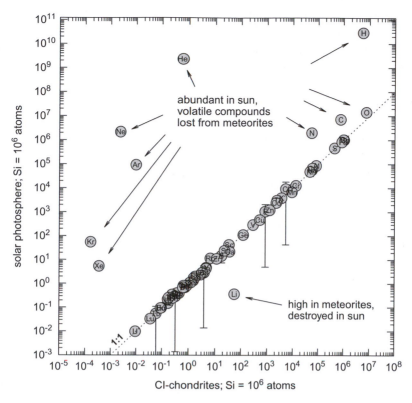

Figure 1.6 Elemental abundances in the solar photosphere and CI-chondrites.[18]

When compared to the photospheric abundances, the relative abundances of H, C, N, O and the noble gases are lower in CI-chondrites by several orders of magnitude, which indicates incomplete retention of volatile compounds in meteorites. Figure 1.6 shows that the Ne, Ar, Kr and Xe abundances in CI-chondrites follow a depletion trend that broadly anti-correlates with their atomic masses. Only one element, Li, has a clearly higher relative abundance in CI-chondrites than in the photosphere. The abundance of Li is well determined in both sources, and the only possible explanation is that Li was destroyed in the Sun.

1.2.6 D, ³He, Li, Be and B

Representative solar system abundances for D, Li and, possibly, Be and B cannot be derived from the Sun's photosphere. Although this section focuses on the elemental abundances, we include D here together with Li, Be and B because of their importance in cosmological models.

Deuterium and the nuclides of Li, Be and B are not produced in any significant amounts through stellar nucleosynthesis; instead, they are more likely to be destroyed in stars because of their low nuclear binding energies. Deuterium and ^{7}Li were mainly produced through big-bang nucleosynthesis, and their abundances are decreased through astration (destruction in stars) over time. The ^{7}Li abundance is continuously supplemented by cosmic ray spallation in the interstellar medium, which happens when heavier elements in the interstellar medium encounter bombardments from so-called cosmic rays. These energetic ions [mainly protons and α ($= \mathrm{He}^{2+}$) particles, but also ions of all other elements and secondary neutrons] break-up abundant C, N and O to yield the lighter element as fragments. Cosmic ray spallation is the major production mechanism of ^{6}Li, Be and B. Lithium (^{7}Li) is also produced in certain giant stars, so that Li is an element that has at least three different nucleosynthetic production origins. Another suggested source of the light elements is neutrino-induced nucleosynthesis in core-collapse supernovae.

Destruction of *deuterium* in stars is unavoidable, and happened in the Sun even before H-burning through the proton–proton chain was ignited. Within its about first million years of age, the temperatures and densities in the contracting proto-sun became favorable for D burning, which requires $> 0.6 \times 10^6$ K. The young Sun was fully convective during that time, and the Sun's atmospheric element inventory could be cycled through interior regions where temperatures were sufficiently high to destroy essentially all D within about $1–2 \times 10^5$ yr. Deuterium burning produces ^{3}He, which increased the ^{3}He/^{4}He in the Sun, and thus the current solar ^{3}He/^{4}He as derived from the solar wind is also not representative of the proto-solar ratio.

Deuterium is essentially gone from the photosphere, but D is present in the solar wind, as seen from the analysis of the solar wind that was captured in the Al and Pt metal foils posted on the lunar surface during the Apollo missions. The D in the solar wind is a product of local spallation reactions in solar flares and therefore it is not representative of the original D inventory of the solar system. However, the overall production of D in flares of other stars is relatively small and cannot balance the loss of primordial D in stellar interiors.[19] Thus, the original amount of D from the big-bang can only decrease over time and the question is how to find the D abundance that was available when the solar system formed.

Figure 1.7 shows the D/H ratios that have been determined in various settings. The primordial D/H ($\sim 2.6 \times 10^{-5}$) is well constrained from the cosmic microwave background measurements by the Wilkinson

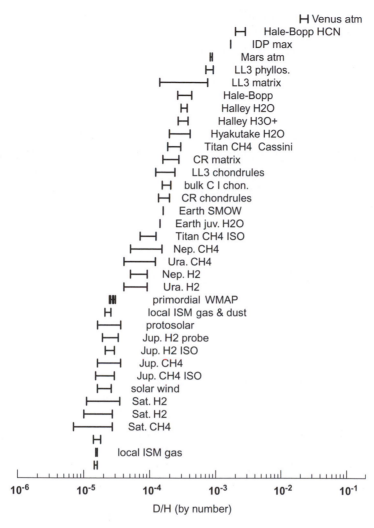

Figure 1.7 D/H abundance ratios in different objects and settings.

microwave anisotropy probe (WMAP)[20] and determinations of the baryon-to-photon ratios and cosmological models.[21–23] The gas in the current local interstellar medium has a low value of $\sim 1.5 \times 10^{-5}$, which either reflects the depletion of D by astration since the formation of the Universe or is simply due to the preferred incorporation of D into organic and icy interstellar dust.[23,24] If the estimate for the D/H in the nearby interstellar gas and dust of $\geq 2.3 \times 10^{-5}$ is correct, the decrease over time was rather modest ($<12\%$). Within these boundaries, the initial solar system D/H should have been between 2.3 and 2.6×10^{-5}.

The D/H ratios measured in the atmospheres of Jupiter and Saturn[25] span this range in D/H (Figure 1.7). Jupiter, the second most massive and, like the Sun, H and He-rich object in the solar system, is probably the least altered source for the original solar system D/H ratio. Since a large portion of Jupiter is gravitationally captured gas from the solar nebula, its atmosphere should have preserved the record of the D/H ratio in the early solar system. In that case, the initial solar system D/H ratio is 2.5×10^{-5}, as given by the average values measured in molecular H_2 by the Galileo spacecraft entry probe (2.6×10^{-5}) and the D/H = 2.4×10^{-5} for Jovian H_2 as determined from spectra with the ISO satellite. Within the larger uncertainties, the same applies to the D/H obtained for Saturn's atmosphere.[25]

The D/H ratios for the outer giant planets, Uranus and Neptune,[25] are larger than the primordial and solar system D/H ratios. These giant planets contain less H and He than Jupiter and Saturn, which shows that they accumulated more solid organics and ices than solar nebula gas. The solids in the outer solar nebula were enriched in D/H, as seen in the D/H measurements from comets,[26] which would explain the larger D/H in Uranus and Neptune. Different meteorite groups and their components[27] are also enriched in D/H. The quite variable D/H ratios in the organic and hydrous phases in meteorites may in part reflect the enrichments in D and/or fractionations in D/H by ion–molecule processes that operated in the solar nebula. Back-reactions of D rich phases with relatively D-poor gas also must be considered to explain the range in meteoritic D/H ratios. The ion–molecule processes that can have occurred in the outer solar nebula are similar to those still operating in the interstellar medium, and especially to those in cold molecular clouds where quite large D/H ratios are observed. The meteorite data give weight to the notion that a lot of D in the interstellar medium today is hidden in dust grains because the ion–molecule reactions favor incorporation of D into solids. In that case, the D/H from gas phase measurements in the ISM are only lower limits to the "true" D/H ratio in the ISM.

The relatively high D/H ratios in water in the atmospheres of Venus ($\sim 2.2 \times 10^{-2}$) and Mars (8.1×10^{-4}), and in the terrestrial oceans (1.56×10^{-4}), cannot be representative for the original D/H ratio of the solar system. Whatever the original D/H in the terrestrial planets may have been, their currently observed D/H ratios are very likely higher than the original values because hydrodynamic escape of H is favored over D escape from all terrestrial planets.

The present-day abundance of *lithium* in the photosphere is ~ 150-times less than that in chondrites. Within the relatively large uncertainties associated with their abundances, the relative abundances of *beryllium* and *boron* are similar in the photosphere and CI-chondrites. The nuclides of the light elements Li, Be and B are relatively fragile like D, but they require 3–8-times higher fusion temperatures than D. The required temperatures for fusion with H are lowest for Li, and highest for B; about 2×10^6 K for ^{6}Li, 2.5×10^6 K for ^{7}Li, 3.5×10^6 K for ^{9}Be and about 5×10^6 K for ^{10}B and ^{11}B. Complete loss of D, Li, Be and B could only occur if the temperatures increased 3–8 times above the temperatures needed for D-burning and if the Sun remained fully convective to allow processing of Li, Be and B in the entire atmosphere at the necessary temperatures. However, the relative solar Li abundance is about 150 times less than in CI-chondrites, but the Be and B abundances are about the same within uncertainties. The reason why these elements are preserved in the photosphere is that the development of a radiative zone over the core prevented full convection of the outer envelope that starts right below the photosphere. The radiative zone formed as temperatures and densities increased in the contracting Sun. As the radiative zone over the core widened, the bottom of the overlying convective zone moved up towards to lower temperatures. As the convective envelope no longer reached down to temperatures necessary for fusion, the Li, Be and B remained in the convective zone. Thus, the photospheric abundances Li, Be and B are useful diagnostics for the depth of the Sun's convective layer. However, to use them as such, the original solar Li, Be and B abundances must be known, which brings us back to the meteoritic abundances from CI-chondrites as standards for the solar system abundances.

1.3   SOLAR SYSTEM ELEMENTAL ABUNDANCES

The average solar system elemental abundances are derived from the analytical data for the solar photospheric and CI-chondrites given in Table 1.2. For the elements that have similar abundances in both sources, one can take either the average or the datum with higher analytical precision. As described above, the abundances of the volatile elements H, C, N, O and noble gases are from solar analysis or other sources than meteorites whereas abundances of Li, Be, B and elements not accessible from the Sun are based on the meteoritic data.

In deriving the proto-solar abundances from the selected meteoritic and photospheric value, one other important aspect has to be considered. Models of the Sun's evolution and interior show that currently observed photospheric abundances relative to H must be lower than those of the proto-sun because He and other heavy elements have settled from the photosphere towards the Sun's interior since the Sun formed ~ 4.6 Ga ago. Therefore, the current photospheric abundances relative to H are not representative of the solar system and only the proto-solar (*i.e.*, un-fractionated with respect to hydrogen) abundances represent the "solar system elemental abundances."

The abundances of elements heavier than He apparently did not fractionate relative to each other because the relative abundances of many rock-forming elements in the photosphere are the same as in CI-chondrites (*i.e.*, the abundances relative to Si are the same). However, the heavy elements are fractionated relative to H and all element/H ratios decreased. Over the Sun's age, the photosphere "lost" $\sim 16\%$ of the heavy elements and the He abundance decreased by $\sim 18\%$ relative to H. This heavy element settling from the photosphere is taken into account when the abundances from Table 1.1 are used to derive the solar system abundances listed in Table 1.2. On the astronomical scale, the proto-solar abundances for elements heavier than He are 0.074 log-units higher than the photospheric values.

However, on the cosmochemical scale by number the solar system abundances of all elements, except for H and He, are the same as the photospheric abundances. This is the obvious outcome for element normalization to $Si = 10^6$ atoms. The proto-solar H abundance is only ~ 0.84 times that of the photosphere, while the respective He abundance is ~ 1.02 times photospheric. The proto-solar He abundance must appear slightly higher than unity on this scale because He settling from the outer layers of the sun was slightly more efficient than that of the heavy elements, including Si, which is used for normalization of the cosmochemical abundance scale. The difference in proto-solar and photospheric abundances thus expresses itself either by a higher heavy element (the astronomers' "metals") content of the proto-sun when the H-normalized astronomical abundance scale is used or by a relative depletion in hydrogen on the Si-normalized cosmochemical scale.

1.4 TRENDS IN SOLAR SYSTEM ELEMENTAL ABUNDANCES AND ORIGINS

1.4.1 Elemental Abundance Trends

The solar system elemental abundances from Table 1.2 as a function of atomic number are shown in Figure 1.8. This diagram with abundances of all stable elements plus Th and U makes an interesting comparison to Figure 1.3 with the limited data set of the element abundances available to Harkins.

Table 1.2 Elemental abundances in the proto-Sun (solar system abundances) (from ref. 18).[a]

	$A(El)_o$	$N(El)_o$		$A(El)_o$	$N(El)_o$		$A(El)_o$	$N(El)_o$
H	≡12	2.59×10^{10}	Ge	3.65±0.06	115	Sm	1.01±0.02	0.265
He	10.986±0.02	2.51×10^{9}	As	2.37±0.04	6.10	*Sm*	*1.01±0.02*	*0.267*
Li	3.33±0.05	55.6	Se	3.42±0.03	67.5	Eu	0.58±0.04	0.0984
Be	1.37±0.03	0.612	Br	2.62±0.06	10.7	Gd	1.14±0.06	0.360
B	2.86±0.04	18.8	Kr	3.33±0.08	55.8	Tb	0.39±0.03	0.0634
C	8.44±0.04	7.19×10^{6}	Rb	2.44±0.06	7.10	Dy	1.19±0.06	0.404
N	7.91±0.12	2.12×10^{6}	*Rb*	*2.45±0.03*	*7.23*	Ho	0.55±0.03	0.0910
O	8.78±0.07	1.57×10^{7}	Sr	2.96±0.03	23.4	Er	1.00±0.06	0.262
F	4.49±0.06	804	*Sr*	*2.95±0.03*	*23.3*	Tm	0.19±0.03	0.0406
Ne	8.10±0.10	3.29×10^{6}	Y	2.25±0.04	4.63	Yb	0.99±0.03	0.256
Na	6.35±0.04	5.77×10^{4}	Zr	2.62±0.04	10.8	Lu	0.17±0.02	0.0380
Mg	7.60±0.06	1.03×10^{6}	Nb	1.48±0.07	0.780	*Lu*	*0.17±0.02*	*0.0380*
Al	6.51±0.07	8.460×10^{4}	Mo	1.99±0.06	2.55	Hf	0.78±0.02	0.156
Si	7.59±0.01	≡1.00×10^{6}	Ru	1.84±0.03	1.78	*Hf*	*0.78±0.02*	*0.156*
P	5.51±0.05	8300	Rh	1.15±0.13	0.370	Ta	−0.09±0.04	0.0210
S	7.21±0.02	4.21×10^{5}	Pd	1.72±0.04	1.36	W	0.72±0.04	0.137
Cl	5.30±0.06	5170	Ag	1.28±0.02	0.489	Re	0.33±0.04	0.0554
Ar	6.55±0.10	9.27×10^{4}	Cd	1.78±0.03	1.57	*Re*	*0.35±0.04*	*0.0581*
K	5.16±0.04	3760	In	0.84±0.03	0.178	Os	1.4w±0.03	0.680
K	*5.16±0.04*	*3760*	Sn	2.14±0.06	3.60	*Os*	*1.42±0.03*	*0.678*
Ca	6.37±0.02	6.04×10^{4}	Sb	1.08±0.06	0.313	Ir	1.41±0.06	0.672
Sc	3.12±0.02	34.4	Te	2.26±0.03	4.69	Pt	1.69±0.03	1.27
Ti	4.98±0.03	2470	I	1.63±0.08	1.10	Au	0.88±0.04	0.195
V	4.04±0.03	286	Xe	2.32±0.08	5.46	Hg	1.25±0.08	0.458
Cr	5.70±0.02	1.31×10^{4}	Cs	1.16±0.02	0.371	Tl	0.85±0.03	0.182
Mn	5.55±0.01	9220	Ba	2.24±0.07	4.47	Pb	2.11±0.03	3.33
Fe	7.51±0.08	8.480×10^{5}	La	1.25±0.02	0.457	*Pb*	*2.11±0.03*	*3.31*
Co	4.96±0.08	2350	Ce	1.66±0.08	1.18	Bi	0.73±0.04	0.138
Ni	6.28±0.04	4.90×10^{4}	Pr	0.82±0.05	0.172	Th	0.13±0.03	0.0351
Cu	4.32±0.04	541	Nd	1.52±0.02	0.857	*Th*	*0.23±0.03*	*0.0440*
Zn	4.70±0.04	1300	*Nd*	*1.52±0.02*	*0.856*	U	−0.46±0.03	8.93×10^{-3}
Ga	3.15±0.02	36.6				*U*	*−0.04±0.03*	*23.8×10⁻³*

[a]Note values in italics refer to abundances 4.57×10^{9} years ago.

The abundance trends already noticed by Harkins are clearly visible in Figure 1.8: hydrogen and He dominate and overall, abundances broadly decrease with atomic number. Major exceptions from this trend are the low abundances of Li, Be and B, and the higher abundances around Fe. Elements with even atomic numbers are more abundant than their odd-numbered neighbors. More detailed trends in the elemental abundance distributions are revealed when the abundances of the even and odd-numbered nuclides of the elements are distinguished and plotted by mass number instead of atomic number (Figure 1.9).

Figures 1.8 and 1.9 show the slight increases in abundances in regions around Ge-Sr; Xe-Ba, the rare earth elements, Os-Pt, and Pb. These abundance peaks are consequences of the larger stabilities of the iso-topes of these elements as they often contain so-called magic numbers of

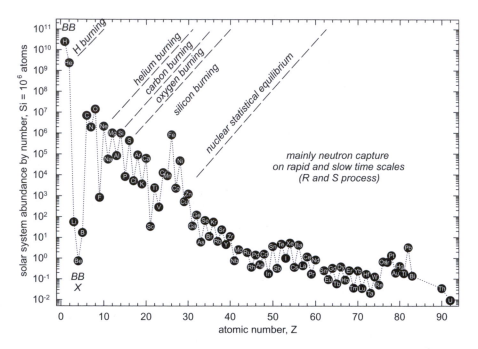

Figure 1.8　Abundances in the solar system as a function of atomic number. Note that abundances are plotted on a logarithmic scale in contrast to the linear scale used in Figure 1.3.

protons and/or neutrons. The magic numbers (2, 8, 20, 28, 50, 82 and 126) correspond to closed nuclear "shells," comparable to the closed electron shell configuration of the noble gases.

The isotopic composition of the elements and a breakdown of the solar system elemental abundances (from Table 1.2) into the isotopic abundances are given in Appendix A. Here we cannot discuss the nuclide abundance distributions as a function with mass number (instead of the elemental abundances as a function of atomic number) in any detail and only point out a few issues about the isotopes that may help to explain the elemental and nuclide abundance distributions. A survey of the 266 stable nuclides of the elements (out of 280 naturally occurring stable and long-lived nuclides in the solar system) shows the following frequency of nuclides with proton number (= atomic number Z) and neutron number N:

Z even, N even: 159 nuclides
Z even, N odd: 53 nuclides
Z odd, N even: 50 nuclides
Z odd, N odd: 4 nuclides: ^2H, ^6Li, ^{10}B, ^{14}N

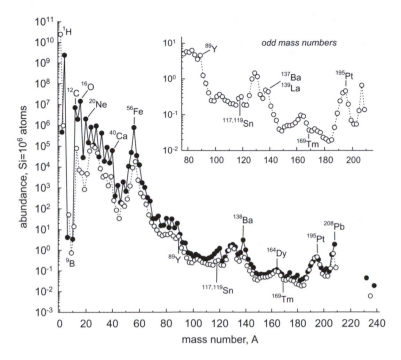

Figure 1.9 Abundances of nuclides as function of mass number. Even numbered nuclides are plotted as filled circles, odd numbered ones by open circles. A few nuclides are labeled for orientation.

The lower number of stable nuclides available among the odd-numbered elements contributes to the fact that these elements are less abundant than the even numbered elements. Among the odd-Z elements, 22 have one stable isotope. Two of them, V and Ta, have an additional very long-lived radioactive isotope. Several of these 22 elements have particularly low abundances, as can be seen from Figures 1.8 and 1.9. The 20 solely mono-isotopic elements are Be, F, Na, Al, P, Sc, Mn, Co, As, Y, Nb, Rh, I, Cs, Pr, Tb, Ho, Tm, Au and Bi. The elements V and Ta each have abundance contributions from a long-lived second isotope (^{50}V with a half-life of 1.4×10^{17} years, and ^{180}Ta with $> 1.2 \times 10^{15}$ years), which for practical purposes can be regarded as stable.

Some of the longer-lived naturally occurring radioactive nuclides (see Chapter 3), for example, ^{40}K (Table 1.3), have odd proton and neutron numbers, most of them have odd Z and even N, but none of these long-lived radionuclides are combinations of even numbers of protons and even numbers of neutrons.

Conversely, stable nuclides with even proton and neutron numbers are favored in frequency and abundance. Elements with even atomic numbers

Table 1.3 Naturally occurring long-lived radionuclides in the solar system (in order of increasing half-life).

Nuclide	Half-life (years)	Decay mechanism[a]	Stable decay product(s)
^{235}U	7.04×10^8	α	^{207}Pb
^{40}K	1.27×10^9	β^-, β^+	^{40}Ca, ^{40}Ar
^{238}U	4.47×10^9	α	^{206}Pb
^{232}Th	1.40×10^{10}	α	^{208}Pb
^{176}Lu	3.78×10^{10}	β^-	^{176}Hf
^{187}Re	4.22×10^{10}	β^-	^{187}Os
^{87}Rb	4.88×10^{10}	β^-	^{87}Sr
^{138}La	1.03×10^{11}	ec, β^-	^{138}Ba, ^{138}Ce
^{147}Sm	1.06×10^{11}	α	^{143}Nd
^{190}Pt	4.50×10^{11}	α	^{186}Os
^{123}Te	1.24×10^{13}	ec	^{123}Sb
^{152}Gd	1.1×10^{14}	α	^{148}Sm
^{115}In	4.4×10^{14}	β^-	^{115}Sn
^{186}Os	2.0×10^{15}	α	^{182}W
^{180}Ta	$> 1.2 \times 10^{15}$	ec, β^+	^{180}Hf
^{174}Hf	2.0×10^{15}	α	^{170}Yb
^{144}Nd	2.1×10^{15}	α	^{140}Ce
^{148}Sm	7×10^{15}	α	^{144}Nd
^{113}Cd	9×10^{15}	β^-	^{113}In
^{50}V	1.4×10^{17}	ec, β^-	^{50}Ti, ^{50}Cr

[a]Decay mechanisms: α = alpha particle (He^{2+}) emission; β^- = electron emission; β^+ = positron emission; ec = electron capture.

typically have more than one isotope. The extreme case is Sn with ten stable isotopes (Sn also has a magic number of $Z = 50$ protons). The occurrence of several isotopes for even-Z elements contributes to the larger abundances of the even numbered elements seen in Figures 1.8 and 1.9.

These simple comparisons show that nuclear properties influence the abundances of the elements. Detailed studies of the abundances of the nuclides as a function of mass number have led to the establishment of detailed rules and explanations for the regularities in the abundance distributions of the stable elements.[9,28,29]

The issue here is that the nuclear make-up and the stabilities of the nuclides, as well as timing of solar system formation, controlled the overall abundances of the elements in the solar system. The overall abundances are not controlled by the elements' electron shell structures, which determine their chemical behavior and position in the periodic table.

The reason for the nuclear control of the elemental abundances is that nuclear properties ultimately determine the yields during stellar element synthesis. The principal nucleosynthesis origins of the elements are indicated in Figure 1.8 by the dashed lines. Briefly, for the lightest elements, "BB" indicates primordial production through the big-bang and "X" indicates production through spallation reactions (see discussion for D, Li, Be and B above). Hydrogen fusion to He through the proton–proton chain and CNO cycle is the main occupation of stars for most of their lives. (About 10 billion years for the Sun, which has converted about 0.3% of its H into He in the past $\sim$4.6 billion years.) In low and intermediate mass stars (up to $\sim$8 solar masses), the fusion of three ^{4}He yields ^{12}C, and reaction of ^{12}C with He can produce elements up to Ne in later stages of stellar evolution. The fusion of 3 He with atomic number 2 to ^{12}C with atomic number 6 circumvents the production of the elements with atomic numbers 3–5 (Li, Be and B), which is the reason why these elements are so low in abundance. In contrast, elements with isotopes that have proton and neutron numbers equal to multiples of the ^{4}He nucleus (^{12}C, ^{16}O, ^{20}Ne, ^{24}Mg, ^{28}Si, ^{32}S, ^{36}Ar, ^{40}Ca) are particularly abundant (Figure 1.9), a fact that was already noticed in 1914 by the Italian chemist Guiseppe Oddo (1865–1954). Several of these nuclides have "doubly magic" compositions (^{16}O with $Z = N = 8$ and ^{40}Ca with $Z = N = 20$). The production of elements beyond C up to Fe through successive C, O and Si burning requires stars with more than 8 solar masses. The reactions leading to the Fe-peak elements involve disintegration of Si and other lighter nuclei to alpha particles and re-assembly of the Fe peak nuclei in a steady state setting of nuclear reactions (nuclear statistical equilibrium).

The *net* fusion reactions of lighter elements to heavy elements up to Fe are exothermic. This is because the average binding energies of the protons and neutrons in nuclei up to mass numbers 56 (isotopes of Fe, Ni) increases. The most abundant nucleus in the iron peak region produced by supernova nucleosynthesis is actually ^{56}Ni with equal "magic" numbers of 28 neutrons and 28 protons. However, ^{56}Ni is unstable with a half-life is 6.1 days and decays to ^{56}Fe, which is why Fe is much more abundant than Ni. The reason why the doubly magic ^{56}Ni is not a stable nucleus can be understood from the nuclear shell model. Strong spin–orbit coupling occurs and destabilizes nuclei with doubly magic numbers of 28 and above, which is not the case for nuclei with the lower doubly magic numbers.

Beyond the region of the Fe group elements, nuclear binding energies decrease with increasing mass number and fusion reactions of the "lighter" elements such as Fe or Ni to form heavier ones up to Th and U require energy. The stability of the Fe-group nuclei puts a natural block on fusion yields from lighter elements, and this can explain why the Fe and Ni abundances spike above the overall abundance trends given by other elements or nuclides (Figures 1.8 and 1.9).

The elements beyond the iron group are mainly built through neutron capture by pre-existing, lighter elements (mainly Fe itself). Figure 1.10 shows the relative contributions from different processes to each element beyond Fe. There are two main neutron capture processes, called S and R processes.[11,12] One occurs on a slow timescale (hence S process) so that unstable nuclides produced through neutron capture may undergo beta-decay to a stable nuclide before another neutron can be captured. The S process operates in low and intermediate mass stars that have reached the giant star stage. It was in these types of stars where Merrill found Tc, which is naturally made by the S process. The S process also operates in massive stars that eventually blow up as supernovae, but the products and relative yields are different (Figure 1.10). The S process in massive stars is called the "weak S" process[30] to distinguish it from the "main S" process in giant stars that produces a much wider mass range of the heavy nuclides that contribute to the elements.[31]

During the R process, nuclides can capture neutrons rapidly (hence R process) before they undergo beta decay. This builds up unstable, neutron rich nuclides. After neutron exposure stops, these nuclides go through a chain of beta decays until a heavy stable or long-lived nucleus is produced. Thorium and U can only be produced this way. Larger contributions of isotopes by the R process occur for elements such as

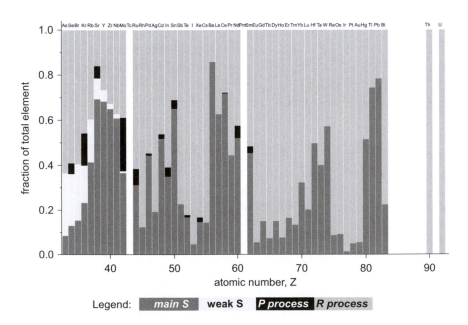

Figure 1.10   Contributions from different nucleosynthesis processes to the abundances of the elements beyond the Fe-group elements.

Te–Xe, Ir–Pt and the heavy rare earth elements (Figure 1.10). The R process requires high neutron densities, which may be realized in supernovae, but the exact site where the R process takes place is still under debate.

In addition to the S and R processes that produce neutron rich isotopes, a so-called- P-process is needed to explain the production of relatively proton-rich isotopes that contribute smaller fractions to a few elements (Figure 1.10). These proton-rich nuclides cannot be obtained through neutron capture processes and subsequent radioactive decay. One possibility is photodisintegration and "boil-off" of neutrons from more neutron-rich isotopes. The P process is believed to operate in supernovae, but appears to be the least understood process for heavy element production.

Since the different nucleosynthesis processes operate in different types of stars, it follows that the element inventory of the solar system is the result of nucleosynthesis in many different types of stars from different stellar generations. On a Galactic scale, the composition of the solar system is a snapshot of the composition at our location in the Galaxy 4.6 billion years ago. More importantly, it is the composition that, once concentrated to the solar system, through chemical and physical processing gave rise to the diversity of planetary objects that we have in our solar system today.

FURTHER READING

D. Arnett, *Supernovae and Nucleosynthesis*, Princeton University Press, Princeton, 1996, 598 pp.

B. E. J. Pagel, *Nucleosynthesis and Chemical Evolution of Galaxies*, Cambridge University Press, Cambridge, 1997, 378 pp.

N. Prantzos, Origin and evolution of the light nuclides, *Space Sci. Rev.*, 2007, **130**, 27.

G. Wallerstein, Synthesis of the elements in stars: forty years of progress, *Rev. Mod. Phys.*, 1997, **69**, 995.

S. E. Woosley, A. Heger, and T. A. Weaver, The evolution and explosion of massive stars, *Rev. Mod. Phys.*, 2002, **74**, 1015.

REFERENCES

1. Élie de Beaumont, Note sur les emanations volcaniques et metalliferes, *Bull. Soc. Geol. Fr., IIe*, 1847, Ser., 4, 1249.

2. D. I. Mendeleev, On the relationship of the properties of the elements to their atomic weights, *Zh. Russ. Fiziko-Khim. Obshch.* (St. Petersburg), 1869, **1**, 60; German abstract in *Z. Chem.*, 1869, **12**, 405.
3. P. W. Merrill, *Astrophys. J.*, 1952, **116**, 21.
4. M. F. Aller and C. R. Cowley, *Astrophys. J.*, 1970, **L145**, 1970; V. Fivet, P. Quinet, É. Biémont, A. Jorissen and A. V. Yushchenko, *Monthly Notices R. Astron. Soc.*, 2007, **380**, 771.
5. H. A. Bethe, *Rev. Mod. Phys.*, 1936, **8**, 82.
6. C. F. von Weizsäcker, *Phys. Z.*, 1937, **38**, 176; 1938, **39**, 33.
7. H. N. Russell, *Astrophys. J.*, 1929, **70**, 11.
8. V. M. Goldschmidt, *Geochemische Verteilungsgesetzte der Elemente. IX. Die Mengenverhältnisse der Elemente und der Atom-Arten*, Skrifter Norske Videnskaps Akademi I. Mathematisk-Naturvidenskapelig klasse, No. 4, Oslo 1937/1938, 148; V. M. Goldschmidt, *J. Chem. Soc.*, 1937, 655–673.
9. H. E. Suess and H. C. Urey, *Rev. Mod. Phys.*, 1956, **28**, 53.
10. R. A. Alpher, H. A. Bethe and G. Gamow, *Phys. Rev.*, 1948, **73**, 803; D. ter Haar, *Science*, 1949, **109**, 81; E. E. Salpeter, *Phys. Rev.*, 1952, **88**, 547.
11. E. M. Burbidge, G. R. Burbidge, W. A. Fowler and F. Hoyle, *Rev. Mod. Phys.*, 1957, **29**, 547.
12. A. G. W. Cameron, *Chalk River Report, CRL-41*, 1957.
13. I. A. Kleiber, *Zh. Russ. Fiziko-Khim. Obshch.* (St. Petersburg), 1885, **17**, 147–171.
14. F. W. Clarke, *Bull. Phil. Soc. Washington*, 1889, **11**, 131.
15. O. C. Farrington, *Meteorites*, Washington (published by the author), 1915 203 pp.
16. W. D. Harkins, *J. Am. Chem. Soc.*, 1917, **39**, 856.
17. G. P. Merrill, *Am. J. Sci.*, 1909, **27**, 469.
18. K. Lodders, *Astrophys. J.*, 2003, **591**, 1247; K. Lodders, H. Palme and H. Gail, in *Landolt-Börnstein, New Series, Section IV, Astronomy & Astrophysics*, Springer Verlag, Berlin, 2009, 560–630.
19. D. J. Mullan and J. L. Linsky, *Astrophys. J.*, 1998, **511**, 502.
20. D. N. Spergel, L. Verde, H. V. Peiris, E. Komatsu, M. R. Nolta, C. L. Bennett, M. Halpern, G. Hinshaw, N. Jarosik, A. Kogut, M. Limon, S. S. Meyer, L. Page, G. S. Tucker, J. L. Weiland, E. Wollack and E. L. Wright, *Astrophys. J.*, 2003, **148**(Suppl), 17.
21. G. Steigman, D. Romano and M. Tosi, *Monthly Notices R. Astron. Soc.*, 2007, **378**, 576.
22. K. H. Sembach, B. P. Wakker and T. M. Tripp *et al.*, *Astrophys. J.*, 2004, **150**(Suppl), 387.

23. B. D. Savage, N. Lehner, A. Fox, B. Wakker and K. Sembach, *Astrophys. J.*, 2007, **659**, 1222.

24. J. L. Linsky, B. T. Draine, H. W. Moos, E. B. Jenkins, B. E. Wood, C. Oliveira, W. P. Blair, S. D. Friedman, C. Gry, D. Knauth, J. W. Kruk, S. Lacour, N. Lehner, S. Redfield, J. M. Shull, G. Sonneborn and G. M. Williger, *Astrophys. J.*, 2006, **647**, 1106.

25. T. Fouchet, B. Bézard and T. Encrenaz, *Space Sci. Rev.*, 2005, **119**, 123.

26. W. F. Huebner, *Earth, Moon, Planets*, 2002, **89**, 179; W. F. Huebner, *Physics and Chemistry of Comets*, Springer, New York, 2000, 376 pp.

27. F. Robert, *Space Sci. Rev.*, 2003, **106**, 87.

28. J. H. D. Jensen and H. E. Suess, *Naturwissenschaften*, 1944, **32**, 374; R. A. Alpher and R. C. Herman, *Rev. Mod. Phys.*, 1950, **22**, 153.

29. E. Anders and N. Grevesse, *Geochim. Cosmochim. Acta*, 1989, **37**, 1435; H. Palme and H. Beer, in *Landolt Bornstein Group VI, Astronomy and Astrophysics*, ed. H. H. Voigt, Vol. 2A, Springer, Berlin, 1993, p. 196.

30. C. M. Raiteri, R. Gallino, M. Busso, D. Neuberger and F. Käppeler, *Astrophys. J.*, 1993, **419**, 207; C. Travaglio, R. Gallino, E. Arnone, J. Cowan, F. Jordan and C. Sneden, *Astrophys. J.*, 2004, **601**, 864.

31. C. Arlandini, F. Käppeler, K. Wisshak, R. Gallino, M. Lugaro, M. Busso and O. Straniero, *Astrophys. J.*, 1999, **525**, 886.

CHAPTER 2

Meteorites

" . . . the study of meteorites has an importance far beyond that
which they [others] have hitherto attributed to it; that it is of importance alike to the physicist, astronomer and philosopher; that
without it no rational conception of the constitution of this universe is possible and that even now no progressive geologist, mineralogist, chemist or teacher of natural history – in short no one
who pretends to a scientific education – can afford to ignore it."

H. Hensoldt, 1889

2.1 INTRODUCTION

Meteorites fall everywhere on Earth, and did so in the past. They inspired legends, myths, superstition and explanations of divine intervention. The Greeks, Romans, Japanese and North American Native
Indians worshipped and kept the stones that fell from heaven on altars.
In other cultures, meteorites were pounded to powder as this was the
only way to deal with these missiles that must have come from an of-
fended deity. In medieval Europe, falling stones were occasionally recognized as special and were kept chained to church ceilings (at least in
one instance, a church collapsed as a result). With the advent of modern
science, stories about rocks falling from the sky were looked upon as
superstitious beliefs. Around the end of the eighteenth century, many
scientific authorities regarded it as impossible that rocks could be falling
out of the sky. Sworn and certified reports of such events were ridiculed

Chemistry of the Solar System
By Katharina Lodders and Bruce Fegley, Jr.
© K. Lodders and B. Fegley, Jr. 2011
Published by the Royal Society of Chemistry, www.rsc.org

and rejected as humbug. At the height of disbelief, many meteorites were removed from museum collections and ended up as vandalized fatalities of enlightenment.

However, the overwhelming evidence for the fact that rocks *do* fall out of the sky did not escape the trained eye of a lawyer. By 1794, E. F. Chaldni[1] had found enough evidence in the library records to claim the extraterrestrial origin of meteorites, even though he had never witnessed such a *corpus delicti* to fall. Timely and well-recorded meteorite falls subsequently helped in the acceptance of meteorites as extraterrestrial objects. By 1933, F.C. Leonard pushed for the establishment of a scientific meteorite research association (which is now known as the Meteoritical Society). He remarked that the significance of meteorites for scientific research is that they are the "only tangible objects of astronomical inquiry." This remained true until samples were returned from the Moon in 1969. Compared to the ongoing history of terrestrial rocks, most meteoritic rocks have brief formation histories restricted to the first 100 million years or less of the solar system. However, the compositions of meteorites are complex and not easy to understand. These space rocks are to space scientists what ruins and ancient scrolls are to archeologists: the only records that tell the stories of the distant past; and with meteorites it happens to be the stories about the processes in the early solar system.

The term "meteorite" itself reflects the long-standing confusion about the origin of these stones. The word "meteorite" borrowed from the Greek means "originating in the air," which is no better word than aërolite ("air-stone") considering that meteorites come from space. The word "thunderstone" is occasionally used in the old literature and is somewhat better fitting as it correctly relates to the sound often accompanying meteorite falls, but as thunder is usually associated with lightning this term is also not suitable. The term "asterolithology" suggested by Edward Howard, who did the first chemical meteorite analysis in the early 1800s, literally means "rocks of the stars." This is as misleading as "asteroid" for the small objects orbiting the Sun between Mars and Jupiter, which are definitively not akin to the stars. In any case, Howard's suggestion never caught on. Currently, some derivative of the word "meteorite" is used in most western languages to denote the stones that fell out of the sky and landed on the Earth's surface.

There is also the technical distinction between meteorite, meteoroid and meteor. A meteoroid is a smaller, solid natural object (in contrast to rocket or space-craft debris) that travels in space and eventually enters the Earth's atmosphere. If all of it or parts of it survive atmospheric passage and lands, it is a meteorite. In many cases, a falling meteoroid was first seen as a brilliant fireball in the sky, but not every fireball yields a meteorite. If a small objects burns up in the atmosphere it may emit an incandescent light trail to be seen as a meteor ("falling star").

2.2   METEORITE CLASSIFICATION

Meteorites are divided by their principal components into stones, stony-irons and irons. Within these three groups, sub-groups are recognized based on chemical and mineralogical composition (Table 2.1).

The first group in Table 2.1 is the iron meteorites, which are made primarily of Fe-Ni metal alloys. The second major meteorite group in Table 2.1 are the stony-irons (pallasites and mesosiderites), containing about equal amounts of metal and silicates. The third group is the stony meteorites, whose silicate minerals make up more than 50% of the meteorite. Stony meteorites are subdivided into two major groups. One group is the chondrites, which mainly contain silicates plus significant percentages of metal and sulfides. The major chondrite subgroups are ordinary, enstatite and carbonaceous chondrites and are designated by a letter system. Chondrites are the most abundant kinds of meteorites and supply most of our information about chemical processes in the early solar system. The achondrites, dominated by silicate minerals, make up the other group of stony meteorites. The SNC and EHD achondrites resemble terrestrial rocks such as basalts, and these achondrites belong to the "planetary" achondrites, as do lunar meteorites. In contrast, the "primitive achondrites" such as acapulcoites (plus lodranites), angrites, aubrites, brachinites, ureilites and winonaites more closely resemble chondritic meteorites that have experienced larger degrees of melting.

In meteoritics, there is the relatively simple (but not necessarily correct) first-order distinction that chondrites are undifferentiated and more primitive than the differentiated achondrites, stony irons and iron meteorites. Chondrites contain an assortment of smaller grains composed mainly of silicates, metal and sulfides. The idea is that these minerals are the closest remains of the original minerals that were dispersed in the solar nebula – the gas cloud surrounding the young, growing Sun – from which larger objects formed. The minerals in the solar nebula accreted to small planet-like objects ("planetesimals"). Some of these planetesimals still exist as asteroids, and are the parent bodies from which most meteorites originate. The material that accreted to the planetesimals of ordinary and enstatite chondrites was compacted, and in several cases mineralogically and chemically altered by mild heating (thermal metamorphism). On other parent bodies, notably those of the carbonaceous chondrites, aqueous alteration of the originally accreted mineral

Table 2.1 Classification of meteorites.

Irons (siderites) Mainly Fe-Ni metal	Stony-irons (siderolites) Metal : silicates ~ 50 : 50	Stones (aërolites) Silicates > 50%		
		Achondrites		Chondrites
		"Planetary"	"Primitive"	
Octahedrites (two FeNi alloys)	Mesosiderites (fine FeNi grains)	EHD group: *Eucrites* *Howardites* *Diogenites*	Acapulcoites + lodranites Angrites Aubrites Brachinites Ureilites Winonaites	Ordinary: H (*Bronzite*) L (*Hypersthene*) LL (*Amphoterite*)
Hexahedrites (low Ni)	Pallasites (coarse FeNi network)	SNC group: *Shergottites* *Nakhlites* *Chassignites*		Carbonaceous: CI, CM, CO, CV, CK, CR, CH, CB
Ataxites (high Ni)		Lunar		Enstatite: EH, EL Rumuruti group R Kakangari group K

assortment took place. Although thermal metamorphism or aqueous alteration affected the originally accreted mineralogy in most chondrites, these processes did not significantly redistribute the chemical elements. Thus, undifferentiated meteorite parent bodies remained more or less homogeneous throughout in chemical composition. The least altered chondrites are also referred to as "primitive chondrites."

On the other hand, the occurrence of compact masses of metal such as the iron meteorites or igneous silicates such as planetary achondrites suggests material processing and density separation. The rocks of the differentiated meteorites are thought to result from substantial melting, re-crystallization and/or density separation of silicates, sulfides and metal. If finer grained silicates, sulfides and metal assembled into a larger planetesimal that was heated enough to produce silicate and/or metal (+ sulfide) melts, the metal could pool and gravitationally separate from the less dense silicates. In other words, some planetesimals differentiated into a metallic core and a silicate layer, analogous to what happened on Earth on a much larger scale. The chemical and mineralogical composition of differentiated planetesimals is heterogeneous, as the chemical elements redistribute between silicates and metal according to their geochemical affinities. If such planetesimals are broken up, silicate fragments provide differentiated planetary achondrites and core fragments yield certain types of iron meteorites. Neither a given achondrite nor iron meteorite group can then be representative of their parent planetesimals' bulk compositions. However, the overall composition of the planetesimal can be reconstructed from the observed compositions of achondrites in a similar manner as the rocks from the Earth's silicate mantle can be used to derive the entire composition of the Earth (Chapter 4).

The notion that iron meteorites are samples of planetesimal cores may have a catch: there are ~ 17 different recognized iron meteorite groups representing up to 17 different "planetary cores" so one would expect an equal number of different achondrites representing the silicate portions of such shattered planetesimals. However, only three known groups of planetary achondrites are known, and two of these achondrite groups are meteorites from the moon and from Mars, which are definitely not shattered down to their cores.

This distinction of differentiated and undifferentiated meteorites implies that "differentiated" meteorites formed from material resembling that of undifferentiated meteorites. Whether this is truly so for all differentiated meteorites has not yet been proven. In the earlier days of meteoritics, it was also thought that various meteorite groups might have genetic relationships, such that, for example, some chondrite groups may have evolved from others through mass exchange of metal and silicate and/or changes in oxidation state imposed by the gas in the solar nebula. However, it became clear that meteorite groups cannot easily be related through simple physical fractionations and chemical exchange reactions once more chemical and mineralogical analyses of different meteorite groups became available. The complete fallacy of this view came when the O-isotopic systematics of the different meteorite groups showed that the major chondrite groups are not related, and that most of the differentiated meteorites as well as planets could not have been derived by processing of one kind of undifferentiated material that is represented by the known chondrites groups.

2.2.1 Oxygen Isotopes

In the early 1970s, Robert N. Clayton and co-workers discovered that the three stable isotopes of oxygen occur in variable proportions in the different meteorite groups and certain meteorite components.[2] In addition to the implications of these variations with respect to the previously assumed homogeneity of the materials in the solar system, measurements of the three O-isotopes are most useful to test genetic relationships among the meteorite groups. We introduce the O-isotopes in this section because they are a valuable tool for meteorite classification. Figures 2.1 and 2.2 show the variations observed in terrestrial, lunar (coinciding with terrestrial rocks) and meteoritic materials.

Oxygen is ubiquitous and is the most abundant element by number in planetary rocks and stony meteorites. Oxygen has three stable isotopes, ^{16}O, ^{17}O and ^{18}O, of which ^{16}O is the most abundant one on Earth. The atom percentages in standard mean ocean water (SMOW) are $^{16}O = 99.76\%$, $^{17}O = 0.04\%$ and $^{18}O = 0.20\%$. It is more practical to work with the atomic abundance ratios $^{17}O/^{16}O$ and $^{18}O/^{16}O$ of the O-isotopes instead of their fractional abundances.

Physical fractionations and chemical reactions involving O exchange lead to mass-dependent isotope fractionations in the O-isotope ratios. For example, during distillation of water, ^{16}O increases in the vapor relative to the liquid, which becomes richer in ^{17}O and ^{18}O. Exchange of O in reactions between water and silicate rocks enriches the heavier ^{17}O and ^{18}O in the silicate at the expense of the ^{17}O and ^{18}O in water, and water becomes isotopically lighter. Mass-dependent fractionations generally occur on per-mil levels because the O-isotopes are still of relatively low mass. The extent of these fractionations among different samples is revealed when the measured isotope ratios are compared to a standard composition such as SMOW. The deviation in per-mil of the $^{17}O/^{16}O$ in a sample from the $^{17}O/^{16}O$ in SMOW is given by:

$$\delta^{17}O(‰) = \left(\frac{(^{17}O/^{16}O)_{sample}}{(^{17}O/^{16}O)_{SMOW}} - 1 \right) \times 1000 \qquad (2.1)$$

Similarly, for the deviation of the $^{18}O/^{16}O$ from the standard ratio:

$$\delta^{18}O(‰) = \left(\frac{(^{18}O/^{16}O)_{sample}}{(^{18}O/^{16}O)_{SMOW}} - 1 \right) \times 1000 \qquad (2.2)$$

The mass difference between ^{18}O and ^{16}O is two atomic mass units whereas that between ^{17}O and ^{16}O is only one, and the mass-dependent fractionations in $^{17}O/^{16}O$ are approximately only about half those in $^{18}O/^{16}O$. Detailed discussions about O-isotope systematics are given by Clayton and co-workers,[2–4] Miller[5] and Thiemens.[6]

In terrestrial water and rock samples,[7,8] the chemical and physical fractionations in the $^{17}O/^{16}O$ ratios relative to $^{17}O/^{16}O$ of SMOW correlate with the $^{18}O/^{16}O$ ratios relative to $^{18}O/^{16}O$ of SMOW as:

$$\frac{(^{17}O/^{16}O)_{sample}}{(^{17}O/^{16}O)_{SMOW}} = \left(\frac{(^{18}O/^{16}O)_{sample}}{(^{18}O/^{16}O)_{SMOW}} \right)^{\lambda} = \alpha_{17/16} = \alpha_{18/16}^{\lambda} \qquad (2.3)$$

Often the notation α is introduced to avoid writing the detailed ratio terms. The mass-dependence of the fractionations enters through the exponent λ, which is proportional to the ratio of the differences of the inverse masses of the compounds containing the different O-isotopes:

$$\lambda = \frac{\dfrac{1}{m_{16}} - \dfrac{1}{m_{17}}}{\dfrac{1}{m_{16}} - \dfrac{1}{m_{18}}} \qquad (2.4)$$

For example, for O_2, in which one O atom is ^{16}O and the other is ^{16}O, ^{17}O or ^{18}O, λ is approximately $(1/32 - 1/33)/(1/32 - 1/34) = 0.515$. The degree of fractionation thus depends on the masses of molecules involved in the fractionation reactions; however, λ is typically around 0.50–0.53 for most exchange reactions of interest. For the water–quartz system, $\lambda = 0.52$ is adopted, and for reactions of terrestrial rocks and water[4,6] $\lambda = 0.5247$.

With the delta notation defined above, the fractionation relation becomes:

$$\frac{(^{17}O/^{16}O)_{sample}}{(^{17}O/^{16}O)_{SMOW}} = \left(\frac{(^{18}O/^{16}O)_{sample}}{(^{18}O/^{16}O)_{SMOW}}\right)^{\lambda} = \frac{\delta^{17}O}{1000} + 1 = \left(\frac{\delta^{18}O}{1000} + 1\right)^{\lambda} \quad (2.5)$$

Taking the natural logarithm gives:

$$\ln\left(\frac{\delta^{17}O}{1000} + 1\right) = \lambda \ln\left(\frac{\delta^{18}O}{1000} + 1\right) \quad (2.6)$$

The logarithmic expressions can be expanded into a series and truncated: $\ln(x+1) = x - \frac{1}{2}x^2 + \frac{1}{3}x^3 - \frac{1}{4}x^4 + - \cdots \approx x$ for $-1 < x \leq 1$, which applies as long as the fractionations are relatively small ($\delta^{17}O$ and $\delta^{18}O < 10‰$). This gives the widely known relation:

$$\delta^{17}O = \lambda\delta^{18}O \quad (2.7)$$

for mass-dependent O-isotope fractionation of terrestrial materials. In a diagram with $\delta^{18}O$ as the abscissa and $\delta^{17}O$ as the ordinate, terrestrial rock samples plot along a line with a slope of $\lambda \sim 0.52$ over a restricted range in $\delta^{17}O$ and $\delta^{18}O$. This slope for the "terrestrial fractionation line" is often used for reference when meteorite O-isotopic compositions are discussed. One should keep in mind that this "line" is actually a curve, and here we use the "terrestrial fractionation curve" from the full logarithmic expression with $\lambda = 0.5247$ from Miller:[5]

$$\delta^{17}O = \left[\left(\frac{\delta^{18}O}{1000} + 1\right)^{0.5247} - 1\right] \times 1000 \quad \text{(terrestrial fractionation curve)} \quad (2.8)$$

As mentioned above, this curve is defined by terrestrial rock and water samples; however, not all terrestrial samples plot along this curve. Stratospheric ozone is a notable exception. There are also rock samples that have exchanged oxygen with anomalous air samples.[6] These samples plot on a "line" with a slope of unity, which indicates that their isotopic fractionations are independent of the atomic masses of the isotopes involved. The causes for such mass-independent isotope fractionations are not completely understood.[6] Mass-independent fractionations in oxygen isotopes are observed in O_3, CO_2 and N_2O in the terrestrial atmosphere, and in certain meteorite samples, where this type of fractionation was discovered first.[2]

This brings us to the O-isotope systematics in meteorites. Contrary to what one may have expected for samples that have a common origin in the early solar system, the O-isotope compositions in meteorites do not follow that of terrestrial samples. Instead, they plot in distinct regions in the three-O-isotope diagram. Figure 2.1 shows characteristic averages for chondrites and the Earth and Moon. The terrestrial fractionation curve is shown for comparison. The enstatite chondrites (EH and EL) plot on or very close to the terrestrial fractionation curve. Ordinary chondrites (H, L and LL) and R chondrites (R) plot above, and the various groups of carbonaceous chondrites (except CI) and the K chondrites plot below the terrestrial fractionation curve.

There is yet another curve labeled "CAI mixing, CCAM," which has a slope of about unity. This curve (which is approximately linear for small variations in $\delta^{17}O$ and $\delta^{18}O$) is mainly defined by Ca-Al-rich refractory inclusions that are preferentially found in carbonaceous chondrites. These inclusions can show large relative ^{16}O enrichments, equivalent to small $\delta^{18}O$ and $\delta^{17}O$ values reaching down to about $-40‰$. Other

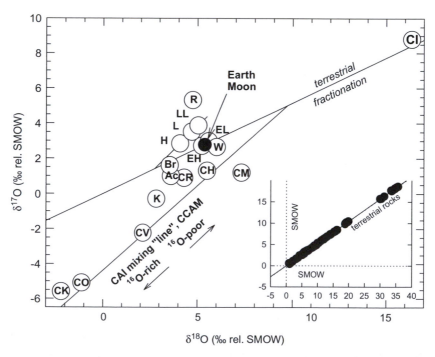

Figure 2.1   Oxygen isotope systematics in terrestrial samples, in chondrites (see text) and some achondrites (Ac = acapulcoites plus lodranites, Br = brachinites, W = winonaites); SMOW = standard mean ocean water. Data from refs 2–8 and references therein.

anhydrous components of carbonaceous chondrites also plot along this curve and therefore it is customary to refer to this curve as CCAM for "carbonaceous chondrite anhydrous minerals." Other "mixing curves" with slopes between 0.5 and 1 are defined by the minerals in CM and CO chondrites that are ascribed to mixing of "wet" and "dry" components.[4]

It was originally thought that the compositions of samples along this curve result from mixing of ^{16}O-poor and ^{16}O-rich components. However, the distinct slope of unity suggests that some still poorly understood, mass-independent process caused the correlation of the O-isotope ratios in these samples.[6]

Figure 2.2 shows the O-isotope systematics for achondrites. The position of the average compositions of chondrites groups, the terrestrial fractionation curve and the CCAM curve are shown for comparison. The O-isotopic compositions in ordinary chondrites define a line with a slope close to unity, which is also indicated here. The compositions of planetary achondrites follow curves that run parallel to the terrestrial fractionation curve, and, over small intervals, these curves define lines with slope of ~ 0.5 as seen in terrestrial rocks.

The SNC meteorites plot above, and the EHD meteorites (triangles) plot below, the terrestrial fractionation curve. The aubrites, or enstatite

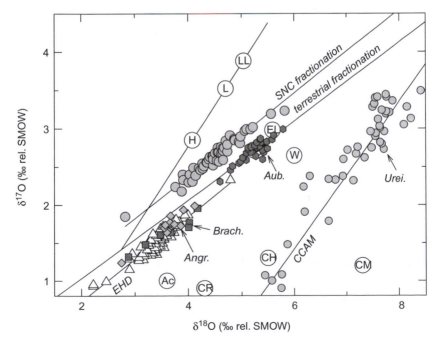

Figure 2.2 Oxygen isotope systematics in achondrites.

achondrites, are highly reduced and their O-isotopic compositions plot along the terrestrial fractionation curve like those of enstatite chondrites. The O-isotopic compositions of angrites coincide with those of the EHD meteorites and brachinites. The mesosiderites and the main group of pallasites also fall into this array, but these stony irons are not plotted for clarity. The ureilites are interesting because their O-isotopic compositions scatter around the CCAM curve with slope of about unity.

The offsets of the meteoritic curves from the terrestrial fractionation curves are expressed as $\Delta^{17}O$, which is characteristic for each achondrite group. The fractionation expression in Equation (2.3) has to include a term $(1 + k)$ to describe the O-isotopic fractionations in a system whose O-isotopic compositions do not fall onto the terrestrial fractionation curve but whose fractionations are measured relative to a terrestrial standard:[5]

$$\frac{(^{17}O/^{16}O)_{sample}}{(^{17}O/^{16}O)_{SMOW}} = (1 + k)\left(\frac{(^{18}O/^{16}O)_{sample}}{(^{18}O/^{16}O)_{SMOW}}\right)^{\lambda} \tag{2.9}$$

Here k is the parallel offset of the sample curve from the terrestrial fractionation curve. Introducing the frequently used δ-notation [Equations (2.1) and (2.2)] instead of the ratios, as done above, gives:

$$\frac{\delta^{17}O}{1000} + 1 = (1 + k)\left(\frac{\delta^{18}O}{1000} + 1\right)^{\lambda} \tag{2.10}$$

Taking the natural logarithm, re-arranging and multiplying by 1000 provides the formal mathematical definition of the offset from the terrestrial fractionation curve in per-mil units:

$$\begin{aligned}\Delta^{17}O &\equiv 1000 \ln(1 + k) \\ &= 1000 \ln\left(\frac{\delta^{17}O}{1000} + 1\right) - \lambda 1000 \ln\left(\frac{\delta^{18}O}{1000} + 1\right)\end{aligned} \tag{2.11}$$

$\times$ (offset from terrestrial fractionation curve)

If the deviations of $\delta^{17}O$ and $\delta^{18}O$ are small relative to the chosen standard, and, similarly, the offset k is small relative to the terrestrial fractionation curve ($k \ll 1$), the logarithms can again be approximated by a truncated series, to give:

$$\Delta^{17}O \cong \delta^{17}O - \lambda\delta^{18}O \tag{2.12}$$

Most of meteoritic and terrestrial data comparisons in the literature use $\lambda = 0.52$, which is sufficient for quick comparisons. However, detailed evaluations of data with a wide range in $\delta^{17}O$ and $\delta^{18}O$ should be carried out using the full logarithmic expression for $\Delta^{17}O$. A plot using $1000 \ln(\delta^{18}O/1000 + 1)$ as abscissa and $1000 \ln(\delta^{17}O/1000 + 1)$ as ordinate gives the offset from the terrestrial fractionation curve, $\Delta^{17}O$, from the intercept, and λ from the slope when a linear regression is performed. This procedure removes the uncertainties in selecting a λ value for comparison. The O-isotopic fractionations in the meteorite may have been governed by fractionation reactions other than those that relate terrestrial rocks and water. Then the value for λ derived from the slope may even shed some light on the type of fractionation since λ depends on the molecular weights of the species involved in isotope exchange as mentioned above.

2.2.2 Meteorite Falls and Finds

Meteorites are usually named after the locations where they are recovered. Most known meteorites were not observed during their fall but were found instead. The distinction between a meteorite "fall" and "find" is important because meteorites weather if they are unprotected from the eroding forces of rain, ice and wind. Meteorites are recognized by their blackened fusion crust and often larger densities and magnetic properties, which distinguishes them from many rocks usually encountered on the Earth's surface. Figure 2.3 shows a small stony meteorite with its fusion crust. Figure 2.4 shows two iron meteorites from the Sikhote-Alin meteorite shower. These pieces broke off from the main meteoroid during atmospheric passage, and their deformed shape and surface "thumbs" and "pits" are in part caused by heating and spattering of surface material during their fall. The shaping of these pieces must have happened before or while their surfaces were oxidized as both pieces are completely covered by a dark, shiny fusion crust. Figure 2.5 is a rare find – a monomict eucrite that still has spattered melt adhering to its fusion crust. The fusion crust on meteorites is typically up to a few mm thick, and the interiors of stony meteorites are not heated during the short flight through the atmosphere. The higher conductivity of metallic iron meteorites permits heating of ~ 1–2 cm below their surfaces where their characteristic structures (see below) may become erased. Iron in meteorites, either metallic or in ferrous silicates, oxidizes to magnetite, which makes up a large part of the fusion crust.

However, not every rock with a dark crust is a meteorite. For example, lumps of industrial slags from ore processing can easily mimic

Figure 2.3   A stone from the Gao-Guenie, Burkina Faso, meteorite shower in 1960
when ~4.3 kg fell as many smaller stones. The meteorite is surrounded by
a flaky, blackened fusion crust from passage though the atmosphere. A
small chip of the fusion crust is missing and reveals the lighter-grey interior of this H5 ordinary chondrite. (Photograph courtesy of R.
Korotev.)

the appearance and properties of iron meteorites. For the identification
of a potential meteorite several other tests, including a chemical analysis,
should be carried out.

The annual meteorite flux to Earth is about 10 metric tons per year,
but this is only a tiny fraction of the estimated 10 000–30 000 metric tons
of cosmic and interplanetary debris that enter the Earth's atmosphere
every year. Most of this is fine dust a few 100 μm in size. Over the past
two centuries, about 1000 meteorite falls have been recorded for which
recovered meteorite material still exists. Most of these falls were obviously observed in populated areas. About 2600 meteorite finds from
settled areas add to the meteorite collections. Since the 1970s, meteorite
hunting expeditions to arid regions increased the number of meteorites
to ~45 000. Most of these meteorites were retrieved in Antarctica by
Japanese and US expeditions since 1969, where the fusion encrusted
meteorites are easily recognizable on the ice. Other places where meteorites accumulated over time are the hot North African, Arabian and
Australian deserts from where large numbers of meteorites have been

Figure 2.4 Two pieces from the Sikhote-Alin iron meteorite shower. The meteorite on the right is *ca.* 6 cm long. A larger iron meteoroid broke up during atmospheric passage and the surfaces of the individual FeNi pieces oxidized to magnetite. The smooth surfaces are the result of melting. The spattering of melted regions caused the thumbprint-like surface depressions. Photograph by the authors.

Figure 2.5 An approximately 7 cm wide monomict eucrite, Dhofar 182, from the desert in Oman, covered with fusion crust. Melted fragments shaped during atmospheric passage still adhere to the fusion crust. (The photograph is from the electronic database of meteorites, MetBase, © 2005, Jörn Koblitz, Bremen, Germany.)

collected in the past decade. The only drawback for scientific work with meteorites from cold and hot deserts is that their terrestrial weathering can be severe, and contamination always poses a hazard for the interpretation of analytical results. For example, Antarctic meteorites are notoriously contaminated with halogens, and meteorites collected from hot deserts often show contamination by sulfates of alkaline earth elements, in particular Sr and Ba. However, such drawbacks cannot outweigh the value of these meteorites since many meteorites from rare groups have been discovered among them. Especially noteworthy is that all known lunar meteorites were recovered in Antarctica and in the deserts, as were the majority of the known SNC meteorites believed to be from Mars.

A comparison of the frequency of the different meteorite groups among observed falls and finds is interesting to see if there was any change in the meteorite influx population over time. Figure 2.6 shows the percent distribution of the different meteorite types among the classified falls and finds. The different meteorite groups are indicated by their abbreviations.

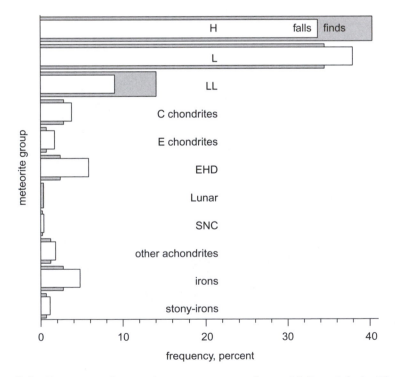

Figure 2.6   Frequency of meteorite groups among observed falls and finds. The falls are shown by white bars, and worldwide meteorite finds by grey bars. The finds include meteorites collected in deserts and in Antarctica.

The ordinary H, L and LL chondrites are the most abundant meteorites, hence the name "ordinary." They comprise $\sim 80\%$ of all falls and $\sim 90\%$ of all finds. In the H and LL chondrite groups, finds are more frequent than falls whereas among L chondrites, falls are more frequent than finds. There are no obvious reasons for this. The carbonaceous (C) and enstatite (E) chondrites, the EHD achondrites, iron meteorites, stony irons and several achondrite groups also have slightly higher percentages among the falls. Notable exceptions are the SNC and lunar meteorites, which are mainly known as finds as already mentioned.

Before the meteorite finds from Antarctica and the desert regions became available, a comparison of the statistics of meteorite falls and finds looked quite different. The frequency of iron meteorites among falls is $\sim 4\%$ (based on ~ 1000 falls); however, iron meteorites make up $\sim 26\%$ of the finds from more settled areas (based on 2600 finds). This distortion towards iron meteorites is primarily due to the fact that an unusually heavy (dense) and often rusty iron meteorite is easier to recognize than a stony meteorite. Taken at face value, the comparison of meteorite finds from more settled areas with that of falls may suggest that the iron meteorites, made almost entirely of Fe-Ni alloys, are more resistant to erosion than stony meteorites. However, such a conclusion is an urban myth as the statistics of Antarctic and desert finds shows. Figure 2.7 compares the frequency of iron meteorites per 100 ordinary

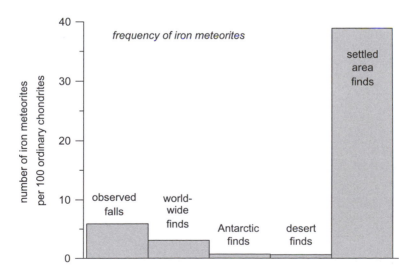

Figure 2.7   Frequency of iron meteorites per 100 ordinary chondrites among observed meteorite falls and finds. The desert finds are for meteorites mainly collected in Africa and Arabia.

(H, L and LL) chondrites for falls and finds from different locations. The number of irons per 100 ordinary chondrites is ~ 14 for falls, ~ 39 for finds from settled areas, ~ 0.6 for finds from the African and Arabian deserts and essentially the same, ~ 0.7, for the Antarctic finds. Compared to the falls, iron meteorites are less abundant among finds from Antarctica and the hot deserts. This would suggest that irons are not very well preserved compared to chondrites, just the opposite conclusion drawn from the statistics for finds from other regions. However, in addition to the human factor in recognizing meteorite finds, there is another caveat in comparing these statistics. The observed falls are mainly from the past two centuries whereas the meteorites in Antarctica and the deserts have accumulated over thousands to hundred-thousands of years. To decipher if the differences in the iron meteorite frequencies are real and reflect changes in the arriving meteorite population over time, one needs to know the residence time of meteorite falls on the Earth's surface. On the other hand, if these differences are caused by weathering, the stability of iron meteorites against corrosion in the deserts and the Antarctic ice needs to be known.

2.3   CHONDRITES

Chondrites comprise $\sim 90\%$ of the meteoritic stones or aeroliths. They are called chondrites because they usually contain abundant silicate spheroidal or ellipsoidal objects called chondrules (Figures 2.8 and 2.9, and Section 2.3.1). Chondritic meteorites are divided into three major chemically and mineralogically different groups (Table 2.1). The most common chondritic meteorites (Figure 2.6) are the ordinary chondrites (H, L, LL), less abundant groups are the enstatite chondrites (EH, EL) and the various subgroups that make up the carbonaceous chondrites. The recently recognized chondrite groups, "grouplets" or "clans" like the R and K chondrites comprise only a few known meteorites.

Table 2.2 summarizes the major components in different chondrites. Most chondrites consist of silicate chondrules, FeNi-metal and sulfides, which are held together by a fine grained, silicate matrix. Objects called CAI for Ca-Al-rich inclusions and AOA for amoeboid olivine aggregates are present in considerable quantities in some chondrites, preferentially in carbonaceous ones. Several primitive chondrites contain small amounts of presolar grains (Table 2.2). Presolar grains condensed around stars, and their chemical and isotopic compositions have remained unchanged since they formed. More about this stardust is found in Chapter 3.

Figure 2.8 A cut specimen of the Bjurböle L/LL4 ordinary chondrite. The meteorite piece is ∼3.8 cm long at its widest dimension. Several mm-size chondrules appear as dark, round objects in the picture. Photograph kindly provided by Randy Korotev.

The carbonaceous, ordinary and enstatite chondrite groups are revealed through the characteristic differences in major and trace element concentrations and oxidation states. Appendix B gives average compositions of the different chondrite groups. Figure 2.10 compares several element/Si concentration ratios in chondrite groups. Carbonaceous chondrites generally have higher Mg/Si ratios than ordinary and enstatite chondrites (the unusual CB chondrites being one exception). In addition, each carbonaceous subgroup has its own characteristic Mg/Si ratio. Similarly, the Ca/Si ratios are higher in carbonaceous chondrites than in other groups, but among carbonaceous chondrites the Ca/Si ratios vary considerably. The same applies to Al/Si and Ti/Si ratios because Al and Ti usually follow Ca. The Mg/Si and Ca/Si ratios for the H, L and LL ordinary chondrites are essentially the same, and other element/Si ratios such as Fe/Si must be used to tell them apart.

Abundance ratios of Na, Cr, Mn and S are particularly characteristic for the carbonaceous chondrite subgroups. These elements have similar chemical behavior during condensation or evaporation, which explains why the element/Si ratios vary mostly in concert across the suite of chondrite groups (of course, there are always exceptions from general

Figure 2.9   A piece of the Allende carbonaceous CV3 chondrite, *ca.* 3.5 cm tall. Grey spherical objects are chondrules, and light, irregular shaped objects are mainly refractory inclusions. Notice the large refractory inclusion on the top of the specimen and flaky parts of the fusion crust on the left-hand side. Photograph by the authors.

trends). The C chondrites have high C/Si ratios; however, in some subgroups the C/Si ratios are not much higher than found in ordinary and enstatite chondrites. Interestingly, the C/Si ratio in carbonaceous chondrites does not vary in a similar fashion as Na and S. Considering that carbon is more volatile than Na or S, combined loss (evaporation) or gain (condensation) of volatile elements may not be the complete answer to explain such abundance trends.

Chondrites have characteristic Fe/Si ratios. The large spike for the CH chondrites is notable and, in CH chondrites, most Fe is in metallic form. However, in chondrites, Fe occurs in metal, sulfide and in silicates, and the bulk Fe/Si ratio alone is not a measure of the metal to silicate ratio of a chondrite group. In contrast, sulfur is in sulfides in most chondrites and, in that case, the S/Si ratio is a useful proxy for the relative proportions of sulfides and silicates.

Another way to compare abundances in the different chondrite groups is to plot elemental concentrations against Si concentrations (or some other element of choice) (Figure 2.11). The diagrams in Figure 2.11

Table 2.2 Chondrite components and overall density.

Group	Chondrules vol.%	Mean diameter (mm)	Matrix (vol.%)	FeNi metal & sulfide (vol.%)	CAI and AOA[a] (vol.%)	Presolar diamonds (ppm by mass)	Density (g cm^{-3})
CI	≪1	–	>99, incl. 10% Fe$_3$O$_4$	<0.01	<0.01	≤1400	1.6–2.2
CM	12–20	0.2–0.4	75–90	0.1–1	2–8	≤750	2.6–2.8
CK	15	0.75	75	<0.01	4		3.4
CO	35–40	0.15–0.30	30–40	1–6	10–18	≤520	3.63
CV	35–50	0.27–1.0	40–45	0–5	3–12	≤620	3.42
CR	50–60	0.7	30–50	4–8	0.5–2	≤400	3.27
CH	~70	0.02–0.09	5	~20	0.1	≤90	4.2
CB	30–40	0.1–5	<5	60–70	<0.1		3.67
EH	70–80	0.2	<5	20	<0.1	≤70	3.58
EL	75–85	0.6	<0.1	15	<0.1		3.8
H	65–75	0.3	10–15	8	0.01–0.2	≤40	3.8
L	70–80	0.5–0.7	10–15	4	<0.1	≤70	3.6
LL	75–85	0.6–0.9	10–15	2	<0.1	≤130	3.55
R	40–65	0.4	35	<0.1	≪0.1		3.6
K	10–30	0.6	70–80	6–9	<0.1		3.7

[a]CAI = Ca-Al-rich inclusions, AOA = amoeboid olivine aggregates.

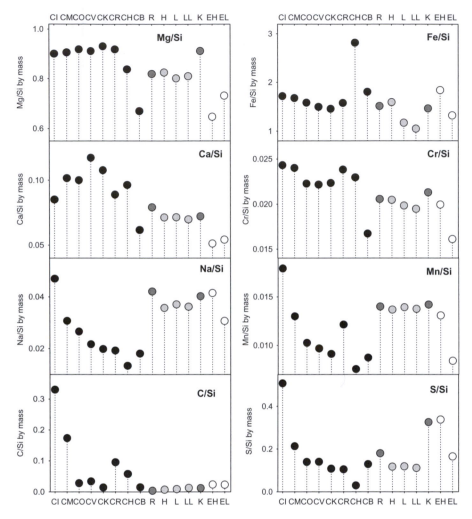

Figure 2.10   Element/Si concentration ratios distinguish the different groups of
chondritic meteorites.

contain the same information as those in Figure 2.10, but do not
convey information on absolute element contents. The dotted lines in
Figure 2.11 indicate positions of constant element/Si ratios as found in
CI chondrites, used for reference because they contain the rock-forming
elements in the same proportions as the Sun (see Chapter 1 and below).
Chondrites with abundance ratios as in CI chondrites plot on or along
the CI chondrite abundance line. For example, the Mg and Si in carbonaceous (CI, CM, CO, CV, CK, CR) and K chondrites roughly follow the CI abundance line. Although their Mg/Si ratios are not exactly

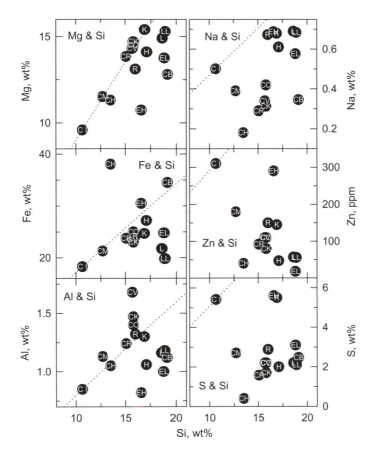

Figure 2.11 Element concentrations in chondrite groups *versus* their Si concentrations. Dotted lines mark the position of constant element/Si ratios as found in CI chondrites.

equal to CI chondrites (Figure 2.10) it is notable that the C chondrites follow one trend whereas H, L, LL and R chondrites with lower Mg/Si ratios follow a separate Mg–Si trend in Figure 2.11. This indicates that these chondrites either lost Mg or gained Si relative to CI chondrites by reactions involving forsterite, Mg_2SiO_4. (Element concentrations described as "enriched" or "depleted" are usually in reference to CI chondrites or solar abundances.)

Compared to CI-chondrites the CM, CV, CO and CK chondrites are enriched in the refractory element Al, whereas the ordinary and enstatite chondrites are depleted in Al. On the other hand, volatile elements like Zn and S are depleted in most chondrite groups relative to CI chondrites. A comparison of the elemental abundances in the different chondrite groups to abundances in the solar photosphere shows that CI chondrites have the least fractionations and agree best with solar abundances, which is the reason why CI chondrites are singled out for use as the solar system abundance standard for rocky elements.

Some chondrite groups are described below after discussing chondrules, the principal constituents of chondrites; also discussed are metamorphism and aqueous alteration, which affected meteorites in all chondrite groups to varying degrees.

2.3.1 Chondrules

Often the silicates in stony meteorites occur in the form of tenths of mm to mm-size spherules called chondrules (from Greek χονδρος, *chondros*, for little sphere). In 1863, the German mineralogist Gustav Rose (1798–1873) suggested the name "chondrites" for meteorites that contain them. Chondrules are easily spotted in a chondrite without a microscope (Figures 2.8 and 2.9). Large chondrules reaching the size of a hazelnut are quite rare. On the other hand, abundant microchondrules occur in some chondrites.

Metallic or sulfide chondrules are occasionally observed, but silicate chondrules are much more frequent. The chondrules are embedded in a fine grained matrix that binds the meteoritic phases together. Some chondrites are almost completely made of chondrules whereas others contain only minor chondrules and more matrix (Table 2.2). The CI chondrites are essentially chondrule-free but they are grouped with the chondrites because they were never processed at high temperatures and because of their chemical compositions. The volume of chondrules and their size distribution (Table 2.2) is also characteristic for each chondrite group.

The diversity of chondrule types, even within one small thin-section of a given meteorite, can be quite amazing (Figure 2.12). The occurrence of different chondrule types within the same meteorite shows that chondrules must have a separate origin from the entire meteorite assemblage.

Most chondrules are crystalline Mg-Fe-silicate minerals such as olivine and pyroxene embedded in a glassy to fully crystallized mineral matrix (mesostasis) and may contain metal and sulfides. Rarely, chondrules rich in plagioclase are found. The existence of chondrules, or fragments of chondrules, within other chondrules suggests that there were multiple chondrule formation events. Many chondrules are surrounded with an assortment of small grains that accreted onto them after the chondrules formed but before the chondrules became part of the meteorite rock. Thus, these rims are called "accretionary rims." The minerals in chondrules create various textures that can be seen in

Figure 2.12   The chondrule variety in the unequilibrated Tieschitz H/L3.6 ordinary chondrite is seen in a thin section through an optical microscope. The field of view is about 5–6 mm wide. (Photograph kindly provided by Herbert Palme and Frank Wlotzka.)

polished thin sections (Figures 2.12 and 2.13). These textures are a means for chondrule classification and may shed some light on chondrule origins.

In barred olivine chondrules, olivine crystallized as parallel lamellae that are now in a glassy or fine crystalline matrix, and often the entire sphere is surrounded by an olivine shell. Chondrules with excentro-radial pyroxene in a glassy or fine grained ground mass appear reminiscent of a "fibrous fan" (Figure 2.13). Other chondrule types contain granular, up to tens of micron size olivine and/or pyroxene grains, others have grains that are only a few microns in size. Some chondrules contain relict grains of clearly different composition than similar minerals that crystallized from the precursor chondrule melt. The "porphyritic chondrules" formed from incomplete melts where small crystals served as nucleation seeds.

Since the time of their discovery about 200 years ago, chondrule formation has been debated. Researchers have now agreed that these spheres were once (partially) molten droplets that were rapidly cooled in a gas. Processes that are ruled out include the ludicrous suggestion from the 1880s that chondrules are fossilized plant and animal remains, and

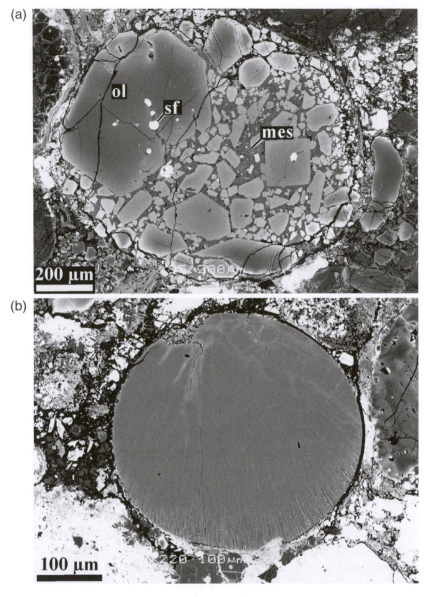

Figure 2.13 Backscattered electron images of individual chondrules from the Tie-schitz H/L3.6 ordinary chondrite taken with a scanning electron microscope. (a) A "porphyritic chondrule" with large FeO-bearing olivine phenocrysts (ol) in a dark mesostasis (mes). The olivine contains sulfide (sf) droplets that appear bright in the picture. (b) Crystallization of thin pyroxene crystals produced a radial, fibrous texture in this "radial pyroxene chondrule." (These photographs were kindly provided by Sasha Krot.)

more serious ideas that chondrules are the result of mechanical processing (rounding by oscillation and attrition), droplets of volcanic origins or meteor ablations.

In 1841, Armand Dufrénoy (1792–1857) concluded that "these globules are evidently the product of fusion." The English geologist and metallurgist Henry Sorby (1826–1908) invented polished thin sections and used them to study terrestrial rocks and meteorites. By 1864–1877[9] Sorby advocated that chondrules formed as free-falling molten objects "like drops of fiery rain," which is still the simplest and most elegant explanation of chondrules that one can use without getting into heated debates about chondrule origins.

Some of the early thoughts on chondrule formation contain the basic ideas still under discussion today. For example, Tschermak (1875)[10] proposed that chondrules formed by solidification of melted rock entrained in a gas eddy; one may come to think of chondrules as products of high temperature "dust devils." Gümbel (1878),[11] inspired by the physical process for making hailstones, described chondrule formation as follows:

"The material of chondrites was formed through disturbed crystallization and scattering as a result of explosive processes in a space filled with mineral forming vapors and H_2 gas that prevented the oxidation of meteoritic iron. The spherules formed through accumulation of mineral masses onto a seed or nucleus during continuous fall or motion in the mineral producing vapors. This explains the eccentric fibrous structure and ellipsoidal structure through localized increase or addition of the material in flight direction, like it happens during the formation of certain hailstones or sleet."

The detailed thoughts of Wahl (1910)[12] on chondrule formation include several concepts that remain relevant today, which are briefly summarized as follows. A free floating, completely molten sphere loses heat from its surface and, in the absence of convection, its surface is cooler than the inside. Hence, crystallization is initiated from the surface layer. If a crystallization seed such as a dust particle touches spots on the periphery, it can initiate the crystallization of eccentric radially oriented crystals typical for many chondrules.

In the absence of a crystallization seed, the outer layer solidifies and crystallizes first, and subsequent internal crystallization leads to more growth connected to the outer layer. This yields lamellar structures, which, when cut perpendicular to their long axis, give the appearance of "bars" such as seen in the barred olivine chondrules. The residual quenched melt from which the olivine crystallized remains as a glassy or a fine crystalline mesostasis, depending on cooling times.

The porphyritic chondrules result from incompletely molten silicates that may not have been heated high enough, which allowed crystals, mostly olivine, to remain while the droplets were dispersed. Upon cooling, these crystals are overgrown by precipitation of crystals from the droplet melt, depending on cooling rates. The types of chondrules thus depend on the composition of the precursor, its degree of melting, the presence of nucleation seeds and the cooling times.

Experimental studies on chondrule formation[13,14] constrain the conditions necessary to obtain the different chondrule textures. The textures of chondrules depend on the number of dust grains that become melted, their extent of melting, the temperature that the precursor material encounters and to some extent on the precursor composition and mineralogy. Experiments by Connolly and Hewins[13] reproduced prominent chondrules types such as the barred olivine and radial type chondrules by letting completely molten droplets collide with mineral dust to initiate nucleation and crystal growth. One important factor on the chondrule outcome is the amount of dust that a completely molten droplet encounters and the temperature of the dust that collides with the droplet. Figure 2.14 gives an overview of the conditions. In the absence of dust grains, a molten sphere quenches to a glass. If the molten drop encounters small amounts of moderately heated dust grains (less than about 1000 °C), these dust grains assemble as a coarse coating on the droplet surface, akin to the accretionary rims around the chondrules seen in thin sections. Hot dust (less than about 1400 °C) induces formation of barred chondrules and eccentro-radial textured chondrules, with large amounts of dust favoring formation of radial-type structures. This is understandable because more potential nucleation seeds can initiate crystallization on several sites on the droplet surface. If the dust encountering a molten droplet is preheated to temperatures above about 1400 °C, small amounts of dust lead to barred structured chondrules. Large amounts of dust induce formation of porphyritic structures, possibly because hot dust is capable of penetrating the droplet to create nucleation sites within it.

Judging from the experimental studies, typical chondrule melts probably had temperatures around 1800 °C and may have reached peak temperatures of up to $\sim$2000 °C. The cooling rates were between 100 and 3000 °C h^{-1}. Thus, chondrules heated to 2000 °C took from about an hour to greater than a day to cool down and crystallize. This cooling is about a magnitude slower than that of silicate liquids ejected during

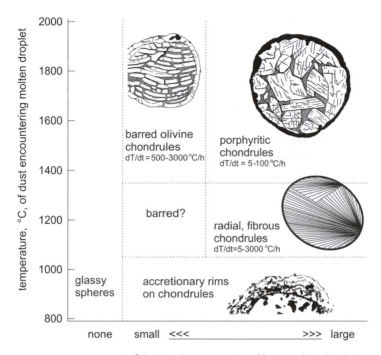

y-axis: temperature, °C, of dust encountering molten droplet (2000, 1800, 1600, 1400, 1200, 1000, 800)

barred olivine chondrules dT/dt = 500-3000 °C/h

porphyritic chondrules dT/dt = 5-100 °C/h

barred?

radial, fibrous chondrules dT/dt=5-3000 °C/h

glassy spheres

accretionary rims on chondrules

x-axis: none   small <<<   >>> large

amount of dust grains encountered by a molten droplet

Figure 2.14 Chondrule formation conditions inferred from experiments by Connolly and Hewins.[13] Cooling rates (dT/dt) from Connolly and Desch.[14]

volcanic eruptions in air. This and the different textures of volcanic spherules are two reasons why a volcanic origin of chondrules was finally ruled out.

The importance of chondrule formation with respect to chemistry is that volatile elements such as Na and K are lost from the precursor materials. Some re-condensation of volatile elements onto the forming chondrules occurs but, overall, chondrule formation seems to result in a net loss of volatiles. Given these element fractionations, parent bodies that accreted volatile-depleted chondrules in larger quantities should contain less volatile elements than present in the precursor materials from which chondrules formed. A related issue is the nature of the chondrule precursor material, which we will discuss in Chapter 3.

As to the *cause* of producing the liquids necessary for chondrule formation to begin with, no firm consensus has been reached – chondrule formation probably will remain a perpetual debate in meteoritics. Since the early studies, chondrules were either thought to be direct condensates from a H_2-rich gas, or products of pre-existing solids that were melted by some mechanism. Both cases can lead to free-floating silicate melt drops dispersed in a gas, and the chondrule diversity provides supporting arguments for both. However, most chondrules are best explained by processing of solid precursors.

The following, while not necessarily complete, is a list of suggested processes, divided into "direct condensation" and "processed precursor" models. The literature on chondrules is quite extensive, and the proceedings of two scientific conferences on chondrules provide a good starting point.[15,16] Detailed references to recent papers by some of the authors listed below, as well as some complete references, can be found on the Astronomical Data Service (ADS) website (see General References and Resources).

Direct condensation models for chondrule formation:

- Following G. Tschermak and W. v. Gümbel, S. Meunier (1883) suggested chondrule formation by fast condensation from a turbulent gas of comparable composition to that of the solar atmosphere; chondrules as possible fossils of "cyclones photosphérique;"
- direct condensation of chondrules from the solar nebula – *e.g.*, H. Suess (1949), J. A. Wood (1962, 1963), M. Blander and J. L. Katz (1967), H. Y. McSween (1977);
- condensation of liquids out of gas at high total pressures, either in the pre-collapse protoplanetary atmospheres of gas giant planets, M. Podolak and A. G. W. Cameron (1974), or at high pressures resulting from shock waves, A. Galy *et al.* (2000);
- condensation of liquids out of the gas of the solar nebula that was enriched in rocky elements (high dust to gas ratio) – J. A. Wood and H. Y. McSween (1977), D. Ebel and L. Grossman (2000) and A. Engler *et al.* (2007).

Processing of pre-existing solids for chondrule formation:

- Chondrules as "molten rain" resulting from asteroid impacts and collisions – H. C. Urey (1952), T. Ringwood (1966), E. A. King (1972) and S. W. Kieffer (1975);
- collisions of small, cm-size planetesimals in space, F. Whipple (1972);
- processing of amorphous dust condensates that collected in a cool dust cloud from which chondrules crystallized during accretion onto small planetesimals – B. Y. Levin and G. L. Slonimsky (1958);
- flash heating induced through lightning in the solar nebula, F. Whipple (1966), A. G. W. Cameron (1966), J. T. Wasson and

K. L. Rasmussen (1982, 1994), W. Phillip *et al.* (1998) and S. Desch and J. N. Cuzzi (2000);

- processing of pre-existing solids by chemical energy (exothermic reactions); D. D. Clayton (1980);
- drag-heating and gas-grain collisions in the shock waves of the collapsing solar nebula – F. Wlotzka (1969), J. A. Wood (1983, 1984), L. L. Hood and M. Horanyi (1991), T. V. Ruzmaikina and W. H. Ip (1995), H. C. Connolly and S. G. Love (1998), J. A. Wood (1996) and J. N. Cuzzi and C. M. O'D Alexander (2006);
- processing in the vicinity of the sun during its T Tauri stage (Chapter 3) – J. A. Wood (1963), J. A. Wood and H. McSween (1977);
- processing in the vicinity of the sun while it may have gone through the FU Orionis stage (Chapter 3), L. Hartman and S. J. Kenyon (1985);
- X-ray melting induced by gamma-ray bursts, P. Duggan *et al.* (2003);
- processing of trapped solids from the solar nebula in bipolar outflows from the young sun (X-wind model, see Chapter 3), W. R. Skinner (1990) and F. Shu *et al.* (1996, 2001); a variant of this model is ablation of small planetary objects in such bipolar outflows, K. Liffman (1992).

2.3.2 Petrologic Types of Chondrites

The petrologic type of chondrites is used to classify the amount of thermal metamorphism and/or alteration by water that the mineral assemblages of chondrites have experienced on their meteorite parent bodies. Metamorphic processes alter the originally accreted mineral assemblages though sub-solidus heating (*i.e.*, heating that stays below the melting point). The other change is through reactions with water, known as aqueous alteration. Chondrites either show signs of aqueous alteration or thermal metamorphism, but rarely both. Van Schmus and Wood (1967)[17] devised a classification that relates the type and extent of alteration to a numbering system from 1 to 6 (number 7 was added later). Table 2.3 summarizes the criteria for assigning the numbers to the different petrologic types. The numbers for the petrologic type are added after the letter designation of the chondrite groups (see Table 2.1), for example, CI1, CM2, H3, LL5, *etc.*

The least altered chondrites have the petrologic type number 3. This may seem odd at first but lower numbers describe the degree of aqueous alteration (1 indicates the most extreme case for this) and higher

Table 2.3 Petrological classification of chondrites (modified from ref. 17).

Petrographic type	1	2	3	4	5	6	7
Texture	No chondrules	Very clearly defined chondrules		Well-defined chondrules	Chondrules can be recognized	Poorly recognizable chondrules	Relict chondrules
Matrix	Fine grained opaque	Chiefly fine, opaque	Clastic and minor opaque	Coarse grained transparent, recrystallized, coarsening from type 4 to 7			
Homogeneity of olivine and pyroxene; Fe, Mg content	–	>5% mean deviation of Fe		0–5%	Homogeneous		
Low-Ca-pyroxene polymorph	–	Mainly cpx, monoclinic		cpx Abundant, monoclinic		Orthorhombic	
				>20%	<20%		
					$CaO<1\,wt\%$		$CaO >1\,wt\%$
Feldspar	–	Primary only; minor and calcic crystalline, secondary feldspar absent		Secondary feldspar very fine grained <2 µm	Fine grained, secondary feldspar <50 µm	Grains clearly visible, coarsening from type 5 to 7, grains >50 µm	
Glass in chondrules	–	Clear and isotropic		Turbid, devitrified	No glass		
Metal, maximum Ni content	–	Taenite minor or absent, <20 wt% Ni		Kamacite and taenite (>20 wt% Ni) in exsolution			
Sulfides, mean Ni content	~1 wt%	>0.5 wt%	<0.5 wt%				
H₂O content (wt%)	18–20	2–16	0.3–3	<2			
Carbon content (wt%)	3–5	1.5–2.8	0.1–1.1	<0.2			
	Hydrous alteration ⇐ ⇐ ⇐		not thermally equilibrated ⇐ ⇐ ⇐ ⇒ ⇒ ⇒ thermally equilibrated				
Metamorphic temperatures (°C)			400–600	600–700	700–750	750–950	>950

numbers (4–7) the degree of thermal metamorphism. Chondrites of type 1 and 2 only occur among the carbonaceous chondrites, and notably the CI chondrites are of type 1 (CI1) with large degree of aqueous alteration. With increasing numbers from 4 to 7, chondrite mineral assemblages have experienced more change through heating at higher temperatures and/or for longer times. Petrologic types 4–7 are found among ordinary and enstatite chondrite groups, and some rarer groups of carbonaceous chondrites.

One consequence of mildly heating a loosely piled mineral mixture to a few hundred °C is that it becomes more compact. Mineral boundaries and surfaces merge to minimize surface energies, which is the well-known process of sintering. This process becomes even more effective if the mineral mixture is mechanically pre-compacted before heating. In the laboratory, one can use pressed pills of material, and on meteorite parent bodies material buried at greater depths will naturally experience some higher lithostatic pressures that compresses it, and shockwaves from meteorite impacts may further assist compaction. With mechanical compaction and mild heating, a former mineral powder changes to a consolidated, lithified mass much faster at a given temperature. Experimentally, a pressed ordinary chondrite powder can be converted into a firm mass within a few hours heating at $\sim 1200\,°C$ in an inert atmosphere.

The increase in the material strengths of chondrites from low to high petrologic type is easily noticed if one tries to take an ordinary chondrite apart. Ordinary chondrites of petrologic type 3 can often be crushed or crumbled between the fingers, whereas breaking apart chondrites of petrological types 5 and 6 requires work with a hammer (however, neither method is recommended to dismantle a meteorite).

During heating, an initially "random" mixture of minerals strives to reach the state where the mineral phases and their compositions are at equilibrium, that is, a state of minimized Gibbs free energy. This drives diffusion, which leads to changes in mineral compositions, or even the disappearance of some minerals and formation of new ones. Metamorphic temperatures are generally low ($<950\,°C$, see Table 2.3). The duration of thermal metamorphism may have been too short to allow diffusion of the elements among the minerals to reach complete chemical equilibrium; nevertheless, chondrites of higher petrologic types 4-6 are called "equilibrated chondrites." The term "equilibrated chondrite" does not imply that all mineral phases in the chondrite are in chemical (thermodynamic) equilibrium; it is used to contrast the more homogeneous mineral compositions of these chondrites to the more heterogeneous mineral compositions of "unequilibrated chondrites" of type 3. For example, the range in $Fe/(Fe+Mg)$ ratios of olivine and pyroxene compositions can vary widely (~ 0 to 30 mol.%) in H3 chondrites whereas on average, type H6 chondrites have narrower $Fe/(Fe+Mg)$ ratios of about 17–19% in olivine and about 15–16% in pyroxene. The reason equilibration through metamorphism was cut short is that the radioactive heat sources on meteorite parent bodies were only available within the first few million years after the first solids accumulated in the solar system (Chapter 3).

The formation of feldspar and phosphates are good examples of the complexity of the metamorphic exchange reactions in a heated ordinary chondrite parent asteroid. Phosphates are not observed in low grade petrologic type chondrites but appear in higher petrologic types 5 and 6. The formation of the phosphates apatite, $Ca_5(PO_4)_3(OH,F,Cl)$, and whitlockite, $Ca_3(PO_4)_2$, required redistribution of the elements and oxidation of P. In type 3 chondrites, P is in the phosphide schreibersite, $(Fe,Ni)_3P$, and Ca is in several silicates, such as pyroxene, $Ca_x(Fe,Mg)_{1-x}SiO_3$. Type 3 chondrites contain small amounts of water and some contain halite (NaCl). An Fe-Ni alloy with varying Ni content, called kamacite, is present in all ordinary chondrites. One possible net reaction for whitlockite formation is:

$$2x(Fe_yNi_{1-y})_3P\,(\text{schreiberstite}) + 3Ca_x(Fe,Mg)_{1-x}SiO_3(\text{Ca-pyroxene})$$
$$+ 8xH_2O(\text{water}) = xCa_3(PO_4)_2(\text{whitlockite})$$
$$+ 3Fe_x(Fe,Mg)_{1-x}SiO_3(\text{Mg-Fe pyroxene})$$
$$+ 3xFe_{(2y-1)}Ni_{2(1-y)}(\text{kamacite}) + 8xH_2$$

$$(2.13)$$

where $x \approx 0.01$ and y is between ~ 0.8 and ~ 0.9 for ordinary chondrites.

Another possible reaction involves reduction of Fe^{2+} in silicates to metal instead of oxidation by water. Whitlockite is common to equilibrated H chondrites, whereas apatite occurs more commonly in L and LL chondrites. A possible net formation reaction of hydroxyl apatite is:

$$3x(Fe_yNi_{1-y})_3P\,(\text{schreiberstite}) + 5Ca_x(Fe,Mg)_{1-x}SiO_3(\text{Ca-pyroxene})$$
$$+ 13xH_2O\,(\text{water}) = xCa_5(PO_4)_3OH\,(\text{apatite})$$
$$+ 5Fe_x(Fe,Mg)_{1-x}SiO_3\,(\text{Mg-Fe pyroxene})$$
$$+ 9xFe_{y-5/9}Ni_{1-y}\,(\text{kamacite}) + 12.5xH_2$$

$$(2.14)$$

where x and y are similar to Equation (2.13).

During metamorphic reactions, trace elements such as the rare earth elements (REE), U and Th are enriched in phosphates. Because radioactive U and Th can be used as chronometers, the phosphates are good mineral phases to determine the timing of metamorphic processes on chondrite parent bodies (Chapter 3).

To summarize, the major features of the petrological types are:

1. **Type 1** designates chondrites that have experienced a high degree of aqueous alteration. The rocks consist almost entirely of fine grained matrix. Hydrated minerals are abundant. Most primary minerals have been replaced by secondary phases and chondrules are extremely rare or absent. Presolar grains are present.

2. **Type 2** chondrites contain abundant hydrated minerals and abundant fine-grained matrix. They have chondrules, and their sulfides contain Ni. Presolar grains are present.

3. **Type 3** chondrites show abundant recognizable chondrules. Chondrites of this type are regarded as the most primitive chondrites with respect to changes of the original mineral assemblages that accreted to their parent bodies. The presence of glass in chondrules makes a chondrite type 3. The "primary igneous glass" in chondrules is quenched glass from the chondrule formation process and this glass is in contrast to shock produced glass that can be all over in a meteorite, not only in chondrules. The composition of olivines and pyroxenes in type 3 chondrites is not homogeneous. Presolar grains are present. The type 3 chondrites are subdivided further to indicate their degree of metamorphism. The type number 3.0 designates the least metamorphosed meteorites, and a type number of 3.9 indicates that the alterations approach the metamorphic type 4.

4. **Type 4** chondrites have abundant recognizable chondrules as in type 3, but the primary igneous glass in chondrules is gone. Olivine compositions are homogenized and the fine-grained matrix is recrystallized.

5. **Type 5** chondrites have chondrules with blurred outlines. Olivine and pyroxene compositions are homogenized through metamorphism. The low-Ca pyroxene seen in types 3 and 4 is converted into orthopyroxene, and the Ca moved into secondary minerals such as small ($<2\,\mu$m) feldspar crystals.

6. **Type 6** chondrites are those in which chondrule outlines have almost vanished and all minerals have homogeneous compositions. Secondary feldspar has grown to sizes $\geq 50\,\mu$m. Heating has not led to melting.

7. **Type 7** chondrites only have a few relic chondrules left, and these meteorites were exposed to temperatures $>950\,°C$. Chondrites of this type are relatively rare and may represent transitional types to primitive achondrites (see Table 2.1 and below).

2.3.3   Ordinary Chondrites

The *ordinary chondrites* mainly contain silicate chondrules and fine silicate matrix, FeNi metal and sulfides. The H, L and LL ordinary chondrites are distinguished by their overall Fe content and the distribution of Fe between silicate minerals, metal and sulfide. The H-chondrites have high (H) total Fe contents, L-chondrites low (L) total Fe and LL-chondrites low total Fe and low metal (LL). Chondrule sizes and O-isotopes provide clear distinctions between the H, L and LL chondrites. All ordinary chondrite groups have members of petrologic types from 3 to 6.

H chondrites have high concentrations of total Fe ($\sim 27\,mass\%$). Metal contains $\sim 8\,atom\%$ Ni. Their chondrules are relatively small ($\sim 0.3\,mm$).

L chondrites are one of the low-iron bearing groups of ordinary chondrites. Their total Fe is $\sim 22\,mass\%$ and metal contains $\sim 13\,atom\%$ Ni. Their chondrules are of moderate size ($\sim 0.7\,mm$).

LL chondrites have an overall low total Fe iron content ($\sim 21\,mass\%$), and, by comparison to the L chondrites, a low metal content. Their metal has $\sim 22\,atom\%$ Ni. On average, they contain larger chondrules ($\sim 0.9\,mm$) than H and L chondrites.

The existence of the three distinct groups of H, L and LL chondrites was revealed with advanced chemical analysis. Most chemical analyses on meteorites up to the 1950s were done by classical wet-chemical methods, and larger uncertainties in the analytical results were to be expected. The wet-chemical data on meteorites suggested the total iron content (in silicates, metal plus sulfide) of ordinary chondrites was similar and that the different proportions of silicates and metal in chondrites and the FeO content of their silicates was the result of oxidation–reduction processes. In 1916, Prior postulated two rules: first, chondrites with low metal contents have high Ni concentrations in the metal, and second, chondrites with low metal content have silicates with correspondingly higher FeO contents. However, the ordinary chondrite groups are not a continuous series with Fe metal and silicate FeO contents solely controlled by changes in redox state. In 1953, Urey and Craig[18] discovered that the ordinary chondrites fall into two separate chemical groups. These are revealed when the metallic Fe content (Fe in the metal plus Fe in sulfide) is plotted *versus* the Fe content in the silicates. Such a plot, known as a "Urey–Craig diagram," is shown in Figure 2.15 for a large sample of ordinary chondrites.

From Prior's rules one would expect that all chondrites plot along a single mixing line in the Urey–Craig diagram if the source material of chondrites had a uniform overall Fe content. The available Fe would be distributed according to the redox equilibria:

$$\text{Fe (in FeNi metal)} + H_2O(\text{gas}) = \text{FeO (in silicates)} + H_2(\text{gas}) \qquad (2.15)$$

$$\text{Fe (in FeNi metal)} + H_2S(\text{gas}) = \text{FeS (in sulfide)} + H_2(\text{gas}) \qquad (2.16)$$

where the partial pressures of H_2S (the major S-bearing gas in a solar composition mixture) and of water (the main oxidizing agent in the solar nebula, see below) in the chondrite-forming region determines the degree of oxidation or reduction of the metal, sulfide and silicates. Ignoring the sulfides for now, one end-member group of chondrites would have higher concentrations of FeO in the silicates and less iron metal because more of the total Fe was oxidized. The oxidation of Fe from a FeNi alloy increases the relative Ni concentration in the metal as Ni is nobler than Fe and requires higher water or oxygen partial pressures for oxidation. The other end-member group of chondrites would contain FeO-poor silicates and more Fe metal because they may have formed under more reducing conditions. However, in a multi-phase system with fixed total Fe, the Fe content of the silicates and metal (plus sulfide) must plot along a mixing curve defined by the end-member compositions.

Urey and Craig found that the ordinary chondrites clustered along two different trends, indicating that there are at least two different groups with different total Fe content, which is clearly seen in Figure 2.15. They named the group with high total Fe content the "H" group of chondrites and the other with low Fe content the "L" chondrites. The two lines in the diagram indicate the position for chondrites that have total Fe contents of 22 and 27 wt%, respectively, which are characteristic averages for L and H chondrites.

The recognition of the LL chondrites as a third separate group of ordinary chondrites came through the analysis of the FeO content in the two major silicate minerals, olivine and pyroxene. Before describing this, we give a few general notes about olivine and pyroxene formulas and naming conventions, which are summarized in Table 2.4.

Olivine has the general formula X_2SiO_4, where X stands for Mg and/or Fe, but other divalent elements such as Ca, Mn and Ni with

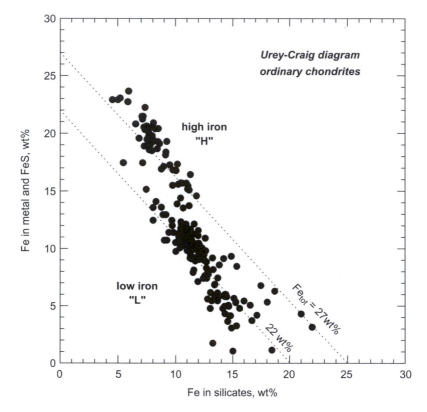

Figure 2.15 Concentration of Fe in metal and sulfide *versus* the concentration of Fe in silicates from individual ordinary chondrites. A chondrite with a total Fe content of 22 or 27 wt%, respectively, can only have Fe concentrations in the metal plus sulfide and silicate phases that are plotted along the respective dotted lines. The position on the lines is controlled by the degree of oxidation of the chondrite. However, the ordinary chondrite data do not follow a single concentration trend, and Urey and Craig (1953) concluded that at least two different groups of ordinary chondrites exist, which they called H (for high Fe) and L (for low Fe).

compatible ionic radii can substitute for Mg and Fe. Olivine is the solid-solution of the Mg-rich end-member forsterite, Mg_2SiO_4, and the Fe-rich Fe_2SiO_4, fayalite. Olivine is often indicated by writing $(Mg,Fe)_2SiO_4$. Instead of writing the entire chemical formula for an olivine with known mole fractions of forsterite and fayalite, such as, $(Mg_{0.9}Fe_{0.1})_2SiO_4$, it is more convenient to use the abbreviation $Fo_{0.9}Fa_{0.1}$, or simply $Fa_{0.1}$, for an olivine that has 90 mol.% forsterite (Fo) and 10 mol.% fayalite (Fa).

Pyroxenes have the general formula $X_{1-x}(Y,Z)_{1+x}Si_2O_6$ where $X = Ca$ or Na, $Y = Mg$, Fe^{2+} or Mn and $Z = Fe^{3+}$ or Al. The major

Table 2.4 Mineral solid solutions.

Mineral system	Formula
Olivine: end-members	
Forsterite = Fo	Mg_2SiO_4
Fayalite = Fa	Fe_2SiO_4
Pyroxene: end-members	
Enstatite = En	$MgSiO_3$
Ferrosilite = Fs	$FeSiO_3$
Wollastonite = Wo	$CaSiO_3$
Diopside	$Ca_{0.5}Mg_{0.5}SiO_3$
Hedenbergite	$Ca_{0.5}Fe_{0.5}SiO_3$
Pyroxene: named solid solutions	
Orthopyroxene	$(Mg,Fe)SiO_3$
Enstatite	$Mg_{0.88-1.00}Fe_{0.12-0.00}SiO_3$
Bronzite	$Mg_{0.70-0.88}Fe_{0.30-0.12}SiO_3$
Hypersthenes	$Mg_{0.50-0.70}Fe_{0.50-0.30}SiO_3$
Clinopyroxene	$(Ca,Mg,Fe)SiO_3$
Pigeonite	$Ca_x(Mg,Fe)_{1-x}SiO_3$ with $x = {\sim}0.1$
Augite	$(Ca,Mg,Fe^{2+},Al)_2(Si,Al)_2O_6$
Feldspar: end-members	
Anorthite = An	$CaAl_2Si_2O_8$
Albite = Ab	$NaAlSi_3O_8$
Orthoclase = Or	$KAlSi_3O_8$
Feldspar: named solid solutions	
Plagioclase	$CaAl_2Si_2O_8$-$NaAlSi_3O_8$
Albite	$An_{0.0-0.1}Ab_{1.0-0.9}$
Oligoclase	$An_{0.1-0.3}Ab_{0.9-0.7}$
Andesine	$An_{0.3-0.5}Ab_{0.7-0.5}$
Labradorite	$An_{0.5-0.7}Ab_{0.5-0.3}$
Bytownite	$An_{0.7-0.9}Ab_{0.3-0.1}$
Anorthite	$An_{0.9-1.0}Ab_{0.1-0.0}$

divalent elements in the pyroxenes are Mg, Ca and Fe and their end-member minerals are called enstatite ($MgSiO_3$), ferrosilite ($FeSiO_3$) and wollastonite ($CaSiO_3$); often abbreviated as En, Fs and Wo, respectively.

In contrast to olivine, pyroxenes form several different solid solutions that are frequently encountered in rocks. The pyroxene solid solutions have separate mineralogical names, which can be quite confusing (see Mason 1960[19]). The Mg-Fe pyroxene solid solutions are orthopyroxenes. Pyroxene with 0–12 mol.% Fs is called enstatite like the pure Mg end-member. Pyroxene with 12–30% Fs is called bronzite, and pyroxene with 30–50% is hypersthene. The pure Fe end-member ferrosilite is not stable, and only occurs in solid solution with $MgSiO_3$ or $CaSiO_3$. Orthopyroxenes may accommodate small amounts of $CaSiO_3$ and are sometimes called low-Ca pyroxenes. Pyroxenes containing the Ca end-member wollastonite belong to the clinopyroxenes, with the general formula $(Ca,Mg,Fe)SiO_3$. The pyroxene with the stoichiometric (or near stoichiometric) composition

$Ca_{0.5}Mg_{0.5}SiO_3$ is diopside, and $Ca_{0.5}Fe_{0.5}SiO_3$ is hedenbergite. Pigeonite is the solid solution $Ca_x(Mg,Fe)_{1-x}SiO_3$ with $x \sim 0.1$, sometimes referred to as low-Ca pyroxene. The high Ca pyroxene augite $(Ca,Mg,Fe^{2+},Al)_2(Si,Al)_2O_6$ is an Al-rich member of the hedenbergite–diopside solid solution series. The pyroxenes do not form continuous solid solution series, and the miscibility gap between orthopyroxenes and clinopyroxenes is responsible for the coexistence of different pyroxenes seen in many rocks, including meteorites.

Originally, the ordinary chondrite groups were distinguished by their frequent pyroxenes. The H chondrites correspond to the older classification of bronzite-olivine chondrites and the L chondrites to the hypersthene-olivine chondrites. These group names (introduced by Prior in the early 1900s) should be avoided because the pyroxenes used in defining the "bronzite" and "hypersthene" chondrites do not cover the same compositional range that is now used in mineralogy to define these pyroxenes (Table 2.4).

One of the first electron microprobe analyses on meteoritic material was performed on the ferromagnesian silicates in ordinary chondrites. The study by Keil and Fredriksson in 1964[20] revealed that the FeO content of olivine and pyroxene separates the ordinary chondrites into three groups, H, L and LL, instead of the two that were distinguished by Urey and Craig about a decade earlier. Figure 2.16 shows a plot of the ferrosilite content in pyroxene *versus* the fayalite content in olivines of equilibrated chondrites. The data for the LL chondrites form a separate cluster at high Fs and Fa contents beyond the L chondrites. The distinction between L and LL chondrites is not well resolved in the Urey–Craig diagram (Figure 2.15) because the overall Fe content of L and LL chondrites is very similar. The LL chondrites are actually included in Figure 2.15 and they plot at the bottom end (highest Fe content in silicates) of the L chondrite line. However, without the information of the chemical composition of the olivine and pyroxenes the identification of the LL chondrites as a distinct group is not easy.

The elemental compositions of L and LL chondrites are generally quite similar and not that much different from H chondrites when compared to other chondrite groups (see, *e.g.*, Figures 2.10 and 2.11). A clear distinguishing feature of the ordinary chondrites is their O-isotopic compositions, where LL chondrites have the highest concentrations in ^{17}O and ^{18}O and H chondrites the lowest (Figure 2.1). The O-isotope composition of individual chondrules from all ordinary groups plot along a mixing line with a slope of ~ 1 that passes through the mean composition of the H, L and LL chondrites (Figures 2.1 and 2.2). This is consistent with the idea that these meteorites received chondrules from a

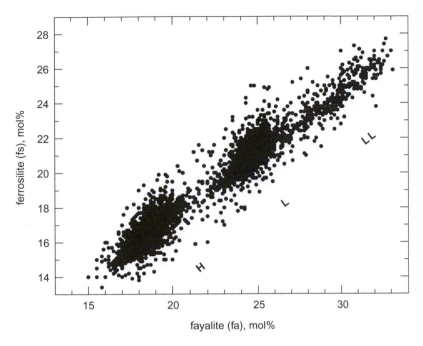

Figure 2.16 Ferrosilite content in pyroxene and fayalite content in olivine of ordinary chondrites correlate within each group and separate the ordinary chondrites into types H, L and LL.

common reservoir. The variation in O-isotopic compositions (Figure 2.1) mimics the trend for the Fa and Fs contents in H, L and LL chondrites (Figure 2.16), suggesting that O-isotopic compositions and degree of Fe oxidation are linked. It remains to be understood why the correlation of the O-isotopes is mass-independent instead of mass-dependent as expected for many physical and chemical processes, including redox reactions. Whatever the reasons may be, if there was a common chondrule reservoir for the H, L and LL chondrites, one further needs to account for the different amounts of chondrules in H, L and LL chondrites, as well as their different chondrule size distributions. Among the suggested mechanisms for this are aerodynamic size-sorting in the solar nebula gas and gravitational size-sorting during planetesimal accretion.

2.3.4 Enstatite Chondrites

Enstatite (E) chondrites[21] are named after their most abundant silicate mineral enstatite. In contrast to other chondrites, pyroxene and olivine in E chondrites are very FeO-poor, which indicates that they must have formed under very reducing conditions. The high state of reduction is apparent from the presence of 1–3 wt% Si in the metal phase, which is not common in other chondrites. Several elements that exclusively occur in oxide form in other chondrites appear in sulfides (*e.g.*, MgS, CaS), nitrides (osbornite, TiN) or oxynitrides (sinoite, Si_2N_2O) in enstatite chondrites as well as in enstatite achondrites (aubrites) listed below.

EH chondrites contain high (H) total contents of Fe, most of which is in an FeNi alloy, which has ~3 wt% Si in solution, and occupies ~10 vol.% of the meteorites. Sulfides (10 vol.%) are niningerite, (Mg,Fe)S, and oldhamite, CaS, and Ti-bearing troilite, FeS. The FeNi silicide-phosphide perryite $(Ni,Fe)_5(Si,P)_2$ and several other exotic phases have been reported. Most of the volume of EH chondrites is filled with relatively small chondrules (0.2 mm average diameter), and matrix is rare (Table 2.2). Most known EH chondrites are of petrologic type 3 and 4.

EL chondrites are the low (L) iron-bearing enstatite chondrites. The chondrules are larger in EL chondrites (0.6 mm) than in EH chondrites. Their metal contains somewhat less Si (~1 wt%) and the ferroan alabandite, (Fe,Mn)S, is frequently encountered instead of niningerite. Most known EL chondrites are of petrologic type 5 and 6.

Characteristic differences in chemical composition between EH and EL chondrites are the higher Na, S, Mn, Zn and Au concentrations in EH chondrites (Figures 2.10 and 2.11). Among all chondrites, the enstatite chondrites have the lowest Mg/Si ratios (aside from the unusual CB chondrites, see Figure 2.10). The elemental fractionations in the highly reduced enstatite chondrites can be expected to be different from that of ordinary and carbonaceous chondrites because the volatility of the elements depends on the oxidation state of the overall system.

Curiously, the oxygen isotopic composition of the enstatite chondrites (as well as that of the enstatite achondrites below) coincides with that of rocks from the Earth and the Moon (Figures 2.1 and 2.2). Thus, the O-isotopes could suggest a genetic connection between these meteorites and the Earth plus Moon. However, considering the very different oxidation states of the enstatite meteorites and the terrestrial mantle rocks (*e.g.*, essentially FeO-free silicates in enstatite chondrites *versus* terrestrial mantle rocks with ~10% FeO), the meaning of the similar O-isotopic compositions of these meteorites and the Earth–Moon system remains puzzling.

2.3.5 R and K Chondrite Groups

The K and R chondrites are small groups of meteorites that do not fit into the categories of ordinary, enstatite or carbonaceous chondrites.

R chondrites are named after the Rumuruti meteorite. The R chondrites have exceptionally Fe-rich olivine, with 37–40 mol.% Fe_2SiO_4, reflecting their overall highly oxidized mineralogy. The occurrence of NiO in the olivine, usually not observed in olivine of other chondrite groups, further attests to their high oxidation state. They have unique fractionations in their elemental abundances when compared to other chondrite groups, *e.g.*, they have lower Mg/Si and Ca/Si ratios than carbonaceous chondrites, but higher Ca/Si than ordinary chondrites (Figures 2.10 and 2.11). Their O-isotopic compositions plot above the terrestrial fractionation curve and above that of ordinary chondrites (Figure 2.1). The currently known R chondrites have petrological types of 3–4.

K-chondrites are named after the Kakangari meteorite, with which they share mineralogical and chemical resemblances. Judging from the FeO content of their olivine (~ 2 mol.%) these chondrites are more reduced than the ordinary chondrites. Their Mg/Si ratios are more comparable to that of C chondrites (Figures 2.10 and 2.11), and in the three O-isotope diagram they plot closer to C chondrites below the terrestrial fractionation curve (Figure 2.1).

2.3.6 Carbonaceous Chondrites

The carbonaceous (C) chondrites[22] contain large amounts of the volatile elements H, N and C (hence the name carbonaceous) and can contain bound water. The carbonaceous chondrites were originally divided into three groups, but currently there are at least eight known groups of C chondrites. The carbonaceous chondrite groups are named by using "C" plus the initial letter of the name from a characteristic meteorite of a group, *e.g.*, the "CI" chondrites are named after the Ivuna meteorite. An exception from this naming convention is for the more recently identified group of CH chondrites, where the "H" in "CH" stands for "carbonaceous chondrite with high Fe content." The following gives an overview of C chondrites (see Table 2.2):

- **CI chondrites** are named after the Ivuna meteorite fall. Ten CI chondrites are known. Five known CI1 chondrites were observed to fall: Alais (15 March 1806), Ivuna (16 Dec. 1938), Orgueil (14 May 1864), Revelstoke (31 March 1965) and Tonk (22 January 1911). The Tagish Lake meteorite (fall 18 January 2000) may also be a CI chondrite but it has the metamorphic type 2 (CI2). Four CI chondrites were recovered from the Yamato ice fields in Antarctica. The CI1 chondrites have a high degree of hydration and are essentially free of refractory inclusions and chondrules; however, rare micro-chondrules (<0.01 mm) have been described. Anhydrous minerals such as olivine and pyroxene that are so common in other chondrite groups occasionally occur as fragments making up <1 vol.%. The major mass of CI1 chondrites is a fine grained matrix of hydrous sheet silicates (phyllosilicates) such as serpentine and saponite. Other minerals are magnetite, carbonates, chlorides and sulfates. Rare sulfides include pyrrhotite and pentlandite but not troilite (stoichiometric FeS) which is common in non-carbonaceous chondrites. Metal is almost completely absent in CI chondrites. More information about CI chondrites is given below.

- **CM chondrites** contain small chondrules and refractory inclusions (0.3 mm) that are embedded in abundant fine-grained matrix (~ 70 vol.%). Hydrated minerals are abundant. They are named after the Mighei meteorite, and the Murchison meteorite is another well-known member of this group. The CM chondrites are mostly of petrologic type 2 ("CM2").

- **CV chondrites** are named after the Vigarano meteorite fall. The best-studied meteorite in this group, Allende, fell in 1969 (Figure 2.9). The CV chondrites contain large (mm-sized) chondrules that are often surrounded by igneous-appearing rims. Matrix makes up ~ 40 vol.% and mm-size refractory inclusions are prominent. Olivine and pyroxene are relatively Fe-rich ($Fa_{0.4-0.6}$, $Fs_{0.1-0.5}$, $Ws_{0.45-0.5}$). The CV chondrites, notably Allende, contain larger amounts of Ca-Al-rich inclusions (CAI) made of anhydrous refractory minerals (*e.g.*, Ca-aluminates, spinel; see Chapter 3) that experienced some degree of aqueous alteration. Secondary minerals like saponite, Na-phlogopite, Al-rich serpentine, Na-K mica, phosphates, carbonates and ferrihydrite from aqueous processing are found in chondrules and isolated olivine grains. A subdivision into oxidized (Bali-type) and reduced (Vigarano type) subgroups of CV chondrites is sometimes practical. The CV chondrites in the oxidized subgroup contain more secondary minerals from aqueous processing and more magnetite instead of metal than their reduced counterparts. However, overall signs of aqueous alteration are still low enough that CV chondrites are classified as petrologic type 3 ("CV3").

- **CO chondrites**, named after the Ornans meteorite, contain notably small chondrules and small refractory inclusions (<0.2 mm). Chondrules and matrix are about equal in volume. Their olivine ($Fa_{0.3-0.6}$) has comparable Fe-content to that of CV chondrites. Their degree of aqueous alteration is typically low; rarely, serpentine and saponite are found in veins of olivine. The CO chondrites are mainly of petrologic type 3 ("CO3").

- **CK chondrites**, named after Karoonda, the first of the two observed falls in this group, contain abundant fine-grained matrix (~ 75 vol.%). They have a relatively high degree of oxidation and their major minerals are olivine ($Fa_{0.3-0.4}$), high and low-Ca pyroxene, and plagioclase ($An_{0.2-0.9}$). Opaque minerals include magnetite, pyrrhotite and pentlandite. Embedded in the matrix are mm-sized chondrules without igneous rims. Refractory inclusions are not as common as in CV or CO chondrites. The CK chondrites have petrologic type 4 ("CK4") or higher.

- **CR chondrites** are named after the Renazzo meteorite. They contain ~ 50 vol.% of large porphyritic chondrules (0.7 mm) that often have igneous rims (Table 2.2). Refractory inclusions are relatively rare. Metal is relatively abundant (5–8 vol.%). About half of the fine-grained matrix is hydrated, which explains their lower densities of 2.92–3.29 g cm^{-3} compared to CV, CO and CK chondrites. The CR chondrites are mainly of petrologic type 2 ("CR2").

- **CH chondrites** are closely related to CR chondrites. Their metal content of ~ 20 vol.% is relatively high. One of the earliest members discovered of this group is the meteorite Allan Hills 85085 found in Antarctica. CH chondrites contain small chondrules and refractory inclusions (~ 0.02 mm). The sulfide content in CH chondrites is low. Fine-grained matrix is only observed in their xenolithic (foreign) clasts. The volatile elements in CH chondrites are substantially lower than in CI chondrites (Figures 2.10 and 2.11).

- **CB chondrites** are relatively metal-rich. They are named after the Bencubbin meteorite found in Australia. The Bencubbin-type chondrites fall into two subgroups. In the CBa subgroup, metal makes up about half of the meteorite volume and the rest is mainly cm-sized chondrule-like objects. The CBb subgroup has larger amounts of metal (~ 70 vol.%) and contains small chondrules (0.2–1 mm). In all CB chondrites, refractory inclusions are rare, and fine-grained matrix that is present in other carbonaceous chondrites is lacking.

Figure 2.17 compares the elemental abundances in CI, CM and CV chondrites, normalized to Si $= 1$. Elements with the same relative abundances as in CI chondrites plot at unity, shown by the dotted lines in the diagrams. Elements enriched in CM and CV chondrites relative to CI chondrites plot above, elements that are depleted plot below these

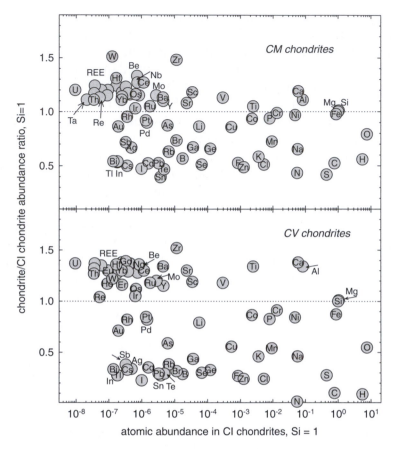

Figure 2.17   Atomic element/Si ratios in CM and CV chondrites relative to CI chondrites plotted against the element/Si abundances in CI chondrites. Elements with abundance ratios as in CI chondrites plot along unity (dotted line).

lines. The CM and CV chondrites have higher relative abundances of refractory elements, *e.g.*, Ca, Al, Ti, rare earth elements (REE) W, Ir and lower abundances of volatile elements, *e.g.*, Na, K, Au In, S, Tl, Pb and C, than CI chondrites. These depletions are also larger for CV chondrites than for CM chondrites. The elemental fractionations are ascribed to condensation and evaporation processes (Chapter 3).

Despite having the most primitive elemental composition, the CI chondrites have experienced more aqueous alteration (petrologic type 1; CI1) than any other (known) chondrites. In the following section, we examine how the aqueous alterations in CI chondrites came about. The CM2 and CR2 chondrites are less aqueously altered and their chemistry provides hints about what may have happened on the CI chondrite parent body. Minor evidence of aqueous activity is found in other carbonaceous chondrites (*e.g.*, CV3), as well as in primitive ordinary chondrites. Aqueous alterations are of course not restricted to meteorites. For example, reactions of olivine and pyroxene with water to form serpentine are important for the destruction of terrestrial mantle rocks (peridotites), and the formation of ore deposits. Formation of secondary minerals such as magnetite, carbonates and sulfates is also important if atmospheric greenhouse gases such as CO_2 or SO_2 become involved. Such reactions apparently also happened on Mars and probably on almost every other solar system object where liquid water once was or still is present (Chapter 4).

2.3.7 Hydrous Silicates and Salts: The Case of CI Chondrites

All CI1 chondrites are soft, friable masses that easily decompose in water. Upon conclusion of their investigation of the Revelstoke CI chondrite, which had landed on a snow field, and of which about 1 g was better preserved, Folinsbee *et al.* (1967)[23] concluded: "... a carbonaceous chondrite...a material as perishable as the alchemist's earth, air, fire, and water and not very stable in any of these media". Indeed, one may wonder how such delicate masses were produced within the first 50 million years, or less, in the solar system.

The mineralogy of CI chondrites resulted from reactions of liquid water with anhydrous precursor minerals such as olivine, pyroxene, metal and troilite that accreted to the CI chondrite parent body.[24,25] These were converted into hydrous silicates, magnetite, pyrrhotite, pentlandite and salts. However, CI chondrites even continue to change in the mineral cabinets in the museums.

Table 2.5 lists some of the abundant hydrous sheet minerals found in carbonaceous chondrites. In CI, CM and CR chondrites, major phyllosilicates are serpentine minerals (named after Latin *serpens* for snake, for the grey-green snake-like mineral colors) and talc minerals called smectites (derived from Greek "*smekhein*" meaning "to clean, to wash off" alluding to the early uses of clay as "soap").

Olivine and pyroxene react with water at low temperatures (*ca.* < 400 °C) to form serpentine. At these temperatures, gas–solid reactions are quite slow and incompatible with the age of the phases in CI chondrites (below and Chapter 3), and all reactions took place in liquid water. We give the standard Gibbs energies of reaction, $\Delta_r G^\circ$, at 298.15 K, which is about the temperature at which aqueous alterations

Table 2.5 Phyllosilicates observed in carbonaceous chondrites.

Serpentines = $M_6^{II}[Si_4O_{10}](OH)_8$	
$Mg_6[Si_4O_{10}](OH)_8$	Chrysotile, lizardite, antigorite
$(Fe^{II},Fe^{III})_{4-6}[Si_4O_{10}](OH)_8$	Greenalite
$Fe_4^{II}[Fe_2^{III}(Si,Fe^{III})_2O_{10}](OH)_8$	Cronstedite
$(Mg, Fe^{II}, Fe^{III})_{4-6}[(Si,Al)_4O_{10}](OH)_8$	Berthierine
$(Mg_2Al)_2[(Si,Al)_2O_{10}](OH)_8$	Amesite
Chlorites = $M_5^{II}Al[(Si_3Al)O_{10}](OH)_8$	
$(Mg,Fe^{II})_5Al[(Si_3Al)O_{10}](OH)_8$	Clinochlore
$(Mg,Fe^{II},Fe^{III})_5Al[(Si_3Al)O_{10}](OH)_8$	Chamosite
Smectite clays = $(A^I)_{0.33}(M^{II},M^{III})_{2-3}[(Si,Al)_4O_{10}](OH)_2 \cdot nH_2O$	
$Mg_3[Si_4O_{10}](OH)_2$	Talc
$(Ca_{0.5},Na)_{0.33}Mg_3[Si_{3.7}Al_{0.33}O_{10}](OH)_2 \cdot nH_2O$	Saponite
$(Ca_{0.5},Na)_{0.33}(Al,Mg,Fe^{II})_2[(Si,Al)_4O_{10}](OH)_2 \cdot nH_2O$	Montmorillonite
Vermiculite clays = $(M^{II})_{0.33}(M^{II},M^{III})_3[(Si,Al)_4O_{10}](OH)_2 \cdot 8H_2O$	
$(Mg,Fe^{II})_{0.33}(Mg,Fe^{II},Al)_3[(Si,Al)_4O_{10}](OH)_2 \cdot 8H_2O$	Vermiculite
Micas = $(A^I)(M^{II},M^{III})_{2-3}[(Si,Al)_4O_{10}](OH,F)_2$	
$CaAl_2[Si_2Al_2O_{10}](OH,F)_2$	Margarite
$K(Mg,Fe^{II})_3[(Si_3Al_2O_{10}](OH,F)_2$	Phlogopite
$Ca(Mg,Al)_3[(Si_{1.25}Al_{2.25}O_{10}](OH,F)_2$	Clintonite

occurred on the CI chondrite parent body. An energetically favorable reaction is olivine with water to form serpentine and hydroxides:

$$4(Mg, Fe)_2SiO_4 \text{ (olivine)} + 6H_2O(liq)$$
$$= (Mg, Fe)_6[Si_4O_{10}](OH)_8 \text{ (serpentine)} + 2(Mg, Fe)(OH)_2 \text{ (hydroxide)}$$

$$(2.17)$$

with $\Delta_rG° = -94\,kJ\,mol^{-1}$ for the reaction of the Mg-end-members forsterite, chrysotile and brucite. Another reaction is olivine and pyroxene to serpentine:

$$2(Mg, Fe)_2SiO_4 + 2(Mg, Fe)SiO_3 + 4H_2O(liq.)$$
$$= (Mg, Fe)_6[Si_4O_{10}](OH)_8$$

$$(2.18)$$

This reaction for the Mg-end-members is also exothermic ($\Delta_rG° = -92\,kJ\,mol^{-1}$). One favorable net reaction for the formation of talc, $Mg_3[Si_4O_{10}](OH)_2$, is:

$$5MgSiO_3 \text{ (enstatite)} + H_2O(liq) = Mg_3[Si_4O_{10}](OH)_2 \text{ (talc)}$$
$$+ Mg_2SiO_4 \text{ (forsterite)}; \Delta_rG° = -46\,kJ\,mol^{-1}$$

$$(2.19)$$

Another favorable net reaction forms talc from chrysotile serpentine:

$$Mg_6[Si_4O_{10}](OH)_8 \text{ (chrysotile)} + 3CO_2(aq) = Mg_3[Si_4O_{10}](OH)_2 \text{ (talc)}$$
$$+ 3MgCO_3 \text{ (magnesite)}; \Delta_rG^\circ = -98\,kJ\,mol^{-1}$$

(2.20)

Reaction (2.20) involves CO_2 dissolved in water, denoted as $CO_2(aq)$.
Reaction (2.20) can occur because water is not the only volatile component that accreted to the CI parent body. CI chondrites contain carbonates as well as ammonia salts. The CO_2 needed to make carbonates can come from reactions of water with abundant organics; or alternatively, CO_2, a minor low-temperature gas or ice in the solar nebula, accreted to the CI parent body. Carbon dioxide is involved in terrestrial serpentinization reactions, and its reactions with olivine to serpentine and talc are extensively studied because they are important for the removal of the greenhouse gas CO_2 from the terrestrial atmosphere (Chapter 5). The simplified reaction for partial carbonation of olivine is:

$$Mg_2SiO_4 \text{ (forsterite)} + CO_2 = MgSiO_3 \text{ (enstatite)} + MgCO_3 \text{ (magnesite)} \quad (2.21)$$

with $\Delta_rG^\circ = -40\,kJ\,mol^{-1}$ for the reaction with CO_2 as gas, and $\Delta_rG^\circ = -48\,kJ\,mol^{-1}$ for the reaction with CO_2 in aqueous solution. Full carbonation of olivine in aqueous solution is:

$$Mg_2SiO_4 \text{ (forsterite)} + 2CO_2(aq)$$
$$= 2MgCO_3 \text{ (magnesite)} + SiO_2 \text{ (quartz)}, \Delta_rG^\circ = -90\,kJ\,mol^{-1}$$

(2.22)

Although the exothermic carbonation reactions are favorable, they may not be relevant to CI chondrites because there is no abundant quartz. However, a reaction of olivine with both water and CO_2 produces serpentine and carbonate, which are both observed in CI chondrites:

$$4(Mg, Fe)_2SiO_4 \text{ (olivine)} + 4H_2O(liq) + 2CO_2(aq)$$
$$= (Mg, Fe)_6[Si_4O_{10}](OH)_8 \text{ (serpentine)} + 2(Mg, Fe)CO_3 \text{ (carbonate)}$$

(2.23)

with $\Delta_r G° = -189 \, \text{kJ mol}^{-1}$ for the Mg-end-member minerals. At higher CO_2 concentrations, the reaction proceeds to talc:

$$4Mg_2SiO_4 \, (\text{forsterite}) + H_2O(\text{liq}) + 5CO_2(\text{aq})$$
$$= Mg_3[Si_4O_{10}](OH)_2 \, (\text{talc}) + 5MgCO_3 \, (\text{magnesite}); \; \Delta_r G° = -283 \, \text{kJ mol}^{-1}$$
(2.24)

The standard Gibbs free energies of the reactions of olivine to serpentine and talc at 298.15 K and 1 bar total pressure are thermodynamically favorable. At higher temperatures (and given pressure) the $\Delta_r G°$ of these reactions decreases, and the reaction equilibria shift to decarbonation. These reactions are also favored at high CO_2 partial pressures, but it is doubtful that CO_2 partial pressures on the CI parent body were ever high enough to drive the reactions towards carbonates. Gas–solid carbonation reactions are very slow and require several hundred thousand years for completion even under terrestrial atmospheric conditions. The rate-limiting step in the reactions is the removal of carbonate, which is more easily achieved in aqueous solution. In the presence of liquid water, carbonic acid forms:

$$H_2O + CO_2 = H^+ + HCO_3^- = 2H^+ + CO_3^{2-}$$
(2.25)

which facilitates the olivine transformation to serpentine by driving Mg^{2+} and Fe^{2+} into solution:

$$4(Mg, Fe)_2SiO_4 \, (\text{olivine}) + 4H^+ + 2H_2O(\text{liq}) + 2HCO_3^-$$
$$= (Mg, Fe)_6[Si_4O_{10}](OH)_8 \, (\text{serpentine}) + 2(Mg, Fe)^{2+} + 2HCO_3^-$$
(2.26)

Carbonates remain in solution in CO_2-rich solutions whereas Fe hydroxides precipitate in CO_2-poor solutions. Divalent Fe^{2+} and Ni^{2+} also remain in solution once NH_4^+ ions are present, which must be considered here. Ammonium salts are present in CI chondrites, hence ammonia ices, like water and CO_2-bearing ices, must have accreted to the CI parent body, which upon heating produced ammoniacal solutions, $NH_3 + H_2O = NH_4^+ + OH^-$. This drives the reaction $NH_4^+ + CO_3^{2-} = NH_3 + HCO_3^-$ and the dissolution of insoluble carbonates to soluble hydrogen carbonates. Similarly, hydroxides can stay in solution when NH_4^+ lowers the OH^- concentration required for precipitation.

2.3.7.1 Hydroxides.
Hydroxides such as brucite, $Mg(OH)_2$, and more abundantly ferrihydrite are present in CI chondrites. Ferrihydrite,[26] occasionally written as $Fe^{2+}(OH)_2$, $Fe_{4-5}(OH,O)_{12}$ or as trivalent

$Fe_2O_3 \cdot 5H_2O$, is closely associated with phyllosilicates and magnetite. Divalent ferrihydrite is a product of olivine serpentinization with water, and hydroxides with intermediate valence between Fe^{2+} and Fe^{3+} result from metal and sulfide oxidation as described below. Round ferrihydrite aggregates ($>0.3\,\mu m$ in diameter) made of ~ 0.008-μm size spheres can resemble the "framboid" morphologies of magnetite.[27] In rare instances limonite, $FeO(OH) \cdot nH_2O$ or simply $Fe(OH)_3$, occurs in CI chondrites. Limonite is the collective name for hydrated iron oxides such as lepidocrocite and goethite, which are well-known terrestrial oxidation products; these minerals may have formed in the meteorites after these landed on Earth.

2.3.7.2 Magnetite. Magnetite occupies $\sim 10\%$ of the volume in CI chondrites and is difficult to separate with a magnet from the fine silicate matrix. In CI, CM and CR chondrites, magnetite forms fine (0.010–0.025 mm) "framboid" aggregates consisting of abundant submicron magnetite spherules.[25] These make up half the magnetite in CI chondrites, the other half are magnetite spheres (up to ~ 0.02 mm in diameter) and plates (plaquettes) that are sometimes within or associated with carbonate grains.[27-29] Some magnetite plaquettes resemble stacks of plates with dislocated spiral growth structures characteristic of crystal growth from vapor. However, crystallographical studies suggest that other external factors may have controlled their growth in CI chondrites. All magnetite crystal shapes in CI chondrites are consistent with magnetite formation in aqueous fluids; however, vapor growth of some fraction of the magnetite cannot be completely ruled out.

Magnetite has different O-isotopic compositions and an older age than the phyllosilicates and carbonates. Metal oxidation before hydrous silicate formation is also necessary to produce a Ni-bearing sulfide-hydroxide that is incorporated in the silicates. The age of magnetite in CI chondrites has been determined from ^{129}Xe measurements of magnetite separates. Magnetite formed when radioactive ^{129}I (half-life 15.7 Ma) was still present in the early solar system. The ^{129}I decayed to ^{129}Xe, which was trapped in the magnetite. The $^{129}I/^{129}Xe$ chronometer indicates that ^{129}I was captured within 2–7 Ma after the oldest known solids formed 4567 Ma ago.[26] In contrast, carbonates formed up to 20 Ma later (see below).

Only a few petrographic studies note kamacite *metal* in the CI chondrites (*e.g.*, Ramdohr[30]). Berzelius[31] observed some white shiny platelets in a magnetic separate of Alais that released H_2 when treated with HCl, indicative of metal. On the other hand, Cloëz[32] noted "not the

smallest bubble of hydrogen was disengaged during attack by HCl"
from his fresh Orgueil sample.

The CI, CM and CR chondrites contain *pyrrhotite*, $Fe_{1-x}S$, or more
specifically, $Fe_{0.87}S$ (or Fe_7S_8), instead of stoichiometric FeS, which is
present in carbonaceous chondrites of petrologic types 3 and 4 (CV, CO,
CK chondrites). Pyrrhotite from CI chondrites contains $\sim 1\%$ Ni, and
$\sim 39.4\%$ S. Pyrrhotite in carbonaceous chondrites is sometimes associated with magnetite and appears to have replaced small metal and
magnetite particles.[30] However, discernable pyrrhotite makes up only
1–3 vol.% in CI chondrites. These strongly magnetic, yellow-bronze
hexagonal prisms up to 50 μm in diameter were already identified as
pyrrhotite in CI chondrites by Daubrée,[33] who described them as
"magnetic pyrite" (the old name for pyrrhotite). In contrast to troilite,
Fe_7S_8 reacts with HCl to release H_2S, and small quantities of H_2S from
whole rock samples or magnetic separates are frequently noted for CI
chondrites. Pyrrhotite in CI chondrites shows signs of breakdown;
sometimes elemental S is found with it; in other cases, phyllosilicates
intruded cracked pyrrhotite.[34] Similarly rare is *pentlandite*, $(Fe,Ni)_9S_8$
with an Fe/Ni ratio of ~ 1, which occurs as individual grains or associated with pyrrhotite.[34] Pyrrhotite and pentlandite grains appear to
be sulfide re-precipitations of original troilite. In CM chondrites,
tochilinite, a product of troilite and pyrrhotite decomposition, is finely
intergrown with phyllosilicates.[35,36] Tochilinites have the general formula $4[(Fe,Ni,Cu)S] \cdot (2.4)[(Mg,Fe,Ca)(OH,CO_3)_{1-2}]$ in which hydroxyl
groups can be partly replaced by carbonate. The Ni-bearing tochilinite
called haapalaite, $4[(Fe,Ni)_{1-x}S] \cdot 3(Mg,Fe^{2+})(OH)_2$, is known for
CM chondrites. This mineral has not been found in CI chondrites,[37] but
a comparable mineral could be suspected. The phyllosilicates in CI
chondrites contain $\sim 1\%$ Ni which may imply that the Ni is present as
NiO. However, analyzing the Orgueil meteorite shortly after its fall, the
French mineral dealer and chemist Felix Pisani (1831–1920) notes:[38]

"I also assured myself that the Ni is combined with the sulfur and
not as oxide in the silicate, where it might be, by treating the stone
with ammonium sulfide which dissolved nickel sulfide."

This agrees with Ni and S correlations found in CI and CM chondrite
matrices, which indicate the presence of a Fe-Ni-S-O phase.[39] In CM
chondrites, this phase was described as tochilinite,[35] or as the Fe-S-O-Ni
phase dubbed "FeSON."[36] However, the nature of this phase in CI
chondrites remained unclear. The varying Ni and S concentrations in
ferrihydrite that is spread through the matrix led to the suggestion that

Ni was adsorbed as sulfate onto ferrihydrite in CI chondrites.[37] In light of the findings of extractable "NiS"[38] in the Orgueil meteorite shortly after its fall, the presence of Ni as sulfate now indicates that oxidation of Ni sulfide occurred during the meteorite's residence time on Earth.

Magnetite forms by oxidation of Fe in metal and troilite. Metal and sulfide oxidation reactions may have occurred at different times, by different oxidizing agents, and/or under different conditions than reactions producing phyllosilicates and their associated carbonates. This is concluded from the different ages and O-isotopic compositions of phyllosilicates, carbonates (see below) and magnetite. The O-isotopes of carbonates and hydrous silicates fall on the same fractionation curve, which has an offset of $\Delta^{17}O = \sim 0.4‰$ from the terrestrial fractionation curve (Figure 2.18), indicating that carbonates precipitated from the same aqueous fluids that are responsible for hydrating the silicates. In contrast, magnetite O-isotopic compositions in CI chondrites define a separate curve with a higher offset ($\Delta^{17}O = +1.7‰$), indicating formation in an environment with different O-isotopic composition.

Metal oxidation to magnetite requires either oxidizing gases or an oxidant in aqueous fluids, which could be water itself. One clue for the reactions is the magnetite composition. Magnetite in CI chondrites is

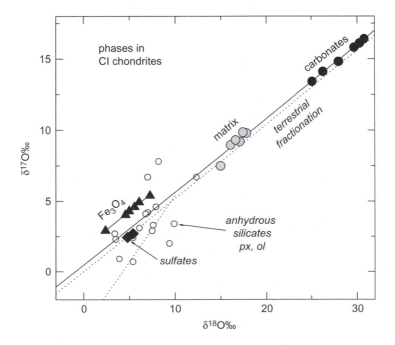

Figure 2.18 Oxygen isotope systematics in CI chondrites. Data from ref. 40.

quite pure; it has only 50 ppm Ni on average.[25] Chondritic metal usually contains several percent Ni and abundant siderophile elements, thus the low Ni in magnetite indicates that magnetite did not form by direct gas–solid oxidation of pre-existing chondritic FeNi metal. Gas–solid oxidation by gases such as CO_2 or H_2O ultimately leads to Ni-rich magnetite:

$$3(Fe, Ni)\,(metal) + 4H_2O(gas) = (Fe, Ni)_3O_4\,(magnetite) + 4H_2(gas) \quad (2.27)$$

$$3(Fe, Ni)\,(metal) + 4CO_2(gas) = (Fe, Ni)_3O_4\,(magnetite) + 4CO(gas) \quad (2.28)$$

Oxidation first removes Fe as Fe_3O_4 to leave more Ni-rich metal behind, but the Fe_3O_4 will have Ni concentrations that are proportional to the Ni-content of co-existing metal. In the end, complete oxidation of FeNi by gases leaves an oxide with the same Ni/Fe ratio as in the initial metal alloy. Starting with chondritic FeNi, Ni-bearing magnetite should result, or magnetite from which trevorite, $NiFe_2O_4$, may have exsolved.

In contrast, oxidation of FeNi metal in aqueous fluids can produce Ni-poor magnetite if oxidation is followed by a selective dissolution or precipitation of Fe. This also provides means to make the Ni available for incorporation into silicates, which contain $\sim 1\%$ Ni in CI chondrites. The oxidation of metal to magnetite in aqueous solutions starts with oxidation to Fe^{2+} and Ni^{2+}, written in the following reactions for Fe:

$$Fe(metal) + 2H_2O(liq) = Fe(OH)_2(solid) + 2H_2(gas) \quad (2.29)$$

In the absence of NH_4^+ salts, the hydroxides precipitate. In $CO_2 -$ bearing solutions, siderite forms:

$$Fe(metal) + 4CO_2(aq) + H_2O(liq) = FeCO_3(siderite) + 2CO(gas) \quad (2.30)$$

Siderite precipitates as long as CO_2 concentrations do not favor formation of aqueous HCO_3^- ions. However, the subsequent oxidation to magnetite on the CI chondrite parent body most likely proceeded through the same "green rust" stages that are observed during metal oxidation in O_2-free, neutral to basic solutions with controlled oxidants. Green rust is the collective name for intermediate Fe metal corrosion products that are not stable in the presence of O_2. The natural mineral, fougerite $(Fe^{2+},Mg)_6Fe_2^{3+}(OH)_{18} \cdot 4H_2O$, resembles green rusts and, just like them, it decays rapidly to brown limonite when fully exposed to air. Green rusts are often encountered in anoxic aquatic systems as intermediates of metal and sulfide corrosion, and the strong reducing action of Fe^{2+} in green rusts plays an important role in the remediation of organic contaminants, such as chlorinated hydrocarbons.[41–43]

Green rust contains Fe^{2+} and Fe^{3+} in hydroxide layers that are stabilized by interlayers made of negatively charged anions. The general green rust formula is:

$$[Fe^{2+}_{6-x}Fe^{3+}_x(OH)_{12}]^{x+}[(A^{n-})_{x/n} \cdot yH_2O]^{x-} \qquad (2.31)$$

where $x = 0.9$–4.2. $(A^{n-})_{x/n}$ represents a n-valent anion, mostly Cl^-, SO_4^{2-} or CO_3^{2-}, but also OH^-, HCO_3^-, SO_3^{2-}, Br^- or I^-. The S^{2-} has not yet been reported as a stabilizing anion. The number of interlayer water molecules incorporated in the green rust structures is typically $y = 2$–4. One can regard green rusts as oxidation products with intermediate valence states between Fe^{2+} and Fe^{3+}. The end-member for $x = 0$ is "$6Fe(OH)_2 \cdot yH_2O$," which is the first expected oxidation product from Fe metal in aqueous solution.

In solution with carbonates, $FeCO_3$ formation from metal competes with the formation of $Fe(OH)_2$. However, subsequent oxidation to magnetite is aided by green rust intermediates that facilitate Fe^{2+} to Fe^{3+} oxidation.[43] In magnetite, the stoichiometric Fe^{2+} to Fe^{3+} ratio is $1 : 2$; thus at least two Fe^{3+} must be present per formula of green rust to make one Fe_3O_4. This corresponds to $x = 2$ in the green rust formula in (2.31):

$$[Fe^{2+}_4Fe^{3+}_2(OH)_{12}]^{2+}[CO_3 \cdot yH_2O]^{2-} \text{ (green rust)}$$
$$= Fe_3O_4 \text{ (magnetite)} + 3Fe(OH)_2 \text{(solid)} + CO_2\text{(aq)} + (3+y)H_2O\text{(liq)}$$
$$(2.32)$$

Oxidation of all initial Fe-metal to magnetite requires a nominal $x = 4$, close to the known limit of x in the green rust formula:

$$[Fe^{2+}_2Fe^{3+}_4(OH)_{12}]^{4+}[2CO_3 \cdot (2.4)H_2O]^{4-} \text{ (green rust)}$$
$$= 2\,Fe_3O_4 \text{ (magnetite)} + 2CO_2\text{(aq)} + (6+y)H_2O\text{(liq)} \qquad (2.33)$$

Green rust formation provides an interesting mechanism to explain how the short-lived radioactive [129]I was captured, whose decay product [129]Xe is now found in magnetite of CI chondrites to attest to its old age. As mentioned above, green rusts contain stabilizing anions such as Cl^- but also can take Br^- and I^-. Hence there is a plausible way to incorporate [129]I into the precursor of magnetite in CI chondrites. "Amorphous" ferrihydrite $Fe_2O_3 \cdot 5H_2O$ has good adsorption properties because it tends to precipitate as small, 0.005–0.01 micron size, spheres, which gives a large surface area.[29] This morphology is reminiscent of the magnetite and ferrihydrite framboids found in CI chondrites. Green rusts have similar large surface areas and adsorption properties, which make them useful for removing environmental pollutants.[44] Taking these analogies, the precipitated green rust phases on the CI chondrite parent body could easily capture [129]I and keep the [129]Xe produced from [129]I decay as an adsorption product. In that case, the [129]I-[129]Xe radioactive dating clock could even start in the green rust phases before these were desiccated to give magnetite.

The Ni in the FeNi metal alloy may oxidize to Ni^{2+} like Fe metal oxidizes to Fe^{2+}, but it is unlikely that Ni is further oxidized to Ni^{3+}. Even the oxidation of metallic Ni to Ni^{2+} could be sluggish if the oxidation potential of the aqueous solution is too low. If Ni is not oxidized, the transitional Fe-oxyhydroxides and green rusts will be essentially Ni-free, and lead to the observed Ni-poor magnetite. However, if Ni is oxidized to Ni^{2+}, Ni must be separated to prevent its incorporation into magnetite. A possible way to do this is by NiS formation from troilite, a likely precursor material on the CI parent body. The exchange reaction $FeS + Ni^{2+} = NiS + Fe^{2+}$ would make NiS available to produce tochilinite ($4[(Fe,Ni)_{1-x}S] \cdot 3(Mg,Fe)(OH)_2$) for later incorporation into silicates; an illustrative reaction is:

$$4FeS \text{ (troilite)} + Ni(OH)_2(s) + 2Fe(OH)_2(s)$$
$$= [3FeS + NiS] \cdot 3Fe(OH)2 \text{ (tochilinite)} \tag{2.34}$$

Tochilinite, which nominally only contains Fe^{2+}, is not reported for CI chondrites, but Ni sulfide in CI chondrites is associated with Fe^{2+} and Fe^{3+}-bearing ferrihydrite in the silicates (see above). Thus, the sulfide-bearing ferrihydrite in CI chondrites could resemble a more oxidized form of tochilinite. Since both tochilinite and green rusts have hydroxyl interlayers as structural components, one may speculate if hydroxides in green rusts could be substituted by sulfides and aid oxidation of Fe^{2+} to Fe^{3+}. Considering that pyrrhotite is also a phase with mixed Fe valences ($Fe_7S_8 = "5Fe^{2+}S + Fe_2^{3+}S_3"$), oxidation of Fe^{2+} to Fe^{3+} in sulfide must have happened.

In addition to carbonates, several percent of *salts* such as sulfates and chlorides (*e.g.*, NH_4Cl, KCl, $NaCl$, $MgSO_4$, $CaSO_4$) can be extracted from CI chondrites, and in smaller quantities from CM chondrites. Berzelius[31] found 10.3% soluble salts in Alais, and Clöez[32] found 5.3–6.4 wt% soluble salts in Orgueil, which he suspected "serve as some sort of cement" to hold the rock together. Salts fill veins and cracks that cross though the phyllosilicates in CI chondrites, and several generations of carbonate deposits are recognized.[24,27,28,45]

2.3.7.3 Carbonates. Carbonates of Mg, Fe and Ca are frequently observed in veins and as individual crystals in the matrix in CI chondrites.[28] Carbonates occur in smaller quantities in CM and other aqueously altered carbonaceous chondrites. Pisani[38] found 0.5–3 mm size whitish, clear crystals that reacted with HCl in the Orgueil meteorite, and Des Cloizeaux[46] identified them as breunnerite, $(Mg,Fe)CO_3$. Daubrée[47] concluded that carbonates must be indigenous to the Orgueil meteorite and not of secondary terrestrial origin because they are present inside freshly broken meteorite samples. Dolomite, $(Ca,Mg)CO_3$, was identified about 100 years later.[24] Calcite, $CaCO_3$, often present as larger crystals, is sometimes associated with magnetite. Dolomite-magnetite intergrowths and carbonate associations with sulfide and sulfates suggest that formation conditions were variable.[27] The formation of carbonates is related to that of the phyllosilicates, which is seen by the O-isotopes (Figure 2.18). However, the varying compositions of carbonates precipitated from solution must reflect changes in fluid composition over time. The ^{87}Rb-^{87}Sr chronometer (^{87}Rb half-life 4.88×10^{10} years) dates carbonate formation to within 50 million years;[48] dating with the short-lived ^{53}Mn-^{53}Cr chronometer (^{53}Mn half-life 3.74 Ma) shows breunnerite deposition within 20 Ma after the first solids had formed in the solar system.[49] This is up to ∼10 times later than the magnetite formation ages of ∼2–7 Ma. One important aspect is that as long as ammonium salts are abundant in solution then carbonates can remain dissolved. However, carbonates begin to precipitate when NH_4^+ concentrations drop, which would happen if NH_3 evaporated from the parent body over time.

2.3.7.4 Sulfates. Sulfates of Mg, Ca, Na and K are the major salts in water extracts of CI chondrites.[24,31,32,38] Epsomite, $MgSO_4 \cdot 7H_2O$, is the major sulfate component, followed by Na, Ca and Ni, but Fe is notably absent. The heptahydrate epsomite or the hexahydrate, $MgSO_4 \cdot 6H_2O$, with the mineral name hexahydrite, are abundant in the phyllosilicate veins. The different amount of crystal water is simply due to the temperature and humidity under which the hygroscopic $MgSO_4$ is analyzed. In epsomite and hexahydrite, Mg^{2+} is six-fold co-ordinated, $[Mg(H_2O)_6]^{2+}$, and the seventh H_2O in epsomite is only weakly bound, hence epsomite easily loses one H_2O in dry air.

The hydrated double sulfate mineral bloedite, $Na_2Mg(SO_4)_2 \cdot 4H_2O$, crystallizes from solutions with epsomite between 5 and 60 °C and has been found *in situ* with epsomite in phyllosilicate veins. Mascagnite, the sulfate of ammonia, $(NH_4)_2SO_4$, has not been observed *in situ* in CI chondrites, but it crystallized from water extracts of freshly fallen Alais and Orgueil samples.

However, most of the sulfates appear to be modern precipitations, which was already suspected by Berzelius,[31] and the amount of extractable sulfates seems to be a function of a CI-chondrite's residence time on Earth (see below). In 1834, Berzelius analyzed the Alais meteorite, which had fallen in 1806, and he notes about his sample:[31]

"Its color is black with a touch of grey, with dense, fine, white dots or incrustations. This is not reported in the older descriptions; only in the *Dictionaire des Sciences Naturelles, XXX*, p. 339 [from 1824], it is noted that this meteorite tends to cover itself with some efflorescence, which the authors give as iron vitriol."

[Berzelius shows that the metal-sulfate (= vitriol) of Fe was a misidentification; $FeSO_4$ is never found in any significant quantities in water extracts or in crystalline form.] Sulfate efflorescences on CI chondrites continue to form easily if samples are not stored in tight desiccators, which was more recently emphasized again.[50]

If sulfates are initially present, they only may hydrate when CI chondrites are exposed to the moist terrestrial atmosphere. The volume increase associated with the hydration of $MgSO_4$ cracks and fractures the meteorite. Hygroscopic activity and water migration are apparently able to transport salt deposits into fresh veins and to bring salt efflorescences to the exterior of the rocks.

The hygroscopic nature of lesser hydrated $MgSO_4$ introduces problems for the determination of the original water content of the CI chondrites. This is important because bound water on early planetesimals is one source of water for the terrestrial planets. Pisani[38] dried a fresh Orgueil sample at 110 °C. It lost 9.15% H_2O, but regained 7% water from the air while it sat open on the balance for a few hours. The D/H ratio of the crystal water released from CI chondrites is terrestrial, indicative of water uptake and/or exchange with the air. However, water from the hydrous silicates, only released above 250–300 °C, has a higher D/H ratio and is clearly non-terrestrial,[51] which is also seen from the O-isotopic compositions of the phyllosilicates (Figure 2.18).

One has to ask whether any non-hydrated $MgSO_4$ or the monohydrate kieserite, $MgSO_4 \cdot H_2O$, were already present in CI chondrites before they arrived, or if most or all Mg-sulfate was produced by reactions of phases containing Mg and S in moist air. The O-isotopes of sulfates from water extracts of two CI chondrites fall onto the terrestrial fractionation curve (Figure 2.18), indicating a terrestrial origin of sulfates, but sulfates in the less altered CM chondrites do not, indicating that sulfates can be of pre-terrestrial origin.[52] One cannot completely rule out that some fraction of the sulfates in CI chondrites is also pre-terrestrial. A signature from small amounts of indigenous sulfate with O-isotopic composition such as measured for carbonates is easily diluted by a large contribution from sulfates that were produced with terrestrial oxygen.

If the sulfates are terrestrial weathering products, we can look at CI chondrite analyses done over time and check if there is an increase of the measured sulfate content over time. The Orgueil meteorite fell more than 140 years ago, and is the most studied CI chondrite, so plenty of information should exist. Figure 2.19 shows the analyses in the literature for S in sulfate form.

Figure 2.19 shows several things. First, three independent wet chemical analyses were done within ~2 months after the meteorite fall, and no other wet chemical analyses were done until the onset of the space age ~100 years later. Second, analyses by different groups, methods and sample sizes give more spread in values. Third, the meteorite keeps changing. The first three analyses give 0.8–1.4% S in sulfate form (the total S in all forms was 5.6–6.7%). These amounts of S in sulfate represent the upper limit to the pre-terrestrial sulfate content in CI chondrites. The five analyses done more recently show about twice the original reported amounts of S in sulfate form. A simple linear regression suggests an initial amount of 1% S as sulfate, and a conversion rate of ~0.012 g-S yr^{-1} for the Orgueil meteorite. Other available data

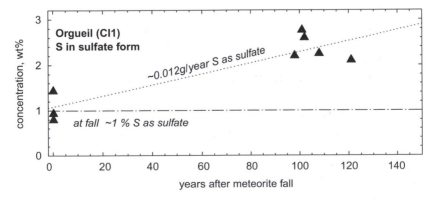

Figure 2.19 Measurements of sulfur in the form of sulfate in the Orgueil CI chondrite as a function of time after its fall, from several sources in the literature. If the analysis date was not given, the year of publication is used.

further suggests that the (overall small) concentration of elemental S increased somewhat, whereas the concentration of S in sulfide form decreased over time. Hence, the sulfide oxidizes to sulfate in air:

$$Fe_7S_8 \text{ (pyrrhotite)} + 14O_2(gas) = 7FeSO_4(solid) + SO_2(gas) \qquad (2.35)$$

If condensed water or crystal water is present, sulfurous acid and sulfuric acid form:

$$SO_2(aq) + H_2O(liq) = H_2SO_3(aq)$$
$$H_2SO_3(aq) + 0.5O_2(gas) = H_2SO_4(aq) \qquad (2.36)$$

If oxidation is by gas–solid reactions, the hygroscopic sulfate may take up water from the air to form mellanterite, $FeSO_4 \cdot 7H_2O$. However, $FeSO_4$ or its hydrated form easily oxidizes to magnetite and further to limonite, which explains why Fe is not found in any significant quantities in the water extracts from CI chondrites:

$$FeSO_4 \cdot 7H_2O \text{ (mellanterite)} + \tfrac{1}{4}O_2(gas) = Fe(OH)SO_4(solid) + 6\tfrac{1}{2}H_2O(liq)$$

$$Fe(OH)SO_4(solid) + 2H_2O(liq) = Fe(OH)_3(solid) + H_2SO_4(aq)$$

Overall:

$$Fe_7S_8 \text{ (pyrrhotite)} + 34.5O_2(gas) + 18.5H_2O(liq) =$$
$$7Fe(OH)_3 \text{ (limonite)} + 8H_2SO_4(aq) \qquad (2.37)$$

The observation of epsomite efflorescence and soluble $MgSO_4$ from CI chondrites also requires a source of Mg. Strong acids such as sulfuric acid facilitate the extraction of Mg^{2+} from hydrous silicates such as serpentine, which contains sheets of SiO_4 tetrahedra alternating with layers of $Mg(OH)_2$ octahedra. Protonization of the O in the Mg–O–Si bonds aids the bond break-up and dissolution of Mg^{2+}:

$$Mg_6[Si_4O_{10}](OH)_8 \text{ (chrysotile)} + 12H^+(aq)$$
$$= 6Mg^{2+}(aq) + 4Si(OH)_4(aq) + 2H_2O(liq) \qquad (2.38)$$

A by-product of this reaction is aqueous silicic acid (H_4SiO_4). Small quantities of SiO_2 have been reported in sulfates analyzed by electron microprobe.[28] If the reported SiO_2 was not from interference of neighboring silicates, the SiO_2 resulting from the acid leach of serpentine may have found its way into the sulfates.

Whatever the detailed formation story of the hydrous minerals, magnetite, and salts may have been on the CI chondrite parent body, at the very end water must have evaporated, leaving behind phyllosilicates cemented together by salt deposits. Only then did the former muddy rock components take the shape of consolidated masses that were ejected from the CI chondrite parent body. Removal of salt-laden aqueous fluids from the rocks did not happen, because there is no evidence for loss of easily soluble compounds (Na, K, Ca, Mg, chlorides, carbonates, sulfates, *etc.*). Thus, aqueous alteration occurred isochemically in a closed system, and the elements were only redistributed among minerals.

2.4 IRON METEORITES

As the name suggests, the principal constituent of iron meteorites is Fe-Ni metal. Depending on the overall Ni content, the metal is the relatively Ni-poor alloy called kamacite (from καμας, "*kamas*", for pole or shaft) and/or the Ni-rich alloy called taenite (from ταινια, "*tainia*," ribbon). Kamacite is α-Fe with Ni < 6% with a body centered cubic (bcc) structure. Taenite or γ-Fe, is the face-centered cubic (fcc) Fe-Ni solid solution with Ni > 6%. Iron meteorites called hexahedrites contain only kamacite, octahedrites contain kamacite bands that are bordered by taenite lamellae, and ataxites contain only taenite. Another component found in octahedrites is plessite, which is not a single phase but a fine mixture of kamacite and taenite that fills the spaces between the intergrown structures of kamacite bands and taenite lamellae, or it is embedded in larger taenite crystals. Inclusions or lamellae of troilite (FeS), schreibersite [$(Fe,Ni)_3P)$], cohenite (Fe_3C), graphite and daubréelite ($FeCr_2S_4$) are accessory minerals in many iron meteorites. In some groups of iron meteorites, silicate and SiO_2 inclusions occur as well.

Nickel concentrations in iron meteorites generally range from 4 to 20%, with very Ni-rich or Ni-poor irons being more exceptional. About ten iron meteorites have Ni contents greater than 20%. This group includes Twin City (30% Ni), Santa Catharina (35% Ni), Dernbach (42% Ni) and Oktibbeha County (58% Ni). We focus on the majority of iron meteorites having up to 20% Ni in our discussion. The relatively high Ni content in iron meteorites is in contrast to the composition of most industrial iron metal and steels, which are essentially Ni-free. Hence, a test for Ni is practical to check whether a found metal mass is indeed meteoritic or some artificial product. The presence of Ni is determined by dissolving a small piece of the metal in dilute HCl, oxidizing the solution with H_2O_2, precipitating $Fe(OH)_3$ with alkaline solution, filtering and neutralizing the solution, and then adding Tschugaeff's reagent, a 1% solution of dimethyl glyoxime in alcohol. Precipitation of red flakes of the Ni-dimethyl glyoxime complex indicates the presence of larger amounts of Ni, while smaller amounts of Ni may only turn the solution pink or red. The Ni test also works for metal of stony-irons and chondritic meteorites, because their metal contains percent-levels of Ni.

The structural classification of iron meteorites into hexahedrites, octahedrites and ataxites uses details in their crystal structures that become visible on freshly polished and etched surfaces. Kamacite is less resistant to acid etching than taenite and taenite stands up in relief, which brings out the regular structures in octahedrites that are known as Widmanstätten figures or patterns (Figures 2.20 and 2.21), named after their discoverer Alois von Widmanstätten (1753–1849) in Vienna in 1808. Widmanstätten did not discover the patterns by etching iron meteorites. He laid down a small polished piece of iron meteorite onto an asbestos plate and then fired it over a Bunsen burner to oxidize it in

Figure 2.20   A polished and etched slice of the Gibeon IVA iron meteorite. The etching reveals the Widmanstätten pattern formed by two different intergrown FeNi alloys. Gibeon is a fine octahedrite with bandwidth of 0.30 mm. The dark spots are mainly troilite inclusions. The piece measures about 13 cm across at the top. Photograph by the authors.

Figure 2.21   A polished and etched surface, *ca.* 10 cm wide, of the Saint Francois
County iron meteorite found in Missouri before 1863. It is a coarse
octahedrite (type IC) with a kamacite bandwidth of 2.7 mm, and it
contains about 6.8% Ni. Photograph by the authors.

air. Alloys with different Fe–Ni content oxidize at different rates and
therefore acquire different thicknesses of oxide tarnishes and films that
appear in contrasting golden-yellow, blue or purple. Such "colored"
Widmanstätten patterns may have some artistic appeal, but etched
surfaces are more suitable for studying the patterns in detail.

The etching is carried out by evenly brushing a freshly polished surface with nital, a solution of 5% nitric acid in alcohol. Dilute aqueous
solutions of HNO_3 also work well on meteoritic metal but will attack
sulfide inclusions if present, which leads to undesirable stains. After
etching to the desired contrast, the samples need to be washed thor-
oughly with clean water, dried with a blower and stored in a desiccator
to prevent rusting. Samples that are intended for decorative purposes
can be coated with polyurethane lacquer to preserve the surfaces, which
may help in humid climates. However, rust spots may appear under the
lacquer if the sample has fine cracks or unprotected surfaces.

Taenite is more resistant to oxidation than kamacite, which leads to interesting results for badly weathered iron meteorites. In some instances, weathering produces Widmanstätten patterns, in other cases kamacite is completely rotted away to limonite, and the intricate structure of the oxidized taenite plates remains.

One factor that determines the crystal structure of iron meteorites is their total Ni content, another is their cooling history together with the low-temperature solid-state crystallization behavior of kamacite and taenite. Irons with less than 6–7% Ni contain only kamacite. If kamacite occurs in cm-size "bars," the iron meteorites are called hexahedrites as kamacite has a cubic (hexahedron) structure; if the metal is a fine aggregate, they are called Ni-poor ataxites (derived from Greek "without structure"). The Ni-rich ataxites, where kamacite and taenite form a fine grained aggregate, have >9% Ni. These ataxites sometimes contain martensite, α_2-FeNi, a metastable phase produced by shock processes.

Octahedrites have 6–18% Ni and they contain both kamacite and taenite that are inter-grown to give the Widmanstätten patterns. The name octahedrite comes from the orientation of the kamacite bands and taenite lamellae that are parallel to octahedral planes. The observed Widmanstätten structure of an octahedrite depends on how the meteorite was cut with respect to the crystallographic structure, which is illustrated in Figure 2.22.

In the late 1800s, Gustav Rose in Berlin, Gustav Tschermak (1836–1927) and Aristides Brezina (1848–1909) in Vienna devised the structural classification of octahedrites as part of their overall work on meteorite classification. For octahedrites, they made use of the kamacite band-widths that range from ~0.03 to >3 mm (Table 2.6). The abbreviations used in this classification may require a little translation. Octahedrites with coarse bands (>1.3 mm) are of type "Og," where the "O" stands for octahedrite, and "g" means "*grob*," which is German for "coarse." Iron meteorites with medium bandwidths (0.5–1.3 mm) are indicated by "Om," where "m" stands for "medium" or German "*mittel*." The "Of" is used for octahedrites with bands <0.5 mm wide, and the "f" stands for "fine" or German "*fein*." The double lettering in "Ogg" for the coarsest octahedrites and "Off" for the finest just stands for twice "*grob*" or "*fein*," respectively.

Etched surfaces of hexahedrites often show fine, long uninterrupted lines in more or less developed form. These are called Neumann lines, after J. Neumann who discovered them in 1848. These lines also occur in the kamacite of octahedrites and are thought to be stress deformations resulting from shock.

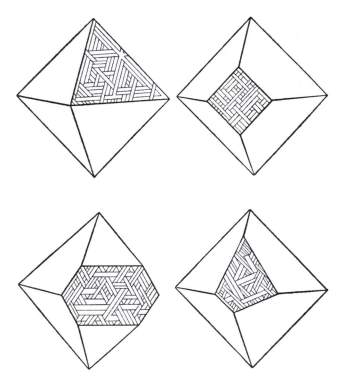

Figure 2.22 The emergence of Widmanstätten patterns depends on how an octahedrite is cut. (After G. Tschermak, *Lehrbuch der Mineralogie*, 1894.)

The kamacite bands in the Widmanstätten patterns are made of long crystals that can reach meters in size. This implies relatively slow cooling so that crystals had ample time to grow. The formation of kamacite and taenite can be understood from the Fe-Ni phase diagram[53] below ~900 °C. Figure 2.23 shows the phase diagram for Ni compositions up to 30% Ni, which covers the metal compositions of most iron meteorites. This is at ambient pressure, but is relevant to iron meteorite parent bodies because the relatively small pressures of a few kilobar inside meteorite parent bodies do not alter the phase equilibria.

Above 900 °C, taenite (γ-FeNi) is the stable face-centered cubic solid solution of Fe and Ni for a wide range of Ni contents. At lower temperatures and Ni concentrations <6–7%, only kamacite (α-FeNi) is stable, whereas at higher Ni concentrations, α-Fe and γ-Fe coexist. Upon cooling of taenite below 900 °C, the boundary to the two-phase stability field is crossed and kamacite precipitates from the taenite. The taenite remaining in equilibrium with the crystallized kamacite is then more Ni-rich, as the original taenite lost Fe to form the kamacite.

Table 2.6 Chemical classification of iron meteorites.

Chem. group	Ni (wt%)	Ga (ppm)	Ge (ppm)	Ir (ppb)	Structure type[a]	Kamacite bandwidth (mm)	Cooling rates (K Ma^{-1})	Frequency (%)	Examples
IA	6.5-8.5	55-100	190-520	10.6-5.5	Om-Ogg	1.0-3.1	1-5	17	Cañyon Diablo, Odessa
IB	8.5-25.0	11-55	25-190	0.3-2.0	D-Om	0.01-1.0	1-5	1.7	Colfax, Four Corners
IC	6.1-6.8	49-55	212-247	0.07-2.1	Anom, Og	<3	3->100	2.1	Bendegó, Etosha
IIA	5.3-5.8	57-62	170-185	2-60	H	>50	2-10	8.1	Chesterville, Hex River
IIB	5.5-6.9	46-59	107-183	0.05-0.46	Ogg	5-15	2-10	2.7	El Burro, Sikhote Alin
IIC	9.0-12	37-39	88-114	4-10	Opl	0.06-0.07	100-500	1.4	Kumerina, Perryville
IID	9.9-11.4	70-83	82-98	3.5-18.5	Of-Om	0.40-0.85	1-2	2.7	Elbogen, Needles
IIE	7.5-9.7	21-28	62-75	1-8	Anom	0.7-2	0.2-400	2.5	Elga, Weekeroo Station
IIF	10.6-14.3	8.9-11.6	99-193	0.75-23	D-Of	0.05-0.21	–	1.0	Dorofeevka, Monahans
IIIA	7.1-8.9	17-23	32-47	0.1-20	Om	0.9-1.3	1-10	24.8	Cape York, Henbury
IIIB	8.6-10.6	16-21	27-46	0.01-1.6	Om	0.6-1.3	1-10	7.5	Bald Eagle, Turtle River
IIIC	10.5-13.3	11-92	8-280	0.07-0.55	Off-Ogg	0.2-0.5	1-5	1.4	Carlton, Havana Tazewell, Wedderburn
IIID	17.0-23.0	1.5-5.2	1.4-4.0	0.02-0.07	D-Off	0.01-0.05	1-5	1.0	
IIIE	8.3-8.8	17-19	34-37	0.05-0.6	Og	1.3-1.6	0.5-2	1.7	Rhine Villa, Willow Creek
IIIF	6.8-8.5	6.3-7.2	0.7-1.1	0.006-7.9	Om-Og	0.5-1.5	5-20	1.0	Moonbi, Nelson County
IVA	7.5-9.5	1.6-2.4	0.09-0.14	0.1-3.5	Of	0.25-0.45	3-200	8.3	Gibeon, Yanhuitlan
IVB	16-18	0.17-0.27	0.03-0.07	13-36	D	0.006-0.03	5-200	2.3	Hoba, Tawallah Valley

^{a}Structural type identifiers: First letter: H = hexahedrite, O = octahedrite, D = ataxites. Subsequent letters: gg = very coarse, g = coarse, m = medium, f = fine, ff = very fine, pl = plessitic. Anom. = anomalous.

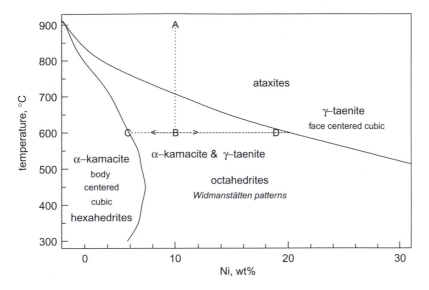

Figure 2.23 The low temperature Fe-Ni phase diagram provides an explanation for the occurrence of hexahedrites, octahedrites and ataxites among iron meteorites. (After ref. 53.)

The total amount of the taenite phase in equilibrium with kamacite must also decrease because mass-balance must be observed. Further cooling leads to more solid state growth of kamacite, which can accommodate a few percent of Ni at lower temperatures. The phase diagram shows that the Ni contents of the coexisting kamacite and taenite phases are a function of temperature, and it gives information on the relative proportions of taenite and kamacite present. This is illustrated by the dotted lines and labeled points (A–D) in the phase diagram. For example, assume a Fe-Ni alloy with 10 wt% Ni is cooled from temperatures above 1000 to 600 °C. At point A at 900 °C, the alloy is the homogeneous Fe-Ni solid solution taenite with 10 wt% Ni. During cooling, the border into the dual stability field of taenite + kamacite is crossed at ∼700 °C and kamacite begins to precipitate. To find the amount of kamacite and taenite present at 600 °C, and the Ni content of these phases, we mark the position of the overall alloy composition as point B. From point B we draw a horizontal line to point C on the curve that separates the two-phase (α-Fe plus γ-Fe) field from the kamacite field. Point C gives the Ni content of the kamacite as 5.9% in equilibrium with taenite at 600 °C. Similarly, point D on the phase boundary between α-Fe + γ-Fe and taenite (γ-Fe) gives the Ni content of taenite coexisting with kamacite as ∼20% Ni at 600 °C. The fractions of kamacite and taenite present follow from the mass balance equation, which requires that the total amount of Ni, C_{tot}, equals the mass fraction of kamacite times the concentration of Ni in kamacite, C_{kam}, plus the mass fraction of taenite, X_{tae}, times the concentration of Ni in taenite, C_{tae}. The sum of the kamacite and taenite mass fractions is unity, so we can write:

$$C_{tot} = X_{kam} C_{kam} + X_{tae} C_{tae} = X_{kam} C_{kam} + (1 - X_{kam})C_{tae} \qquad (2.39)$$

Solving for the fraction of kamacite and comparing the concentrations with the points B, C and D in the phase diagram shows that the fraction of kamacite is proportional to distance between points B and D ($\overline{BD}$) and equal to the absolute value of the distance ratios $\overline{BD}/\overline{CD}$:

$$X_{kam} = \frac{C_{tot} - C_{tae}}{C_{kam} - C_{tae}} = \frac{|\overline{BD}|}{|\overline{CD}|} \qquad (2.40)$$

This is the lever rule. Graphically, the fraction of taenite is obtained from the distance ratios $\overline{BC}/\overline{CD}$.

In our example, the initially homogeneous FeNi alloy with 10% total Ni cooled to 600 °C changed into two phases: the phase with the larger fraction (71%) is kamacite with 5.9% Ni, and the smaller fraction (29%) is taenite, which has a Ni content of 20%.

In principle, the phase diagram and analysis of the Ni content in coexisting taenite and kamacite determine the temperature under which the octahedrites crystallized because the Ni *content* of coexisting taenite and kamacite is independent of the total Ni content in the entire assemblage (for total Ni $\approx >6\%$). At a given temperature, the compositions of kamacite and taenite coexisting at equilibrium are fixed (such as the fixed Ni contents at points C and D in the 600 °C example above). This is true even if assemblages with different total Ni contents were cooled to the same temperatures. However, the proportions of the mass fractions of kamacite and taenite in each assemblage will be different because this changes as a function of total Ni content. For example, if an alloy had a higher Ni content than the 10% as in point A of our example above, cooling to 600 °C leads to precipitation of the same taenite and kamacite compositions (points C and D) as before. However, if point A shifts to the right, so will point B and the distance between points $\overline{CB}$ increases and that of $\overline{BD}$ decreases, which means that the fraction of kamacite decreases and that of taenite increases.

The practical use of this "Ni-composition thermometer," however, has limits. To use it, the kamacite and taenite compositions must be homogeneous, which is never really the case. The low temperature crystallization or exsolution of kamacite from taenite requires that

Fe diffusion rates through taenite into kamacite are faster than the cooling rates. For example, at 1000 °C, diffusion coefficients of Fe and Ni in taenite are on the order of $\sim 10^{-12}\,cm^2\,s^{-1}$, which is equivalent to $\sim 3 \times 10^{-3}\,mm^2\,yr^{-1}$. In a year, Fe can diffuse a distance $x = \sqrt{Dt} \approx 0.05\,mm$, in 10^6 years this is $\sim 50\,mm$. However, the phase diagram shows that exsolution of kamacite only starts below 900 °C and diffusion coefficients decrease exponentially with decreasing temperature. They are about a factor of ten lower for each temperature drop of 100 °C. With a drop from 1000 to 500 °C, diffusion slows down $\sim 10^5$ times, and only allows diffusion over $\sim 0.2\,mm$ per million years at 500 °C. To have diffusion controlled kamacite growth of this size, the temperature has to remain at 500 °C for 1 Ma. However, cooling rates of iron meteorites show that temperatures dropped by 10–100 °C over this period. As diffusion slows down with cooling, exsolution of kamacite and homogenization of taenite eventually stops. The low diffusion rates of Fe and Ni are also the reason why the low-temperature portion of the FeNi phase diagram is very difficult to investigate experimentally.[53]

With slow diffusion, the Ni concentrations are inhomogeneous with Ni concentration gradients in the taenite lamellae. The Ni concentrations are high at the contact to kamacite, and low in the center. Detailed analyses of such concentration gradients, especially at the kamacite and taenite interfaces, can be used to derive the cooling rates.[53] Typical cooling rates of the most abundant iron meteorite groups (IAB, IIAB and IIIAB) as well as pallasites (see below) are in the range of ~ 10–$100\,K\,Ma^{-1}$. These cooling rates are compatible with slow cooling of iron masses in the cores of asteroids of about 10–100 km in diameter. Irons in group IVA have faster cooling rates of $\sim 200\,K\,Ma^{-1}$, corresponding to a differentiated asteroid of ~ 10–15 km in diameter. The IVB irons, which are peculiar in chemistry, have apparent cooling rates that range from a few tens to a few thousand K per Ma. These widely varying cooling rates do not make sense if the IVB irons originated from the core of a single asteroid. The metal core of an asteroid is probably isothermal and all metal in the core is likely to have the same thermal history.

The kamacite band width is useful for the structural classification of iron meteorites. However, chemical analyses of the bulk Ni and trace element concentrations lead to a more reliable classification scheme, which was developed in detail by J. T. Wasson, E. R. D. Scott and colleagues.[54] Plots of the concentrations of Ga, Ge, Ir and Au *versus* Ni from individual iron meteorites give more or less well resolved compositional fields that are used to group the iron meteorites. Figure 2.24 shows examples of such correlations. The chemical classification uses

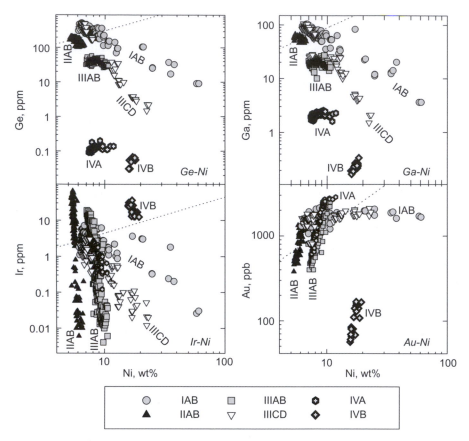

Figure 2.24 Concentrations of Ga, Ge, Ir and Au and Ni separate the different iron meteorite groups. Only data for a few iron meteorite groups are plotted here to avoid overcrowding the diagrams. The dotted line indicates the compositions that correspond to a CI chondrite abundance ratio.

roman numerals and capitalized letters for the different groups of irons. Several of these groups (*e.g.*, IA and IB) may be "end-members" of one group (*e.g.*, IAB) with continuous composition, which only became apparent as more iron meteorites were investigated. Those end-member groups are combined to give the IAB, IIAB, IIIAB and IIICD groups.

The trace elements can be positively correlated with Ni (*i.e.*, concentrations of the trace elements and Ni both increase), negatively correlated with Ni (trace element concentrations decrease as the Ni concentration increases) or uncorrelated. For example, Ga and Ge show positive correlations for all iron groups, whereas correlations of Ge and Ni can be positive, negative or uncorrelated. Some of the observed correlations are shown in Table 2.7.

Table 2.7 Correlations of trace elements.

Pair	Trend	Found in
Ga-Ge	Positive	All groups
Ga-Ni	Positive	IIIA, IVB
	Negative	IA, IB, IIB, IIIB, IIICD
Ge-Ni	Positive	(IIA), IIC, IID, IIIA, IVA, IVB
	Negative	IA, IB, IIB, IIIB, IIIC, IIID
	None	IIE, IIIC, IIIE, IIIF
Ir-Ni	Negative	All groups
W-Ni	Negative	IA, IB, IIA, IIB, IIIA, IIB, IVA

The compositional regions for most iron meteorite groups in Figure 2.24 can be explained by fractional crystallization (group IVB irons are one clear exception). The positions of the regions spanned by the element correlations for iron meteorites depend on the overall content of the elements in the iron meteorite source region and the degree of melting.

The presence of small oxygen-bearing phases, such as silicates, silica, phosphates or chromite, in some iron meteorites provides a check for possible genetic relationships of irons to other meteorite classes through the O-isotopes. The O-isotope composition of silicate inclusions in IAB irons coincides with that of winonaites (Figure 2.2); similarly, the O-isotopes also suggest a link between the IIIAB irons and main group pallasites. The O-isotope compositions in IIE irons are compatible with those in H chondrites, and that in IVA irons with those in L or LL chondrites. These similarities suggest that the iron meteorites and chondrites formed from material that was processed within the same regions of the solar system. It does not prove, and should not even imply, that meteorites with similar O-isotopic compositions come from the same meteorite parent body.

Iron meteorites may also result by condensation and metal accumulation, and the IVB iron meteorites are the most probable candidates for this. Such a possibility can be tested by comparing the observed abundance distribution in iron meteorites to calculated element distributions expected for either condensation or planetary differentiation (Chapters 3 and 4).

2.5 STONY-IRON METEORITES

The stony-irons consist of pallasites and mesosiderites. Iron metal and silicates occur in about equal amounts in the stony irons. Pallasites are named after the type-specimen meteorite, which the German naturalist

Peter Simon Pallas (1751–1811) encountered on his journey through Russia. It was this meteorite that stimulated Chladni's researches on meteorites. Most pallasites contain well crystallized, Mg-rich olivines (~88% forsterite) embedded in a sponge-like metal network, as for example, in the Springwater pallasite (Figure 2.25). The metal in pallasites displays Widmanstätten patterns like the octahedrite iron meteorites, and the main group of pallasites appears to be related to IIIAB irons through their O-isotopic compositions. Another group of pallasites has olivines that are more Fe and Ca-rich than that of main group pallasites, and their metal contains more Ni and Ir, which links them to IIF iron meteorites. The mixture of coarse olivine embedded in metal suggests that pallasites formed at the boundary of a metal core and silicate mantle in a smaller asteroid. A few pallasites contain pyroxene instead of olivine, and differences in the O-isotopic compositions among pallasites indicate that pallasites originated from more than one parent body.

The mesosiderites, after μεσος (mesos) for "being in the middle" and σιδηρος (sideros) for "iron," received this name from G. Rose in 1863 because they contain about equal amounts of silicates and iron metal plus troilite. Thus, the mesosiderites are in the middle between stone and iron meteorites. Mesosiderites are breccias of fine to coarse fragments

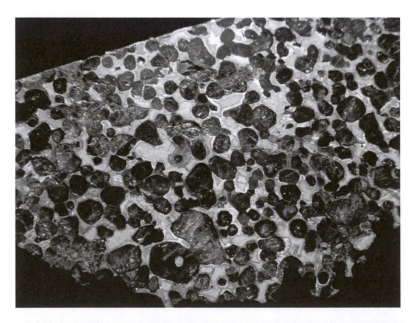

Figure 2.25   The Springwater pallasite contains cm-size olivine grains in a FeNi metal network. The piece is *ca.* 10×15 cm wide. Photograph by the authors.

consisting of igneous silicate rocks typical for differentiated planetary objects (basalts, gabbros, orthopyroxenites). Metal (and sulfide) grains are up to a mm in size and are "sprinkled" in with finer silicates. The metal forms a much finer mesh than that in pallasites, and they have a wider variety of silicate minerals. Several subgroups of mesosiderites have been identified, but their complexity makes mesosiderites one of the least understood meteorite groups.

2.6 ACHONDRITES

Several groups of silicate-dominated meteorites – the achondrites – have been recognized (Table 2.1). Their mineralogy suggests that igneous (melting and crystallization) processes are involved in their formation. Depending on the degree of such processing, achondrites are divided into primitive achondrites and planetary achondrites.

The primitive achondrites include the acapulcoite-lodranite group, angrites, aubrites, brachinites, ureilites and winonaites. A popular view is that the parent bodies of the primitive achondrites experienced some degree of melting that left the achondrite rocks behind as residues. The acapulcoites, brachinites and winonaites may still show chondrule relics. To preserve chondrule textures, melting cannot have been extensive. Localized partial melting is an attractive possibility to rid the achondrite source region of sulfides and metal. However, sulfides and metal are still present in smaller amounts in several primitive achondrites although in lower amounts than seen in chondrites. The extent of metal and sulfide removal from a chondritic source can be evaluated from the concentrations of trace elements that are distributed between silicates, sulfide and metal during differentiation.

Such modeling has been done extensively for the planetary achondrites whose mineral assemblages largely formed by crystallization from magmas. Among the planetary achondrites are the lunar meteorites, the SNC group containing the shergottites, nakhlites and chassignites, and the EHD group with the eucrites, howardites and diogenites.

2.6.1 Acapulcoites and Lodranites

The acapulcoites and lodranites, named after the Acapulco and Lodran meteorite falls, were first described as separate groups of primitive achondrites but their mineral chemistry and overlapping O-isotopic composition warrants treating them as a single group of meteorites. Pyroxene, olivine, pyrrhotite and metal are their principal minerals. The chemical composition of acapulcoites is comparable to that of H chondrites, but the different O-isotopic compositions of these meteorites rule out a direct genetic link. Another difference from H chondrites is that acapulcoites are more gas-rich. In H chondrites, the gas content decreases from low to high petrologic type as expected for more thermally processed materials. The textures of the acapulcoites suggest that they have experienced even more heating than the equilibrated H chondrites of petrologic types 5 and 6, and one should expect that acapulcoites would be even more degassed. However, they contain 3–10 times larger gas volumes (per volume element of meteorite) than H5–6 chondrites.

Both acapulcoites and lodranites seem to be residues from partial melting, with lodranites experiencing more melting and recrystallization. An early study by Meunier of the Lodran meteorite describes the metal texture in these meteorites:

> "if a chip is heated and then suddenly plunged into mercury, the silicates fall to pieces, while the metallic portion is seen to form a very fine network or sponge like mass. This network is the same as, but finer than, that formed by the metal of in the celebrated Pallas meteorite, to which this is allied."[55]

Meunier describes the network of iron as "metallic cement" binding the silicate grains.

The O-isotopes are comparable to the ranges found for the CR and CH chondrites (Figures 2.1 and 2.2). However, it is not clear if partial melting of these carbonaceous chondrites leads to residues that resemble the acapulcoites and lodranites in mineral chemistry.

2.6.2 Brachinites

These primitive achondrites are largely made of coarse olivine with significant percentages of augite pyroxene, plagioclase, sulfide and FeNi metal. These rocks appear to be residues from partial melting, although their elemental compositions are close to those of chondrites. The O-isotopic composition of brachinites (Figure 2.2) coincides with that of the eucrites, howardites and diogenites (EHD group). Whether this is incidental or implies a genetic link between these groups is not yet entirely clear.

2.6.3 Winonaites

The rare winonaites have similar mineralogy and compositions to chondrites, and they may contain relict chondrules. Originally, these primitive achondrites were known as "forsterite chondrites." Their mineral assemblage is more reduced than that of H chondrites. Oxygen isotopes may relate winonaites and silicate inclusions found in IAB and IIICD irons.

2.6.4 Angrites

The angrites, named after Angra dos Reis which fell near Rio de Janeiro in 1869, are very Ca, Al and Ti-rich achondrites. They consist of $\sim 90\%$ Ti-bearing augitic pyroxene and some olivine and troilite. Angrites are extremely depleted in alkali elements when compared to chondrites and even eucrites, but other volatile elements such as Zn and Se are not depleted in a similar manner.

The O-isotopic composition of angrites falls together with that of the planetary EHD meteorites and is close to that of brachinites. However, chemical and mineralogical differences among these meteorites are inconsistent with an origin from a common parent body.

2.6.5 Aubrites

The aubrites (named after the Aubres meteorite fall), or, in the older literature, bustites (after the Bustee meteorite fall), have a highly reduced mineralogy that resembles that of the enstatite chondrites in many respects. They consist mainly of almost FeO-free enstatite with (Fs < 1%), hence they are also called enstatite achondrites. Other minerals include diopside, forsteritic olivine (Fa < 1%), plagioclase (An < 25%), kamacite with low Ni contents (< 4%), metallic Cu, heideite ($FeTi_2S_4$) and Ti-bearing troilite (up to $\sim 10\%$ Ti), daubréelite ($FeCr_2S_4$), djerfisherite [$K_3(Cu,Na,)(Fe,Ni)_{12}(S,Cl)_{14}$], ferroan alabandite (Fe,Mn)S, oldhamite (CaS), schreibersite ($(Fe,Ni)_3P$) and osbornite (TiN). Several of the sulfide minerals are unique to the enstatite chondrites and achondrites.

The exotic mineralogy is a consequence of the low oxygen fugacity under which the minerals were formed. In the absence of sufficient oxygen, elements normally found as oxides in silicates are also found as sulfides (*e.g.*, Ca, Mg, Mn, Na) or nitrides (*e.g.*, Ti). For example, in aubrites Ca resides in diopside as $CaMgSi_2O_6$, in feldspar as $CaAl_2Si_2O_8$ and in oldhamite as CaS.

Most aubrites are brecciated and are samples of their asteroid's surface regolith, which is rich in noble gases implanted by the solar wind. The O-isotopic compositions of aubrites plot along the terrestrial fractionation line like those of the E chondrites. In contrast to the E chondrites, aubrites lost most of their metal and sulfides. The aubrite parent body may have had a similar composition to that of EH chondrites before silicates and metal fractionated. However, the degree of reduction and the similar O-isotopes only allow the conclusion that the aubrites, EH chondrites and EL chondrites formed in the same region of the solar system; it is not proof that they come from the same parent asteroid. The differences in chemistry require at least three different parent bodies for each meteorite group.

2.6.6   Ureilites

Ureilites are the largest group of primitive achondrites. These Ca-poor achondrites chiefly contain coarse olivine (Fo $\approx$ 20%) and pigeonite grains in a fine, carbon-rich matrix. Kamacite containing 2–4% Ni, troilite, graphite and diamonds occur in small amounts. Between 1957 and 1962, diamonds were identified by X-ray diffraction in several ureilites. The formation of these diamonds is ascribed to shock from collisions and impacts after the parent body had formed. Some ureilites contain fragments of material from other meteorite groups, such as CI-chondrites, or angrites, which attests to the bombardment that the ureilites parent asteroid must have endured.

These ultramafic rocks are residues from small degrees of partial melting, which is inferred from the absence of larger amounts of plagioclase. Their O-isotopic compositions are unusual as these scatter along the "CAI mixing line" of the slope unity (Figure 2.2). This is not expected from larger scale processing on a parent body, which should result in mass-dependent fractionations of O-isotopes that follow a slope of ~ 0.5.

2.6.7   Eucrites, Howardites and Diogenites

The eucrites, howardites and diogenites form a larger group of achondrites. Rose (1863) proposed the name eucrites, after ευχριτος (meaning well to determine), for the easy classifiable basaltic rocks in this group. Eucrites are mainly composed of Ca-rich pyroxene (pigeonite) and plagioclase feldspar (An $\sim$ 80–95). Minor minerals include magnetite, chromite, ilmenite, whitlockite, troilite and kamacite, and the three SiO_2 modifications quartz, tridymite and cristobalite. Eucrites show brecciated structures, indicating that their parent body experienced extensive bombardment over time. Several eucrites contain rock fragments from other meteorite types, such as CM chondrites.

The howardites were named by Rose in honor of the British chemist Edward Howard (1774–1816), who performed the first detailed meteorite

analyses. Howardites are mainly orthopyroxene (Fs<40%), and pigeonite (Fs>45%). Less abundant are plagioclase feldspar (An>75%),
olivine of variable composition (Fa≈8–40%), kamacite (~4% Ni), taenite (~40%Ni) and troilite. The howardites can be understood as
products from mixing eucrite and diogenite fragments, which makes
howardites polymict breccias.

The diogenites, described under the names chladnites, rodites or
shalkites in the older literature, are Ca-poor and contain mainly pyroxene of bronzite-hypersthene compositions (Fs = 23–27). Other components are olivine (Fa~28%), plagioclase feldspar (An 85–90%), and
minor chromite, troilite, kamacite (~3% Ni) and glass. Most diogenites
are monomict breccias that contain orthopyroxene rock clasts in fine
to coarse-grained matrices. In some polymict brecciated diogenites,
material from eucrites is mixed in.

The O-isotopes of the EHD meteorites follow a mass-dependent
fractionation curve with a slope of ~0.52 that runs parallel to the
terrestrial fractionation curve with an offset of about − 0.3 per-mil
(Figure 2.2). Compared to other achondrites, their concentration of
volatile elements such as the alkali elements is very low. They also lack
siderophile elements, which is indicative of metal segregation and core
formation. The eucrites have very old crystallization ages and the EHD
parent body seems to have differentiated within a few million years after
the first solids formed in the solar nebula. The EHD meteorites are
thought to come from the differentiated asteroid Vesta; more about the
chemistry of the EHD meteorites is included in Chapter 4.

2.6.8 SNC Meteorites and Lunar Meteorites

The SNC and lunar meteorites are differentiated meteorites of Mars and
the Moon, respectively, which are described in more detail in Chapter 4.
Like the asteroids, the cratered lunar and Martian surfaces bear records
of numerous larger impacts over time. Some of these impacts were
apparently catastrophic and energetic enough to eject rock fragments
with more than the required escape velocities of ~2.4 and ~5 km s^{-1}
from the Moon and Mars, respectively. After several million years of
travel, these samples landed on Earth as gram- to kg-size masses without
causing any catastrophic impact. After they sat in deserts or on the
Antarctic ice for several thousand years, they became targets of pro-
fessionally organized meteorite search expeditions, and now these rocks
are the pride of meteorite collections.

The case for *lunar meteorites* is robust because direct comparisons of
their mineralogy and chemistry with the rocks collected during the

Apollo missions can be done. Given the spread of impact craters over the lunar surface, the locations sampled by the ~ 50 lunar meteorites (totaling $\sim 30\,kg$) may be larger than the six different lunar sites visited by the Apollo missions. However, the Apollo missions brought back $382\,kg$ of soil, fines and drill cores. The automated sample return missions by the former USSR visited three additional lunar sites and sampled $\sim 300\,g$. Since it is known where samples were taken during the missions, they can be interpreted in context with the existing geology. However, for the lunar samples delivered as meteorites, there is no easy way of telling from which exact location they were ejected since sampling is more or less random. As we will see in Chapter 4, the lunar meteorites have some characteristic chemical differences from lunar rocks collected during the Apollo missions.

There are also meteorites that landed on the Moon and were brought back from the Moon in contrast to the lunar meteorites that are materials from the Moon coming to Earth without human intervention. The Bench Crater meteorite, resembling CI chondrite matrix material, was among the lunar rocks collected at the Apollo 12 landing site in 1969, and the Hadley Rille enstatite chondrite was collected in 1971 during Apollo 15.

As with lunar meteorites, there are two kinds of *Martian meteorites*: those that have landed on the surface of Mars, and those that landed on Earth and are believed to have originated from Mars. The IAB iron meteorite discovered by the Mars Rover Opportunity in 2005 at Meridiani Planum certainly is one of the first kind. The iron meteorite type is inferred from its composition ($\sim 7\%$ Ni, Ga $< 70\,ppm$, Ge $\sim 300\,ppm$) determined by the Alpha-Particle X-ray spectrometer (APXS) instrument on board the Mars rover.[56] The second kind of Martian meteorites are those that have reached the Earth's surface after impacts ejected larger rock fragments from Mars. The most likely candidates for a Martian origin are the *shergottites, nakhlites* and *chassignites* (SNC) meteorites.

The SNC meteorites are linked through their O-isotope compositions, which plot on a curve parallel to the terrestrial fractionation curve ($\Delta^{17}O = +0.30$ per-mil; Figure 2.2). The shergottites (named after the Shergotty meteorite) are basaltic achondrites. The Antarctic and desert finds make shergottites a larger group of achondrites. Until the late 1960s, the two shergottites known (Shergotty fell in 1865 and Zagami fell in 1962) were not distinguished from eucrites, with which they share several similarities in mineralogy. However, the occurrence of maskelynite, a glassy form of labradorite feldspar, and the absence of brecciation sets shergottites apart from eucrites. Nakhlites are named after the

1912 Nakhla meteorite shower in Egypt, during which a dog was fatally hit by one of the falling fragments. Chassignites, named after Chassigny, are olivine-rich meteorites that resemble terrestrial mantle rocks in mineralogy. The orthopyroxenite meteorite ALH84001 also belongs to the SNC group but is in its own mineralogical category. More about SNC meteorites and the indirect arguments that SNC meteorites originated on Mars are described in Chapter 4.

2.7 METEORITE HOME WORLDS

The occurrence of iron meteorites and meteorites with mixed metal and silicate mineralogy gave rise to the early view by Chladni and others that meteorites might be fragments of a broken-up celestial object. Chladni called them "Weltenspäne," (shavings of worlds), and he thought of meteorites as fragments of a broken planet or primordial matter that failed to accumulate into larger objects.

The acceptance of meteorites as extraterrestrial objects and the notion that they could be parts of a broken planet gained in popularity when G. Piazzi discovered the first asteroid, Ceres, in 1801. Other asteroid discoveries soon followed. The geometric distance series of planets devised by Titius and Bode had predicted a planet for the region between 2.3 and 3.3 AU (AU stands for "astronomical unit" and 1 AU is the Earth–Sun distance), but instead of one large planet this region was found to be filled with small planetoids (or asteroids). At the time, Olbers thought that asteroids were pieces of a single scattered planet; however, it is now clear that the asteroids are an ancient population of chemically and physically distinct objects.

The low-pressure mineralogy of chondrites is consistent with an origin from small objects like the asteroids (Chapter 4). The mineralogy of achondrites is also consistent with low-pressure igneous processing. The basaltic EHD achondrites must come from a differentiated asteroid, and the best candidate is Vesta (Chapter 4). On the other hand, achondrites may simply have formed on or near the surfaces of their parent bodies where lithostatic pressures are lower. This applies to meteorites from the Moon and Mars. Before lunar and Martian meteorites were identified, it was frequently argued that, on dynamical grounds, it is difficult to bring material ejected from the Moon and Mars to Earth. However, as we know now, the lunar and SNC meteorites apparently overcame these difficulties.

The cosmic-ray exposure ages of meteorites show that they travelled ~ 10–100 million years as smaller objects through space. Short exposure ages imply short travel times, suggesting that there must be source regions of meteorites near Earth. A nearby and long suspected source of meteorites is the Moon. The Willamette iron meteorite had been greeted as "Visitor from the Moon" by the Native Indian Americans. In 1660, Paolo Maria Terzago concluded "the Moon was the cause of the falling of the stones" at Milan in 1650, which, according to legend, killed a Franciscan monk. Olbers in 1795 and Laplace around 1810 thought about the possible lunar origin of meteorites since it was about the time when the possible volcanic and impact origins of lunar craters were debated. Once the relatively young exposure ages of meteorites had been determined in the 1960s, some researchers thought that either ordinary or carbonaceous chondrites might be from the moon. However, the first genuine lunar meteorite was only identified long after the Apollo missions.

The other nearby-sources of meteorites are asteroids with Earth-crossing orbits. This population is replenished by asteroid fragments coming from the main asteroid belt and by extinct comets, which we will visit in Chapter 4.

FURTHER READING AND RESOURCES

Astronomical Data Service: A free, web-based literature reference resource, ideal for searching the primary astronomical, planetary and meteorite literature. It provides abstracts to research papers and links to scientific journals to access the papers in full (subscriptions from libraries or individuals may be required). However, many papers can also be downloaded directly through the links provided in ADS. http://adsabs.harvard.edu/abstract_service.html.

METBASE, a computer data base on meteorites by Jörn Koblitz, Bremen, Germany, is an indispensable tool with extensive information on meteorite chemical compositions and literature; http://www.metbase.de.

The Meteorite Bulletin at: http://tin.er.usgs.gov/meteor/metbull.php is issued by the Meteoritical Society and provides regular updates on the official classification of newly recognized meteorites in a searchable database.

V. F. Buchwald, *Handbook of Iron Meteorites*, vol. 1–3, University of California Press, Berkeley, 1975, 1426 pp.

R. T. Dodd, *Meteorites, a Petrologic-chemical Synthesis*, Cambridge University Press, Cambridge, 1981, 368 pp.

B. Mason, *Meteorites*, John Wiley & Sons, New York, 1962.

J. T. Wasson, *Meteorites*, W. H. Freeman & Co, New York, 1985, 267 pp.

Meteorites and the Early Solar System, ed. J. F. Kerridge and M. S. Matthews, University of Arizona Press, Tucson, AZ, 1988.

Meteorites and the Early Solar System II, ed. D. S. Lauretta, H. Y. McSween, University of Arizona Press, Tucson, AZ, 2006.

Planetary Materials, ed. J. J. Papike, Reviews in Mineralogy, vol. 36, Mineralogical Society of America, 1998.

Treatise on Geochemistry, series eds. H. D. Holland and K. K. Turekian, vol. 1, *Meteorites, Comets, and Planets*, ed. A. M. Davis, Elsevier-Pergamon, Oxford, 2004.

REFERENCES

1. E. F. F. Chladni, *Ueber den Ursprung der von Pallas gefundenen und anderer ihr ähnlicher Eisenmassen*, J. F Hartknoch Publisher, Riga, 1794.
2. R. N. Clayton, L. Grossman and T. K. Mayeda, *Science*, 1973, **182**, 725; R. N. Clayton, N. Onuma, L. Grossman and T. K. Mayeda, *Earth Planet. Sci. Lett.*, 1977, **34**, 209.
3. R. N. Clayton, *Annu. Rev. Earth Planet. Sci.*, 1993, **21**, 115.
4. R. N. Clayton and T. K. Mayeda, *Geochim. Cosmochim. Acta*, 1996, **60**, 1999; *Geochim. Cosmochim. Acta*, 1999, **63**, 2089.
5. M. F. Miller, *Geochim. Cosmochim. Acta*, 2002, **66**, 1881.
6. M. Thiemens, *Annu. Rev. Earth Planet. Sci.*, 2006, **34**, 217.
7. F. Robert, A. Rejou-Michel and M. Javoy, *Earth Planet. Sci. Lett.*, 1992, **108**, 1.
8. I. A. Franchi, I. P. Wright, A. S. Sexton and C. T. Pillinger, *Met. Planet. Sci.*, 1999, **34**, 65.
9. H. C. Sorby, *Philos. Mag.*, 1864, **28**, 157; *Nature*, **15**, 495, 1877.
10. M. Tschermak, *Akad. Wissenschaften*, Wien, 1875, 22.
11. C. W. Gümbel, Sitzungsberichte k. bayrischen, *Akad. Wissenschaften*, Munich, Mathematisch-naturw. Klasse, I. Abt., 1878, 14.
12. W. Wahl, *Z. Anorg. Allg. Chem.*, 1910, **69**, 52.
13. H. C. Connolly and R. H. Hewins, *Geochim. Cosmochim. Acta*, 1995, **59**, 3231.
14. S. Desch and H. C. Connolly, *Meteoritics Planet. Sci.*, 2002, **47**, 183.
15. *Chondrules and their Origins*, E. A. King, (ed.), Lunar and Planetary Institute, Houston, 1983, 377 pp.
16. *Chondrules and the Protoplanetary Disk*, ed. R. H. Hewins, R. H. Jones and E. R. D. Scott, Cambridge University Press, 1996, 346 pp
17. W. R. van Schmus and J. A. Wood, *Geochim. Cosmochim. Acta*, 1967, **31**, 747.
18. H. C. Urey and H. Craig, *Geochim. Cosmochim. Acta*, 1953, **4**, 36.

19. B. Mason, *Principles of Geochemistry*, John Wiley & Sons, New York, 1960.
20. K. Keil and K. Fredriksson, *J. Geophys. Res.*, 1964, **69**, 3487.
21. K. Keil, *J. Geophys. Res.*, 1968, **73**, 6945; *Meteoritics*, 1989, **24**, 195.
22. B. Mason, *Space Sci. Rev.*, 1962, **1**, 621; H. Y. McSween, *Rev. Geophys. Space Phys.*, 1979, **17**, 1059.
23. R. E. Folinsbee, J. A. V. Douglas and J. A. Maxwell, *Geochim. Cosmochim. Acta*, 1967, **31**, 1625.
24. R. E. DuFresne and E. Anders, *Geochim. Cosmochim. Acta*, 1962, **26**, 1085.
25. K. Boström and K. Fredriksson, *Smithsonian Misc. Collection*, Publ. 151, 1966, 1.
26. U. Schwertmann and R. M. Taylor, *Iron oxides*, in *Minerals in Soil Environments*, J. B. Dixon and S. B. Weed, (ed.), Soil Science Society of America, Madison, Wisconsin, 1977, p. 145.
27. K. Fredriksson and J. F. Kerridge, *Meteoritics*, 1988, **23**, 35.
28. G. F. Herzog, A. Anders, E. C. J. Alexander, P. K. Davis and R. S. Lewis, *Science*, 1973, **180**, 489; O. Pravdivtseva, *et al.*, *Lunar Planet. Sci.*, 2003, **34**, 1863.
29. M. Endress and A. Bischoff, *Geochim. Cosmochim. Acta*, 1996, **60**, 489.
30. P. Ramdohr, *J. Geophys. Res.*, 1963, **68**, 2011, P. Ramdohr, *The Opaque Minerals in Stony Meteorites*, Elsevier, Amsterdam, 1963.
31. J. J. Berzelius, *Ann. Phys. Chem.*, 1834, **33**, 113.
32. S. Cloëz, *Comptes Rend. Paris*, 1864, **59**, 37; 1864, **59**, 986.
33. G. A. Daubrée, *Comptes Rend. Paris*, 1864, **59**, 984.
34. J. F. Kerridge, J. D. MacDougall and K. Marti, *Earth Planet. Sci. Lett.*, 1979, **43**, 359.
35. I. D. R. MacKinnon and M. E. Zolensky, *Nature*, 1984, **309**, 240.
36. K. Tomeoka and P.R. Buseck, *Geochim. Cosmochim. Acta*, 1985, **49**, 2149.
37. K. Tomeoka and P. R. Buseck, *Geochim. Cosmochim. Acta*, 1988, **52**, 1627.
38. F. Pisani, *Comptes Rend. Paris*, 1864, **59**, 132.
39. H. Y McSween and S. M. Richardson, *Geochim. Cosmochim. Acta*, 1977, **41**, 1145.
40. R. N. Clayton and T. K. Mayeda, *Earth Planet. Sci. Lett.*, 1984, **67**, 151; R. N. Clayton and T. K. Mayeda, *Geochim. Cosmochim. Acta*, 1999, **63**, 2089; M. W. Rowe, R. N. Clayton and T. K. Mayeda, *Geochim. Cosmochim. Acta*, 1994, **58**, 5341; L. A. Leshin, J. Farquhar, Y. Guan, S. Pizzarello, T. L. Jackson and M. H. Thiemens, *Lunar Planet. Sci. Conf.*, 2001, **32**, 1843.

41. I. R. McGill, B. McEnaney and D. C. Smith, *Nature*, 1976, **259**, 200.
42. E. J. O'Loughlin and D. R. Burris, *Environ. Toxicol. Chem.*, 2004, **23**, 41.
43. L. Legrand, R. Maksoub, G. Sagon, S. Lecomte, J. P. Dallas and A. Chause, *J. Electrochem. Soc.*, 2003, **150**, B45.
44. A. H. Cuttler, D. R. Glasson and V. Man, *Thermochim. Acta*, 1984, **82**, 231.
45. S. M. Richardson, *Meteoritics*, 1978, **13**, 141.
46. A. Des Cloizeaux, *Comptes Rend. Paris*, 1864, **59**, 829.
47. G. A. Daubrée, *Comptes Rend. Paris*, 1864, **59**, 830.
48. J. D. Macdougall, G. W. Lugmair and J. F. Kerridge, *Nature*, 1984, **307**, 249.
49. M. Endress, E. Zinner and A. Bischoff, *Nature*, 1996, **379**, 701.
50. M. Gounelle and M. E. Zolensky, *Meteoritics Planet. Sci.*, 2001, **35**, 1321.
51. F. Robert and S. Epstein, *Geochim. Cosmochim. Acta*, 1982, **46**, 81.
52. S. A. Airieau, J. Farquhar, M. H. Thiemens, L. A. Leshin, H. Bao and E. Young, *Geochim. Cosmochim. Acta*, 2005, **69**, 4166.
53. J. I. Goldstein and R. E. Ogilvie, *Geochim. Cosmochim. Acta*, 1965, **29**, 893.
54. E. R. D. Scott and J. T. Wasson, *Rev. Geophys. Space Phys.*, 1975, **13**, 527.
55. S. Meunier, *quoted in Science*, 1883, **1**, 70.
56. Meteoritical Bulletin No. 90, *Meteoritics Planet. Sci.*, 2006, 1383 (see also the Meteoritical Bulletins online: http://meteoritics.org/).

CHAPTER 3

The Solar Nebula

... As knowledge of the universe increases, set limits between various branches of research break down. Compartmentalized science now survives chiefly in the names of academic departments and the titles of professors. Researchers easily cross the artificial boundaries; physicists, for example, make important discoveries in astronomy, and many astronomers talk somewhat like physicists. Some men are astronomers on Monday and Wednesday, physicists on Tuesday and Thursday, and chemists on Friday and Saturday. What they do on Sunday, I cannot even guess. In the past few years meteorology and geology have made friendly and constructive infiltrations into astronomy. Astronomy cordially welcomes even closer contact with other sciences.

Paul W. Merrill 1951

3.1 INTRODUCTION

The solar nebula – a rotating protoplanetary accretion disk made of gas and dust around the forming Sun – is the first stage that began to resemble the general layout of the solar system. Most of the matter from this disk accreted to the growing Sun, and, at the same time, the planets began to form. The formation of the solar nebula must be understood as part of the general star forming process, and the processes within the

Chemistry of the Solar System
By Katharina Lodders and Bruce Fegley, Jr.
© K. Lodders and B. Fegley, Jr. 2011
Published by the Royal Society of Chemistry, www.rsc.org

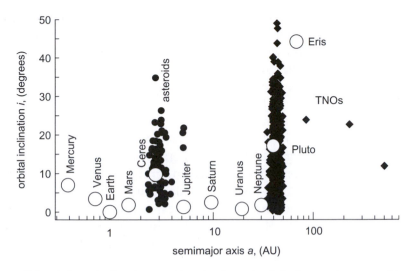

Figure 3.1   Orbital inclinations and semimajor axis of the larger planetary objects in the solar system. Only objects with diameters over 100 km are shown. Planet Mercury and the dwarf planets Ceres, Pluto and Eris have relatively large inclinations compared to the major planets.

resulting solar nebula must explain the following general observations of the solar system's structure (Figure 3.1) and planetary properties:

- There is one central star – the Sun – orbited by eight major planets with near-circular, coplanar orbits, and three or more dwarf planets. The larger objects are spaced at regular radial distances (*i.e.*, with increasing distance from the Sun, each larger object is located at about twice the distance of the previous one; Titius–Bode rule). The Sun's rotation and the orbital revolution of the planets and dwarf planets have the same direction.
- The Sun contains >99% of the solar system's mass but the planets have most (~97%) of the angular momentum, as shown by Kant and Laplace.
- The four small terrestrial planets in the inner solar system – Mercury, Venus, Earth and Mars – are mainly composed of rock and metal. Their diameters, masses and observed densities do not vary as a smooth function of radial distance from the Sun. Mercury and Venus are "moonless," Earth has a large natural satellite (the Moon with ~0.012 Earth masses) and Mars has two small satellites, Phobos and Deimos.
- The four large gas-giant planets in the outer solar system – Jupiter, Saturn, Uranus and Neptune – are rich in H_2 and He but also have

larger quantities of volatile compounds such as water, CH_4 and NH_3. Their diameters, masses and observed densities do not vary as a smooth function with radial distance from the Sun. Uranus and Neptune are smaller and contain larger fractions of water ice, CH_4 and NH_3, and show higher D/H ratios in their atmospheres than Jupiter and Saturn. The rocky materials in the outer planets make up only a few percent of their total mass. Each outer planet is surrounded by a plethora of moons, some of which may be co-genetic with the giant planets, others which may be stray trans-Neptunian objects that were gravitationally captured.

- The asteroid belt at 2–4 AU is populated with thousands of smaller rocky objects. The largest one, Ceres, was recently raised to the status of "dwarf planet" by the International Astronomical Union. Asteroids have variable densities and surface compositions. It is generally believed that most meteorites are samples from asteroids.

- The trans-Neptunian region (> 30 AU) is populated with thousands of "icy" objects and is the source region of comets. Pluto is located within the Kuiper belt. Beyond Pluto are the high-inclination objects from the disk ("scattered objects"), followed by the Oort cloud. The two largest objects in the Kuiper belt are the IAU-certified dwarf planets Pluto at 39 AU (discovered by Tombaugh in 1930), with its moons Charon (discovered 1977), Nix and Hydra (both discovered in 2005), as well as Eris at 97 AU (discovered in 2003 by Mike Brown and colleagues), with its moon Dysnomia. Ices of water, methane, nitrogen, CO_2 and/or organic substances are inferred on their surfaces.

- The surfaces of the Moon, Mercury, Mars and outer planet satellites (*e.g.*, Callisto) are heavily cratered; similarly, craters are found on asteroids and cometary nuclei.

- The orbital inclinations (in degrees) relative to the ecliptic (the Earth's orbital plane) are large for the innermost planet Mercury (7°), and two dwarf planets Pluto (17°) and Eris (44°). The orbit of the Moon is inclined 5.1° relative to the Earth's orbital plane.

- The obliquities (tilt of the planetary rotational axis to the planet's orbital plane) of Venus ($\sim 177°$) and Uranus ($\sim 98°$) are unusually high.

- Terrestrial, lunar and meteoritic materials have characteristic O isotopic compositions. Compared to the magnitude of the differences in O isotopic compositions, the isotopic compositions of other, heavier and abundant elements (Mg, Si, Fe) are the same in terrestrial, lunar and meteoritic rocks.

- Short-lived radioactive nuclides such as ^{26}Al, ^{182}W and ^{129}I, with half-lives from ~ 1 to 100 Ma were present when the solar system formed but are now "extinct." Longer-lived radionuclides such as ^{40}K, ^{87}Rb, ^{235}U, ^{238}U and ^{232}Th are still present.

- Radiometric dating of Earth, Moon and meteorites shows that rocky materials began to assemble within a few million years after the first solids assembled their refractory mineralogy 4567 Ma ago. Thermal and aqueous alterations on asteroid-size bodies occurred within ~ 50 Ma after that, and core formation in terrestrial planets and differentiated asteroids occurred within similar time spans.

- Several chondritic meteorites contain stardust, which had formed around evolved stars and was not destroyed in the interstellar medium during solar system formation.

3.2 STAGES OF SOLAR SYSTEM FORMATION

The stages from the formation of the Sun and its protoplanetary nebula to the formation of planets are sketched in Figure 3.2. The major stages are collapse and fragmentation of an interstellar molecular cloud, formation of the proto-Sun and the solar nebula around it, evaporation/condensation of solids and accumulation into planetesimals, and accretion of planetesimals into planets. We follow the elements and their chemistry as we describe these stages.

3.2.1 Interstellar Medium and Presolar Grains

The interstellar medium (ISM) contains gas and dust. Most material is gas, which, in turn, is mainly H and He. The amount of other elements in the ISM is continuously supplemented by nucleosynthesis from evolved stars such as red giants and supernovae (Chapter 1). In the ejecta of stars, dust can condense, and stellar ejecta add both gases and dust to those already found in the ISM. The dust in the ISM is reprocessed over time but a tiny fraction of the dust that originally formed around stars was preserved as "presolar grains" in chondritic meteorites.

Presolar grains were first isolated from primitive meteorites in 1987 and their study with micro-analytical methods has allowed detailed insights into stellar evolution and nucleosynthesis processes that cannot yet be obtained even with the best telescopes.[1] The major recognized presolar minerals include nano-diamonds, followed by micron to sub-micron size grains of SiC, graphite, Al_2O_3, spinel and silicates.

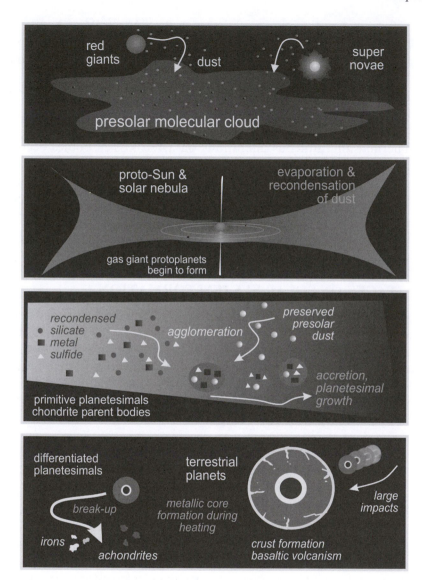

Figure 3.2   Cartoon sketching some of the major stages of solar system formation that led to the formation of terrestrial planets.

Figure 3.3 shows examples of presolar graphite and SiC grains. Small quantities of presolar grains are present in chondrites of petrologic types 1–3, and it is significant that presolar grains are found in ordinary, carbonaceous and also enstatite chondrites. However, C chondrites seem to have retained the largest quantities of presolar grains (see, for example, Table 2.2 for presolar nano-diamond abundances in

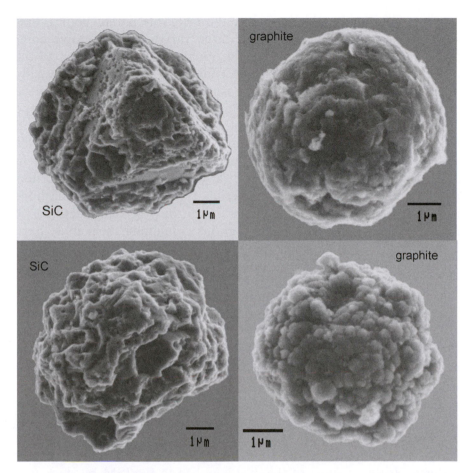

Figure 3.3 Presolar SiC (left) and graphite (right) grains formed in the ejecta of stars and found in primitive chondrites as rare components. Note the different morphologies. Other major presolar minerals include nano-diamonds, corundum, spinel and silicates such as pyroxene and olivine. (Photographs kindly provided by S. Amari.)

C-chondrites). The absence of presolar grains from chondrites of higher petrologic types is caused by grain assimilation into "normal" meteoritic minerals during thermal metamorphism. Aqueous alteration on the parent bodies of type 1 and 2 chondrites apparently did not destroy the entire presolar grain population that was initially present.

Presolar minerals were present in the molecular cloud from which the solar system originated (hence "presolar" grains), and these grains were not significantly thermally processed during the formation of the solar system, which would have lead to their destruction. The grains accumulated together with other solar system solids during accretion of larger planetesimals, which are now the meteorite parent bodies. Presolar grains are identified by their large variations in isotopic compositions of C, N, O, Si and several other elements compared to the isotopic compositions found in terrestrial materials, which are usually taken as reference standards. Large isotopic variations that cannot be explained by mass-dependent fractionations are also called "isotope anomalies." In presolar grains, isotope ratios can vary over several magnitudes compared to the normal terrestrial compositions. This includes variations in O isotopes, which can exceed percent-levels in individual presolar grains, but only vary on per-mil levels among terrestrial and normal meteoritic materials (Figures 2.1, 2.2). The huge isotopic anomalies in presolar minerals can only be manufactured by nucleosynthesis processes, and the grains containing them must have formed near the stellar source that produced the characteristic nuclides. Therefore, formation of these grains must clearly predate the formation of solids within the solar system.

The nuclear signatures of a star become fixed in grains for isotopes of those elements that can form stable solids in the stellar ejecta. However, some elements may not condense or condensation may be incomplete. In such a case, the star's characteristic isotopes of such elements remain in the gas. Since the gas from the stellar ejecta will quickly become diluted with other gas in the ISM, these individual star signatures become lost.

Stardust particles retain their stellar signatures as long as they are not destroyed through processing in the ISM or during star formation. The average residence time of dust in the ISM is estimated as 800–1000 Ma before becoming incorporated into newly forming stars and planetary systems. However, most of the dust supplied to the ISM by stars goes through reprocessing at least once before being incorporated into new stars and planetary systems. Grain destruction mechanisms in the ISM include sputtering and erosion through shockwaves (*e.g.*, from supernovae), and photo-evaporation near luminous stars. Average grain sizes in the ISM are taken as $\sim 0.2\,\mu m$. Estimates for the average grain destruction time scales in the ISM are 60–400 Ma for silicates, ~ 500 Ma for iron metal alloys and ~ 600 Ma for carbonaceous dust.[2] Re-condensation homogenizes the dust composition in the ISM. However, since there is a steady supply of new dust and gas from aged stars, there is always a steady-state amount of "fresh" or unprocessed stardust among the dust population in the ISM.

On a large scale, the elements made and released from different types of stars at different times merge and produce a relatively homogenous elemental composition in the ISM. Therefore, it is not too surprising that the Sun and many stars like it have relatively similar compositions since all stars incorporate a large fraction of homogenized material from the ISM when they form.

A direct comparison of the solar composition with that of the nearby ISM is somewhat complicated because only the elemental composition of the gas phase from the ISM can be determined. The dust in the ISM includes silicates, ices and organic solids, but it is not easy to obtain quantitative elemental compositions for the dust. However, the elements are distributed between gas and dust according to their relative volatility, which can be utilized in a comparison of the solar and the ISM compositions.

Observations of interstellar gas are carried out along lines of sight to young, hot stars like ζ Ophiuchi, which ionize the gas of the ISM in their vicinity. Figure 3.4 shows the atomic composition of the cool ISM component towards ζ Ophiuchi relative to solar abundances as a

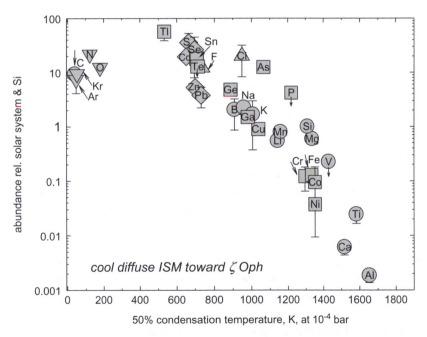

Figure 3.4 Atomic abundances[3] in the gas phase of the cool interstellar medium (ISM) towards ζ Ophiuchi relative to solar abundances as a function of condensation temperature of the elements. The abundances are normalized so that the Si ratio plots at unity. The gas-phase of the ISM is depleted in refractory elements that are in dust. The symbol shape indicates the condensate type for an element: circles are oxides and silicates, squares are for metal alloys, diamonds are for elements with affinity to FeS, normal triangles are halogen-bearing condensates, and inverted triangles are for elements that condense as low-temperature ices such as water.

function of condensation temperatures (see below).[3] The refractory elements, which condense first (or evaporate last) at the highest temperatures, are very low in abundance in the gaseous ISM. This suggests that the refractory elements are in dust. On the other hand, volatile elements that evaporate first (or condense last) at low temperatures are high in abundance relative to solar. Overall, there is a relatively smooth trend of the solar-normalized abundances in the ISM gas with the condensation temperatures of the elements calculated for a solar composition gas. This would not be the case if the overall composition of the ISM (gas plus dust) were drastically different from solar because condensation temperatures are also a function of the overall composition of a system. Hence, the correlation with condensation temperatures calculated for a solar composition system indicates that the overall compositions of the ISM (gas + dust) and the Sun are very similar. Because the composition of the Sun was that of the local ISM around 4600 Ma ago, and measurements are for the present day ISM compositions, the similarity between the past and current compositions also shows that the elements heavier than He (the astronomer's metals) are supplied to the ISM in about the same proportions that they had been 4600 Ma ago (note that the abundances in Figure 3.4 are normalized to Si).

3.2.2 Interstellar Molecular Cloud Collapse and Solar Nebula Formation

The composition of the ISM gives us a good idea of the element distributions between gas and dust in giant molecular clouds. These clouds can extend several tens of light years in diameter and can contain several to hundreds of solar masses of material. Star-forming regions are often associated with molecular clouds, indicating that these clouds are the stellar birthplaces. They are called "molecular clouds" because H is in the form of molecular H_2 and they contain molecules such as CO, whereas other types of interstellar clouds contain H as neutral atoms (*e.g.*, in diffuse clouds). Within giant cloud complexes, "hotter" cloud cores of interstellar gas and dust are present, which have temperatures of ~ 100–200 K and typical densities of $\sim 10^8$ particles cm^{-3} (equivalent to 2×10^{-16} g cm^{-3} for a solar composition mixture). These clouds are stable against self-gravitational collapse as long as turbulent motions in them work against it. Cloud collapse may be triggered through disturbances such as localized density clumping from random motions, or external forces such as supernova shockwaves. The triggering of cloud collapse by supernova shock waves is an attractive mechanism for solar system formation because this would provide a source and a mechanism to introduce some of the short-lived radioactive nuclides (see below) into the forming solar system.[4]

When molecular clouds gravitationally contract and fragment into individual collapsing pieces or globules, more than one star forms, and usually a cluster of new stars with varying stellar masses results. The percentage of stars with large masses (> 10 solar masses) is much smaller than that of intermediate masses, and the largest fraction consists of stars with about a solar mass or less. The time it takes a dense molecular cloud core to collapse through free fall ($\tau_{\text{cloud free fall}}$) is several thousand years, which is estimated from:

$$\tau_{\text{cloud free fall}} = \sqrt{\frac{3\pi}{32G\rho_{\text{ini}}}} \tag{3.1}$$

This estimate assumes the unlikely case that the cloud has an initially uniform density ρ_{ini}, is spherical and non-rotating, and that collapse is isothermal and homologous (*i.e.*, density increases at the same rate throughout the cloud). However, it is useful to obtain first-order estimates of the time scales involved.

Free fall works as long as the pressure within the fragment does not impede it. However, as the density increases, the temperature and pressure in the collapsing cloud also increase, and collapse of the cloud fragment slows down. After about half a million years, the cloud fragment has a size of a few times the current solar system, and a core of about 10 AU in radius begins to accumulate. Gravitational energy freed during contraction starts to heat the material in the cloud fragment. Only about half of this energy is used for heating, the other half is carried away through radiation. (The virial theorem applies here: $2E_K + U = 0$, where E_K is the internal kinetic energy and U is the gravitational potential energy of the cloud.) However, radiation can only escape as long as the density is low and the material is transparent to radiation. Most energy is released by radiation from the center where the mass accumulation is the greatest. Continuing infall of material increases the density and the opacity of the material near the center, which causes more adiabatic heating. Pressure builds up in the developing proto-core and counteracts free fall of material onto it. At this stage, the solar proto-core has a few percent of the final Sun's mass and has shrunken down to the size of a few hundred times the current solar radius. To put this size into perspective, the current Earth–Sun distance ($= 1$ AU) is equivalent to 215 times the solar radius. Within this proto-core, the original dust from the interstellar cloud has long since vaporized, and gravitational heating raises temperatures sufficiently to ionize atoms in the gas. The energy used to dissociate H_2 and to ionize atoms decreases the thermal energy available to maintain the pressure that works against the forces of free-fall collapse.

This leads to a second collapse phase from which the Sun emerges as a protostar of a few solar radii, which happens $\sim 10^5$ years after the start of molecular cloud core collapse. The time it takes for further contraction to a hydrostatic equilibrium structure is governed by how quickly the released gravitational energy can be radiated away, which also influences the thermal evolution of the proto Sun's surroundings. The Kelvin–Helmholtz timescale gives a measure of this. The energy freed by gravitational collapse of $M = 1$ solar mass into a sphere with a radius R of the current radius of the Sun is $E_{grav} \approx 3\,GM^2/(10R) \approx 2 \times 10^{41}$ J (G is the universal constant of gravitation). Assuming that the luminosity ($L = 3.8 \times 10^{26}$ J s^{-1}) of the Sun was about the same as now, the Kelvin–Helmholtz timescale is $\tau_{KH} = E_{grav}/L \approx 10$ Ma. This simple first-order estimate is of the same magnitude as more detailed models of pre-stellar evolution provide. At the end of this second collapse stage, the densities and temperatures in the Sun's center have finally become high enough to sustain H-burning, which makes the Sun a real star. However, during the first million years of the second collapse stage, the fully convective Sun only reaches high enough interior densities and temperatures for deuterium-fusion, which delayed further contraction for a short period. The relatively early capability of the Sun to burn D rules out theories to make the planets from large amounts of material that was processed by thermonuclear reactions in the Sun and was later ejected from the Sun, since planetary, meteoritic and cometary materials still contain D (see also Chapter 1).

The advancement to the D-burning state happens relatively quickly because the Sun accretes most of its mass within about 50 000 years (*i.e.*, at a rate of about $\sim 2 \times 10^{-5}$ solar masses per year) from a rotating disk of gas and dust that had stabilized around the proto-Sun. This rotating disk from which matter accreted onto the growing Sun and from which the planets originated is called the *solar nebula* when the evolution of our solar system is discussed; protoplanetary accretion disks around other stars are referred to as proplyds.

Gas and dust continues to fall onto the disk from the remains of the molecular cloud fragment that surrounds the disk, and material from the disk continues to move onto the Sun. At the same time that material moves toward the growing Sun, angular momentum is transported outward through the geometrically thin disk. The mechanism for transferring the angular momentum from the growing Sun to the disk is not completely understood. It may involve Balbus–Hawley magnetic instabilities, spiral density waves or disk-driven bipolar outflows.

Observations of star forming regions and young protostars can provide information on what the young solar system may have looked like. Several young protostars, known as classical T Tauri stars, are less than ~ 10 Ma old, and are surrounded by optically thick protoplanetary accretion disks of gas and dust. T Tauri stars are named after the young solar-mass star T Tauri in the Taurus star-forming region. These protostars are in the stage of accreting the final few percent of their total masses, which are similar to that of the Sun (0.5 to 3 solar masses).

The absorption spectra of disks reveal the presence of solids such as silicates and ices at far radial distances. The type of solids depends on the temperature and pressure within the disk; ices are only stable in the outermost disk regions whereas silicates remain stable in the hotter, inner regions. Infrared observations of protoplanetary disks reveal that they contain mainly crystalline silicates. This is consistent with the observations of crystalline olivine and pyroxene in comets, which contain preserved material from the outer regions of the solar nebula. However, infrared observations show that the silicates in the ISM are amorphous. This either indicates that dust from the ISM was recrystallized in the outer solar system where cometary materials originated, or that crystalline silicates were mixed from the inner solar nebula to the outer solar nebula. Interestingly, observations of evolved dust-producing stars also show that they already release mainly crystalline silicates into the ISM, and the presolar silicate grain observations are consistent with this. Apparently, the processing and likely homogenization of silicates in the ISM leads to amorphous silicates that may only re-crystallize in protoplanetary disks.

In this context it is interesting to mention the GEMS (glass with embedded metal and sulfides)[5] found as 0.1 to 0.5 μm diameter particles in interplanetary dust particles that are believed to come from comets. It is long suspected that comets may have preserved some of the interstellar material that escaped processing in the solar nebula. The silicates of GEMS show an excess of oxygen compared to the expected stoichiometric amount from Mg- and Ca silicates, and, relative to solar composition, GEMS are depleted in Fe and S. They show "normal" O-isotopic compositions, which would not be surprising if dust that originally had formed in stellar ejecta was processed and homogenized through sputtering and re-condensation in the interstellar medium. After all, we can expect that a large fraction of the solids present in the pre-solar molecular cloud was processed stardust, since pristine stardust only has a limited lifetime against destruction in the interstellar medium.

However, it is not easy to prove whether the GEMS are indeed relics of processed interstellar solids. In favor of such an origin is that they show physical radiation damage that plausibly can be caused by intense galactic cosmic ray radiation. On the other hand, the edge of the solar system is also more prone to cosmic ray exposure, and phases formed in the solar nebula may have experienced radiation damage before entering cometary materials.

During this disk stage, dust may also have evaporated in the region now occupied by the terrestrial planets, but re-condensation of solids from the cooling solar nebula gas also takes place. The types of condensates and the condensation temperatures of the elements are described in the next section, after a brief description of the temperature and density structure of the solar nebula.

Protoplanetary disks around T Tauri stars have estimated masses between ~ 0.001 and ~ 0.1 solar masses within $\sim 100\,\text{AU}$. Such amounts of mass are more than enough to account for the mass of all the planets of ~ 0.0014 solar masses (see Figure 1.5) in our solar system. The minimum mass of the solar nebula required to account for the planets can be used to derive the nebular density structure, and, with additional thermal modeling, the pressure–temperature structure of the solar nebula can be estimated.

The temperature and pressure structure within the solar nebula disk depends on the radial distance from the Sun, the height above the midplane of the disk, accretion rates of material onto the disk and from the disk into the Sun, and the surface area of the disk that allowed radiative energy loss. The highest temperatures and pressures exist for high accretion rates onto the disk, and at close distances to the Sun. At a given distance, temperature and pressures are highest at the disk's midplane where matter concentrates and the least radiation can escape.

The minimum disk mass of the solar nebula required to create the planets can be estimated from the solar abundances of the major elements and approximate compositions for the planets and their total masses. We can estimate the amount of the major elements that must be added to the planetary material to get back the solar composition (solar abundances in Table 1.2 must be converted into mass concentrations for this purpose). The condensation calculations described below tell which elements are found in silicate rock, metal, sulfide, ices and in the gas. Silicates and metal are the principal material of the terrestrial planets, and their contribution to the solar system composition is easily estimated by summing up the oxides of the rock-forming elements Si, Mg, Al, Ca, Ti, Na, K, *etc.*, S in FeS, and metal from total Fe not in sulfide plus Ni. Iron enters silicate rocks as "FeO," but whether some fraction of total Fe is present as metal or oxide does not change this simple estimate by much. Hence, we find that the solar system composition contains $\sim 0.5\%$ silicates, sulfide and metal. Mercury, Venus, Earth and Mars consist of rock and metal, and their masses are equivalent to 5.9×10^{-6} solar masses. Terrestrial planets only retain these 0.5% rocky mass of the solar nebula, and, scaling back to total solar composition, the total mass needed in the terrestrial planets region must have been at least $5.9 \times 10^{-6}/0.5\% = 0.001$ solar masses of solar composition material.

The outer planets are more massive and contain significant amounts of H_2 and He, but in smaller fractions than in the solar system composition. They also contain H_2O, CH_4 and NH_3 in addition to rocky elements. In condensed form, the ices of H_2O, CH_4 and NH_3 make up $\sim 0.6\%$ by mass in the solar system composition, which is essentially the same mass fraction as for rocky material. It is estimated[6] that Jupiter requires between 0 and 11 Earth masses of rock- plus ice-forming elements, corresponding to up to 3.5% of Jupiter's current mass; the rest is H_2 and He. Adopting a middle value of 5.5 Earth masses corresponds to $\sim 2\%$ rock and ice in Jupiter. Saturn contains between 9 and 22 Earth masses of ice- and rock-forming elements, but accreted less H_2 and He. Using an intermediate value of 15 Earth masses then suggests that $\sim 16\%$ of Saturn's total mass is from rocky and icy materials of the solar nebula.

The amounts of rocky and icy materials are estimated as $\sim 86\%$ and $\sim 90\%$ of the current masses of the denser outermost planets Uranus and Neptune, respectively. The mass fraction of ice and rock in the solar system composition is about 1 : 1, and assuming that the giant planets accreted rock and ice in this proportion (which may not be a valid assumption) we have all of the necessary information to complete the estimate for the minimum mass of the solar nebula from the rocky (but ice-free) fraction of the planets (Table 3.1).

Table 3.1 The rocky (but ice-free) fraction of the planets.

	Planet mass, M (in solar masses)	Mass fraction of rocky material in planet(s), f	Required solar nebula mass = $M \times f/0.005$ (in solar masses)
Terrestrial planets	5.94×10^{-6}	1.0	0.001
Jupiter	9.55×10^{-4}	0.01	0.002
Saturn	2.86×10^{-4}	0.08	0.005
Uranus	4.35×10^{-5}	0.43	0.004
Neptune	5.15×10^{-5}	0.44	0.005
			Total = 0.017

From Table 3.1 the estimated minimum mass of solar nebula is 0.017 solar masses.

To get an idea of the mass density distribution in the nebula, one can use these estimates for the masses and "spread" them over an annulus about a planets' orbit. This leads to a surface mass density (σ_{nebula}) variation in the solar nebula that is approximately proportional to $r^{-1.5}$ (r = radial distance). One simple estimate[7] of the total pressure in the midplane is:

$$P(r) = \sigma_{nebula}\left[GMRT/(2\pi\mu r^3)\right]^{0.5}$$
$$\approx 1.5 \left(g\,cm^2\,s^{-2}\,K^{-0.5}\right) T^{0.5}\left(K^{0.5}\right)/\left[r^3\left(cm^3\right)\right]$$

(3.2)

where μ is the mean molecular weight of the solar nebula gas ($\sim 2.4\,g\ mol^{-1}$), M is the system mass, T is the temperature at radial distance r and G and R are the gravitational and gas constants, respectively. We approximate the pressure and temperature conditions at the midplane by the adiabatic relationship:

$$P(r)/P(r_o) = (T/T_o)^{Cp/R}$$

(3.3)

where the subscript "o" denotes some reference pressure and temperature, and C_P is the heat capacity at constant pressure. Figure 3.5 indicates characteristic values for T_o and $P(r_o)$. For the H_2-rich solar nebula gas, C_P/R is about 7/2 for temperatures between ~ 150 and 2000 K. This indicates that the midplane total pressures vary as $\sim r^{-4.1}$ and midplane temperatures as $\sim r^{-1.1}$.

Results of some more elaborate models for characteristic temperature–pressure conditions as a function of radial distance in the solar nebula are also illustrated in Figure 3.5. At 1 AU, the models[8] suggest temperatures of ~ 600 K and total pressures of about 10^{-4} to 10^{-5} bar. Closer to the proto-Sun (~ 0.1 AU), the lower midplane temperatures are more realistic when compared to temperature estimates for observed disks around T Tauri stars, which suggest ~ 300 K at Earth's orbit (1 AU). In the outer solar nebula, which is usually taken as Jupiter's orbit and beyond (> 5.2 AU), temperatures generally are lower (< 200 K); typical values for T Tauri disks are < 50 K at ~ 10 AU. However, one should not forget that the temperature and pressure structure varies with time and depends on the mass accretion rates, which are initially high (leading to higher T) and drop to low values in the T Tauri stage.

During the last few million years of the terminal accumulation stage of the Sun, the accretion rates slow down from the estimated initial rates

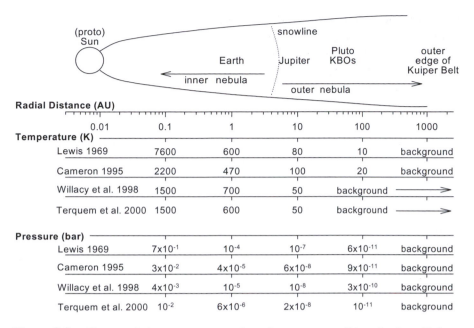

Figure 3.5 Characteristic temperature and total pressure conditions in the midplane of the solar nebula as a function of radial distance.[8]

of $\sim 2\times10^{-5}$ to 10^{-7} to 10^{-9} solar masses per year, which are comparable to the rates inferred for several T Tauri stars. This leads to cooling of the solar nebula, and one can expect that re-condensation of any previously evaporated solids occurs during this stage. Since there is still infall onto the solar nebula disk during this stage, the chances of preserving stardust grains from the molecular cloud also increase at lower temperatures.

The T Tauri stars can show irregular short-term (days) changes in their luminosity; probably caused by the ongoing accretion of material from their disks. This activity can heat material close to the protostars and has been suggested as a mechanism of chondrule formation (see Chapter 2). Several young (1–5 Ma old) T Tauri stars accrete material from their surrounding disks and simultaneously eject hot ionized matter in conically shaped, far-reaching jets of up to several thousand AU from their polar regions. These bipolar jets appear to be collimated by magnetic fields and to be powered through the material accreting onto the star. The jets are an attractive means to remove angular momentum from stars. Processing of inner disk material in bipolar outflows of the early Sun is also another potential mechanism for chondrule formation. In the "X wind" model by F. Shu and colleagues,[9] material from the accretion disk and the solar surface regions merge into a magnetized "X-wind" that launches from the plane of the accretion disk. Any solids swept up this way could become melted or partially evaporated before being ejected from the jets. However, one problem with this chondrule formation mechanism is that chondrules become widely distributed across the nebula, which does not provide chondrule populations that are characteristic in composition for the different chondrite groups. For example, it is not easy to account for the distinct oxidation states and O isotopic compositions of chondrules of carbonaceous, ordinary, and enstatite chondrites. In addition, chondrules in a given meteorite have complementary element abundances to the matrix that binds them, which indicates that the chondrules formed at locations similar to the fine matrix materials. Such complementary element abundances are not expected if chondrules are from materials processed in the jets, X wind or flares of the Sun. However, the X wind model provides a means to melt and partially evaporate refractory Ca-Al-rich inclusions, which are found in many chondritic meteorites (preferentially in carbonaceous chondrites). The mineralogy of these inclusions requires high temperature formation events, and they do not seem to require a localized origin like chondrules and their counterpart matrix. As a group, these inclusions have more chemical and isotopic similarities to each-other than to the different types of chondritic meteorites that they were extracted from (of course secondary alterations of these inclusions on meteorite parent bodies depend on what type of chondritic material they were embedded in).

In addition to dust evaporation and condensation, the dust in the solar nebula began to accumulate into larger planetesimals and proto-planetary objects. Models suggest that Moon to Mars size objects have accretion timescales of a few 10^5 to a few 10^6 years, which is consistent with the observations that dust disappears in 2–3 Ma around young stellar objects. According to theoretical simulations of planetary accretion, it is more difficult to build Earth- and Venus-size planets in non-eccentric and low-inclination orbits. However, larger planetary objects must have formed before the gas in the solar nebula began to dissipate; otherwise, it would be hard to explain how the gas-giant planets acquired their large amounts of H_2 and He. To attract and keep these light gases gravitationally bound, proto-planetary cores of several Earth masses are necessary. On the other hand, planetary objects are in danger of moving inward towards the protostar in a viscous gaseous disk. Such inward migration of planets is clearly indicated by the presence of gas-giant planets in very close orbits (< 0.1 AU) around other stars that have been discovered over the past decade. Another indication that even larger objects accrete onto protostars is the energetic outbursts associated with "FU Orionis"-type protostars (of which the star FU Orionis is the prototype), although other mechanisms for such energetic outbursts have been suggested. In any case, planet formation and preservation of planets is a delicate balance of the timescales of gas and dust accretion, growth of planetesimals and loss of gas from protoplanetary disks.

Observations show that 3–5 Ma (sometimes up to 10 Ma) old protostars have lost or are in the process of losing the gas from their accretion disks. The disks develop central holes as they dissipate outward, and gas loss leads to optically thin disks around the protostar. The dust remains, but clearing of dust from the inner disk proceeds quickly. There are two possible mechanisms to clear the inner disks of dust: accretion of dust onto the protostar, or accretion of dust to build larger planetesimals and planets. These two processes must compete, and it seems that building larger dust aggregates or even planetesimals wins out in many cases. The accretion of dust onto the star can occur because orbital motions of small dust grains (less than a micron in size) around the protostar are destabilized through radiation pressure and the Poynting–Robertson mechanism, which causes dust to spiral onto the proto star within $1000–10^6$ years. On the other hand, dust can grow from micron to mm sizes and up, and metre to km-size planetesimals begin to form, whose orbits are less influenced by radiation pressure forces. Dust must grow to mm-size particles in less than 10^6 years in order to prevent large losses of dust due to spiraling onto the protostar. Recent observations reveal radial density variations within protoplanetary disks, indicative of planetesimal and planet formation.

Collisions of larger particles or planetesimals also replenish fine dust, which is needed to maintain the observable dust in accretion disks around older stars. Since collisions can be quite energetic, such collisions may have played a role in producing the chondrules found in chondritic meteorites. Another indicator of collisions among different objects in the early solar system is the "foreign" fragments and clasts in given types of meteorites that clearly came from a different meteorite group. For example, the achondritic eucrites often contain embedded clasts and fragments of CM chondritic material. However, collisions among planetesimals are not restricted to the early few million years of the solar system, and if such collisions were a major source of chondrules one would expect to find chondrules with ages up to ∼ 10 Ma, which is the maximum age for the solar nebula before it lost its gas. (The gas is needed to get the proper cooling rates for chondrules, see Chapter 2.) It is known that planetesimal and planet formation was quite rapid within the solar nebula ($<$ 10 Ma), and chondrule formation only lasted for a few (perhaps 2–3) Ma after the first solids had formed.

For other planetary systems, it is assumed that planet formation does not take longer than 100 Ma. However, a heavy bombardment period of planetary objects in the solar system went on for 700–800 Ma until $\sim$ 3900 Ma ago, as witnessed by the large impact craters on the Moon, Mars, Venus and other solar system bodies. Huge collisions at earlier stages may well be responsible for the retrograde motion of Venus, and the inclined orbits of Mercury, and the Moon, leaving aside the impact-triggered formation of the Moon itself. The rather long period of collisions also gives a measure of the duration that was necessary for bringing the orbits of smaller bodies into a dynamical steady state. The period of intense bombardment continued after solar nebula gas was lost. Interestingly, there does not seem to be much evidence that meteorite parent bodies formed from any debris resulting from such late collisions, which may indicate that planetesimal growth required the presence of nebular gas.

With respect to dusty debris disks remaining after planets formed, the $\sim$ 800 Ma old system epsilon Eridani,[10] which is among the many discovered stars with planets (the "exoplanets"), could be an interesting analogue to the earlier solar system. The star epsilon Eridani is only slightly smaller than the Sun and a $\sim$ 1.6 Jupiter-mass planet orbits it at a radial distance of 3.4 AU. A clumpy debris disk at $\sim$ 40–60 AU has been discovered by excess infrared radiation from dust, and a second planet is suspected to be present in this disk. The age of this system coincides with the age of our solar system during the heavy bombardment period and, therefore, that system could resemble the solar system during this stage. The denser disk of epsilon Eridani is at a distance similar to Pluto and the Kuiper belt, a region where water ice and other ices were stable within the solar nebula (see next section). This region is also a source region from which comets are dynamically scattered into orbits that pass through the inner solar system. The dynamics of scattering bodies from the outer solar system into the inner solar system at an earlier time may have also led to a larger number of icy planetesimals, which contributed their volatile compounds, such as water, to the terrestrial planets, during the later heavy bombardment.

3.3 CHEMICAL ZONES IN THE SOLAR NEBULA

The overall elemental composition of the solar system (Chapter 1) does not tell us much about the *compounds* that were present at the various stages of solar system formation. If the temperature and pressure conditions are known, we can use chemical thermodynamics to infer the gases and solids that were present. However, the heating and cooling in the solar nebula also depended on the distance from the sun, and was time dependent, hence chemical kinetics become important. In addition to thermochemistry, which played a major role in the denser midplane of the inner solar system, photochemistry was important in close proximity to the young Sun, and in the outer region of the envelope around the accretion disk, which was exposed to ambient UV and X-ray radiation from the interstellar medium and the proto-Sun.

There are at least three chemical zones in the solar nebula. Zone I is the inner region with high temperatures and pressures where thermochemical equilibrium is reached. Zone II is an intermediate zone where thermochemical reactions become kinetically inhibited. Different reactions are inhibited at different temperatures and pressure and, therefore, kinetically inhibited departure from chemical equilibrium is gradual. However, in this intermediate zone, reactive species produced by UV and X-ray driven photochemistry and ion–molecule chemistry do not yet counteract thermochemical equilibria. Zone III is at low temperatures and pressures, where disequilibrium reactions driven by photochemistry and ion–molecule chemistry control the gas (and dust) composition. All zones extend in both radial and vertical directions and, at a given distance from the Sun, two or three zones may be present when moving perpendicularly through the disk from the midplane position.

Local exceptions to this gradual zoned vertical and horizontal disk structure are the circumplanetary subnebulae. These are miniature "solar nebulae" around the gas-giant proto-planets that later become Jupiter, Saturn, Uranus and Neptune with their principal moons. The chemical environs of the outer solar nebula had the characteristics of zone II and/or III. The denser environments around the accreting gas-giant protoplanets are better characterized with conditions as in zone I. Figure 3.6 sketches the relative importance of thermochemical and photochemical reactions for planetary objects in the solar nebula. Similar accretion processes that led to the formation of the terrestrial planets must have been going on in the subnebulae, which is indicated by the similarities in compositional variations and in relative orbital distances of the four terrestrial planets from the Sun to that of the Galilean satellites from Jupiter (similar orbital structuring is found to some extent for the regular satellites of Saturn, Uranus and Neptune).

The preservation of solids in the solar nebula and formation of condensates out of the nebular gas is the first step in forming planets. In the inner solar system, temperatures were high enough to vaporize a large

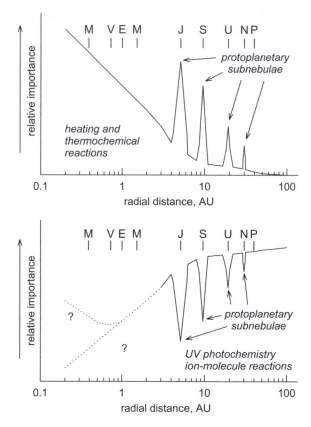

Figure 3.6   Relative importance of thermochemistry (top) and photochemistry (bottom) for planetary objects in the solar nebula. (Modified from ref. 11.)

fraction of the interstellar dust that was initially present in the molecular cloud. The only possible exceptions are the very refractory compounds such as found in Ca-Al-rich inclusions, but even these could undergo partial evaporation and partial melting. Certainly all high temperature processing of solids leads to resetting or disturbances of radioactive chronometers such as K-Ar, U-Pb, Th-Pb and Rb-Sr (see also below), because the ages of meteorites and their components determined with these radiometric systems are all limited to ≤ 4.57 Ga. We would expect older ages for interstellar materials that condensed around other stars or that were processed in the ISM, but no such solids with older ages have been found and confirmed yet. Absolute age determination on small presolar grains using chronometers with long-lived isotopes such as U, Th or [87]Sr are not yet possible, and are also hampered by the problem that the initial amounts of the radioactive decay products, which need to be known for absolute age determinations, are difficult to model.

One simple approach to model the solids available to build a planet at a given heliocentric distance is to look at the gas and condensate chemistry in the solar nebula. The thermodynamically stable gases and solids depend on the temperature and pressure conditions as the nebula cooled. Meteorites provide good justification for such an approach. Already in 1893, Daubrée concluded that rapid, direct condensation out of turbulent gases leads to formation of meteorite components, because even closely associated meteorite components show heterogeneous compositions. These can only be explained if the individual components formed separately and accumulated later into the rock. However, meteorites also show that the solar nebula was not homogeneous in composition, and that the silicates, metal and sulfide in chondrites did not simply result by equilibrium condensation from a solar composition gas. The different oxidation states of the ordinary, enstatite and carbonaceous chondrites suggest that the solar nebula had varying redox states at different locations.[12] The common antiquity of chondrites rules out that the redox conditions changed over a short period at one location. The characteristic O isotopic compositions of chondrites also indicate processing at different locations. Chondrule formation must have been operating over larger distances within the solar nebula because chondrule formation confined to a certain region within the solar nebula and subsequent transport of chondrules to different regions cannot explain why chondrules in the different chondrite groups have very different FeO contents, and O isotopic compositions. The modeling is further complicated by the fact that temperature, density and pressure varied with time at a given location within the solar nebula.

3.4 COSMOCHEMICAL CLASSIFICATION OF THE ELEMENTS

The cosmochemical character of an element describes its distribution between the stable gases and solid phases in a gas of solar elemental composition at given temperature and total pressure. In other words, the cosmochemical classification is a measure of the relative volatility of an element. The ongoing reactions in the solar nebula were condensation reactions of gases to solids, and evaporation reactions of dust to gas, depending on the previous history of the element mixture. Both types of reactions are relevant. During the evolution of the solar nebula, a large fraction of the dust from the molecular cloud must have been heated and evaporated, whereas new dust condensed from the hot gas as the solar nebula cooled. Nebular condensation chemistry is relatively straightforward to model with chemical thermodynamics and kinetics, and models of the temperature and pressure structure of the solar nebula.

3.4.1   Condensation Temperatures

The cosmochemical classification of the elements based on volatility describes how the different elements of a solar composition mixture distribute between gas and different condensed phases of different thermal stability as a function of temperature and total pressure. The *relative* stability sequence of the condensed minerals with varying temperature is not strongly affected by total pressure. A quantitative measure for the volatility of an element is its condensation temperature. This provides a general grouping of the elements into refractory and volatile elements.

The calculations of thermodynamic equilibria have a rich history, starting with the early works by Wildt 1933[13] and Russell in 1934[14] on gas chemistry in the sun and gas and condensation chemistry for cool stars. Lord (1965)[15] expanded on this subject, and many detailed studies have been done since.[15–30]

The solar nebula gas was mainly H_2, which led to a reducing overall oxidation state, which determined which elements are stable as oxides and silicates, and which ones are stable as metals and sulfides in condensed form. Elements that geochemically prefer to be in a metallic phase are called siderophile elements; those that geochemically prefer to be in an oxide or silicate phase are called lithophile elements, and elements with affinities to a sulfide phase are called chalcophile elements. Table 3.2 shows how the elements are grouped in the cosmochemical and geochemical classifications.

Figure 3.7 shows which elements behave as siderophile, lithophile, chalcophile or atmophile during condensation, and lists the abundances of the elements and their 50% condensation temperatures at 10^{-4} bar total pressure.[30] The 50% condensation temperature is the temperature at which half of an element is in the gas phase and the other half is in a condensed phase. Several elements do not form pure phases and instead condense into solid solution, which is indicated by "s" after the condensation temperature in Figure 3.7. The formation of solid solutions is also the reason why 50% condensation temperatures are a more practical measure of relative volatility for all elements.

The "normal" condensation temperatures describe when a condensed phase first becomes stable from a cooling gas. However, these condensation temperatures only apply to the major minerals. Figure 3.8 shows the condensation temperatures of major minerals as a function of total

Table 3.2 A cosmochemical and geochemical classification of the elements.[a]

Cosmochemical character	Geochemical character			
	Lithophile (oxides & silicates)	*Siderophile (metal alloy)*	*Chalcophile (sulfide)*	*Atmophile (gas phase)*
Highly refractory $T_{cond} > 1580$ K	Al, Sc, Y, Ti, Zr, Hf, La, Pr, Nd, Sm, Gd, Tb, Dy, Ho, Er, Tm, Lu	Mo, W, Re, Os, Ir	—	—
Refractory $1400 < T_{cond} < 580$ K	Ca, Be, Sr, Ba, V, Nb, Ta, Ce, Yb, Th, U, Pu	Ru, Pt	—	—
Major condensate	Mg, Si,	Fe, Ni,		
Intermediately volatile $1100 < T_{cond} < 1400$ K	Li, Cr, Mn, Eu	P, Co, Pd, Rh	—	—
Moderately volatile $700 < T_{cond} < 1100$ K	Na, K, Cl, B, Rb, Cs, F, (Ga)	Ga, Ge, Sn, Pb, As, Sb, Bi, Te, Cu, Ag, Au	—	—
Volatile[a] $200 < T_{cond} < 700$ K	Br, I, (Zn)	—	S, Fe, In, Tl, Se, Zn, Cd, Hg, (Pb, Te)	—
Highly volatile $T_{cond} < 200$ K	—	—	—	H, C, N, O, noble gases

[a]Based on 50% condensation temperatures at 10^{-4} bar from a solar composition gas.

Figure 3.7 A cosmochemical periodic table of the elements listing atomic abundances (above element symbol) and 50% condensation temperatures (below element symbol) for solar composition. An "s" after the condensation temperature indicates that the element condenses into solid solution with a major mineral. The shading indicates which type of phase (metal, oxide, sulfide) an element enters during condensation.

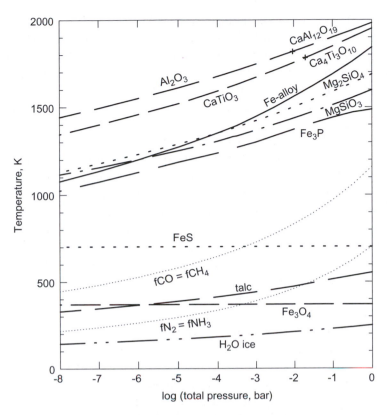

Figure 3.8 Equilibrium condensation temperatures of some major phases as a function of total pressure. The condensation temperatures indicate when a phase first becomes stable upon cooling a gas of solar system composition. The most refractory condensates form directly from the gas whereas some condensates at lower temperatures form by gas–solid reactions of previously condensed phases, *e.g.*, enstatite ($MgSiO_3$) from forsterite (Mg_2SiO_4) or troilite (FeS) and schreibersite (Fe_3P) from Fe-metal alloy. Thin dotted lines show where the major C-bearing gases CH_4 and CO and the major N-bearing gases NH_3 and N_2 have equal fugacities (partial pressures). Low temperature equilibria ($< 1000\,K$) at low total pressures ($P \ll 1$ bar) may not be reached within the lifetime of the solar nebula, especially for those reactions having large activation energies. Typically, reactions involving solid state diffusion and endothermic steps have large activation energies and proceed slowly at low temperatures. Three such examples are the gas-phase reduction of CO to CH_4, of N_2 to NH_3 and hydration of anhydrous silicates by reaction with water vapor near room temperature.

pressure. These major minerals are Ca-Al-oxides, Ca-Ti-oxides, Mg-silicates, FeNi alloy, and FeS, and several low temperature condensates, which are discussed in more detail below. The condensation temperatures for the minerals are higher than the 50% condensation temperatures of individual elements making these phases. For example, forsterite starts to condense at 1354 K (at 10^{-4} bar total pressure), but condensation of 50% of total Mg is only reached at 1336 K, and 50% of Si is only condensed at 1265 K, when enstatite is stable as well.

Condensation of most elements in solid solution requires that the mineral host phase is already present, and formation of solid solution essentially starts when the host phase condenses. However, the amount of a trace element condensing into the host mineral when it starts forming can vary from essentially all of the trace element condensed to only minor fractions, so that the condensation temperature of the host phase does not accurately reflect the relative volatility of the trace elements condensing into it.

Figures 3.7 and 3.8 provide the information needed to classify the elements as refractory, moderately volatile, volatile or highly volatile. The refractory elements either condense first from a hot gas of solar composition or vaporize last from a solid with CI chondritic abundances. Refractory lithophile elements condense into oxides and/or silicates. Aluminium, Ca, Ti and the rare earth elements (REE) fall into this category. Refractory siderophile elements condense into metal and are exemplified by W, Re, Os, Ru, Ir and Mo.

Iron metal and the magnesian silicates (*i.e.*, forsterite Mg_2SiO_4 and enstatite $MgSiO_3$) together constitute most of the rocky material that can form in the solar system. At pressures greater than about 10^{-4} bar iron metal condenses at higher temperatures than forsterite, while the reverse is true at lower pressures. Within the range of total pressures existing in the inner solar nebula, the condensation (or evaporation) temperatures of Mg, Si and Fe are relatively close to each other and are benchmarks that separate the refractory elements from the moderately volatile elements. In turn, the moderately volatile elements are separated from the highly volatile elements by the condensation of troilite, FeS, which starts to condense at the pressure-independent temperature of 704 K. With the possible exception of Hg, the highly volatile elements condense in the temperature interval between troilite (704 K) and magnetite (370 K) condensation. The most volatile elements are the atmophile elements (H, C, N, O and the noble gases). Water condensation as ice or liquid (at sufficiently high total pressures) is another important benchmark temperature because oxygen is the third most abundant element in solar system composition material. Therefore water ice (or liquid water) is expected to form a large fraction of condensate mass at the lowest temperatures. The following gives more details about the cosmochemical categories.

3.4.1.1   Refractory Elements.   The *lithophile* refractory elements condense as oxides at high temperatures (or remain as oxide residues from high temperature evaporation). In Table 3.2, the highly refractory elements are separated from the refractory elements by the temperature where ~ 50% of all Ti is condensed. The elements in the two refractory categories include Al, the alkaline earths, elements in groups 3b (Sc, Y), 4b (Ti, Zr, Hf) and 5b (V, Nb, Ta) of the periodic table, and most of the lanthanides and actinides. The refractory element oxides constitute about 5% by mass of the total rocky material in a solar composition system (Figure 3.9). Extensive studies of the chemical composition of chondrites show that refractory elements behave as a group in most meteorites. This is illustrated by the close correlations between the abundances of pairs of refractory lithophiles such as Ca and Al or Sc and Ti. Another indication of their similar behavior is that the CI chondrite normalized abundances of these elements in different meteorite groups show enrichments or depletions by about the same factor (see also Figures 2.10 and 2.11 in Chapter 2).

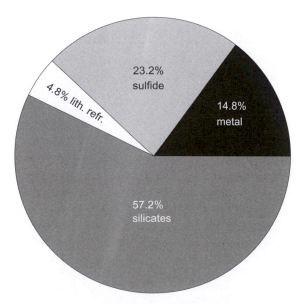

Figure 3.9   Mass distribution of high-temperature condensates from a solar composition gas. This distribution assumes that the silicates are Fe-free and that there are no icy condensates.

The condensation chemistry of the refractory lithophile elements is relatively well known. The first Al-bearing condensate is either corundum (Al_2O_3) or a calcium aluminate such as hibonite ($CaAl_{12}O_{19}$) or grossite ($CaAl_4O_7$), depending on total pressure. However, the initial condensates can continue to equilibrate with the gas with decreasing temperature and convert into other stable condensates at lower temperatures. However, the existence of corundum, grossite, hibonite, melilite, perovskite, spinel and other high temperature minerals in the Ca, Al-rich inclusions in the CV3 carbonaceous chondrite Allende and other meteorites shows that sometimes the high temperature condensates were isolated from the nebular gas and did not continue to react with it at lower temperatures.

Table 3.3 gives the condensation temperatures of some major minerals at 10^{-4} bar and shows the temperatures at which the higher temperature condensates transform into other minerals with decreasing temperature. The condensation temperature is where the mineral becomes thermodynamically stable and is higher than the 50% condensation temperature where half of the major element in the mineral is condensed and half remains in the gas.

Complete condensation (or evaporation) occurs gradually over a range of temperatures. For example, Table 3.3 and Figures 3.8 and 3.10

Table 3.3   Condensation temperatures[a] of some major minerals.

Ideal formula	Mineral name	Condensation temperature (K)
Al_2O_3	Corundum	1677
$CaAl_{12}O_{19}$	Hibonite	1659
$CaAl_4O_7$	Grossite	1542
$Ca_2Al_2SiO_7$	Gehlenite	1529
$CaTiO_3$	Perovskite	1593
$Ca_4Ti_3O_{10}$	Ca-titanate	1578
$Ca_3Ti_2O_7$	Ca-titanate	1539
$Ca_4Ti_3O_{10}$	Ca-titanate	1512
$CaTiO_3$	Perovskite	1441
$MgAl_2O_4$	Spinel	1397
$CaAl_2Si_2O_8$	Anorthite	1387
Mg_2SiO_4	Forsterite	1354
$MgSiO_3$	Enstatite	1316
$CaMgSi_2O_6$	Diopside	1347
Fe	Fe-alloy	1357
Fe_3P	Schreibersite	1248
FeS	Troilite	704
Fe_3O_4	Magnetite	371
H_2O	Water ice	182

[a]Note – at 10^{-4} bar total pressure.

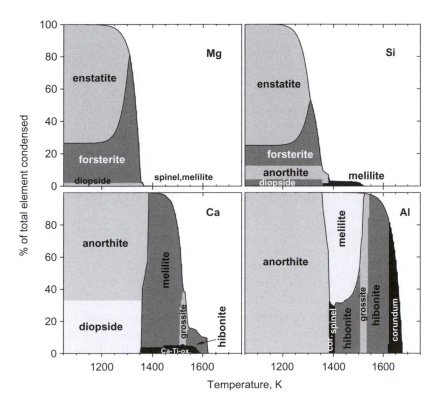

Figure 3.10   Distribution of Mg, Si, Ca and Al between gas and condensates as a function of temperature at 10^{-4} bar total pressure.

show that Al starts to condense as corundum at 1677 K, which is converted into hibonite within a few tens of degrees. With decreasing temperature more Al condenses from the gas into hibonite. Grossite, another Ca aluminate with the formula $CaAl_4O_7$, becomes stable about ~ 100 K below the hibonite condensation temperature. At even lower temperatures, the Ca aluminates react with the nebular gas to form spinel $MgAl_2O_4$, and melilite, which is a solid solution of gehlenite ($Ca_2Al_2SiO_7$) and åkermanite ($Ca_2MgSi_2O_7$). The higher temperature Al-bearing phases become unstable with decreasing temperature and feldspar (mainly anorthite $CaAl_2Si_2O_8$) and Al-bearing clinopyroxene are the major Al minerals (Figure 3.10).

The total abundance of each condensate, *e.g.*, hibonite, is limited by the abundance of the least abundant element in the condensate. Aluminum, Ca and O are the three elements in hibonite. Their abundances in solar composition material are 84 100, 62 870 and 14.13 million atoms, respectively, relative to one million Si atoms. Although Ca is slightly less abundant than Al, hibonite contains one Ca atom per 12 Al atoms and thus the abundance of hibonite is limited by the abundance of aluminum.

Condensation of Ca also proceeds through a procession of stable mineral phases (Table 3.3, Figure 3.10). It starts with hibonite condensation at 1659 K, and continues with grossite and melilite condensation. The 50% condensation temperature of Ca is ~140 K lower than the hibonite condensation temperature. A small amount of Ca also condenses into Ca-titanates, but the amount of Ca that can be removed from the gas by Ti-condensation is limited by the much smaller abundance of Ti. Titanium starts to condense as perovskite ($CaTiO_3$), and converts into other Ca-titanates at lower temperatures before becoming integrated into Ti-bearing clinopyroxene.

The less abundant, refractory lithophile trace elements such as the REEs generally do not form their own minerals because it is energetically more favorable for them to condense into solid solution with Ca-titanates or Ca-aluminates. However, in some cases phases with high concentrations of the most refractory lithophiles, such as Hf or Zr, occur in meteorites. Likewise, refractory metal nuggets containing refractory siderophile trace elements (*e.g.*, W, Re, Os, Ir and Ru) also occur in meteorites (inside the Ca, Al-rich inclusions in Allende and other carbonaceous chondrites). Kinetic limitations to the condensation of pure trace element phases are apparently unimportant in the nebular regions where these phases formed.

The 50% condensation temperatures in Figure 3.7 for Sr, Ba, and the REE assume ideal solid solution in perovskite, or Ca-Al-bearing phases. Perovskite is a major host phase for these elements in chondrites. Perovskite in chondrites also contains the more volatile elements Sr, Ba and Eu, although perovskite (or other Ca-Ti-oxides) will have reacted away by the temperatures where 50% condensation of Sr, Ba and Eu is calculated to occur with melilite and anorthite as condensate hosts. This indicates that fractional condensation or fractional evaporation took place, and that condensed phases were removed from equilibrium with the gas at some point.

Large enrichments of ~20-fold solar elemental abundances on average of the refractory lithophile elements are found in refractory inclusions in the Allende meteorite and other carbonaceous chondrites. These Ca, Al-rich inclusions (or "CAIs") have a mineralogy dominated by Ca-, Al- and Ti-rich minerals such as hibonite melilite, spinel and perovskite $CaTiO_3$. Several of these minerals are among the first expected condensates to appear from a cooling solar composition gas, which is why the CAIs are also called refractory inclusions. However, the CAIs are far more complex than just being condensate assemblages.

In whole-rock chondrites, the refractory lithophiles behave as a coherent group, but CAIs sometimes display large chemical fractionations from one another. These fractionations are consistently explained by the relative volatilities of the different refractory lithophiles.[21,27] The volatility-specific fractionations among the lanthanide abundances are used to define different groups of CAIs.[31]

The counterpart to the refractory lithophile elements are the refractory siderophile elements. These are the Pt-group metals (except Pd), Mo, W and Re. Like the refractory lithophiles, the refractory siderophile elements are enriched up to about 20-times solar elemental abundances (on average) in CAIs. Refractory metal nuggets containing W, Re, Os, Ir, Ru, Rh and Pt occur inside the Ca, Al-rich inclusions in Allende and other carbonaceous chondrites. Less refractory metals such as Fe and Ni also occur in the nuggets. These nuggets and complex multiphase assemblages of metal, oxides and sulfide, known either as Fremdlinge (little strangers) or as opaque assemblages, are the principal host phases for the refractory siderophiles in CAIs.[32] The refractory metal nuggets are generally believed to be condensates from the solar nebula and their compositions are reproduced by equilibrium condensation calculations.[26] However, the origin of the complex multiphase assemblages (Fremdlinge or opaque assemblages) is more controversial and proposed models vary from formation in the meteorite parent bodies,[33] to condensation in the solar nebula,[34] to formation by the partial evaporation and melting of interstellar dust aggregates.[35] As so often, each of these models has some attractive features, but none can easily account for all of the complexities observed in the Fremdlinge/opaque assemblages, and their origin remains somewhat enigmatic.

3.4.1.2 Iron Alloy and Magnesian Silicates. The condensation of the major elements Mg and Si as magnesian silicate and Fe as metallic iron is an important benchmark because it is the point where most of the rocky materials in solar composition material condense. As illustrated in Figure 3.8, at total pressures greater than about 10^{-4} bar iron metal alloy starts to condense at higher temperatures than the magnesian silicate forsterite, whereas at lower pressures the magnesian silicates condense first. In both cases, the separation between the metal and silicate condensation temperatures increases as the pressure is either increased or decreased from the crossover point.

The fraction of Mg and Si condensed is illustrated in Figure 3.10. The condensation of forsterite *via*:

$$2Mg(g) + SiO(g) + 3H_2O = Mg_2SiO_4 \text{ (forsterite)} + 3H_2 \qquad (3.4)$$

leads to a higher 50% condensation temperature of Mg than Si because forsterite accommodates two Mg and one Si per formula unit whereas the solar abundances of Mg and Si are essentially the same. Thus, up to about half of all Si remains in the gas (largely as SiO) until forsterite reacts with it to enstatite:

$$Mg_2SiO_4 \text{ (forsterite)} + SiO(g) + H_2O = 2MgSiO_3 \text{ (enstatite)} + H_2 \quad (3.5)$$

which is responsible for Si removal from the gas with decreasing temperatures.

The magnesian silicates condensing at high temperatures are essentially pure $MgSiO_3$ and Mg_2SiO_4 because the large excess of H_2 in solar gas leads to extremely low oxygen fugacities. (The oxygen fugacity is effectively the same as the oxygen partial pressure.) The temperature dependent oxygen fugacity (fO_2) in solar composition gas is given by:

$$\log fO_2 \text{ (bar)} = 2\log(H_2O/H_2) + 5.67 - 25\,664/T(K) \quad (3.6)$$

This equation depends on the H_2O/H_2 molar ratio and is valid from 300 to 2500 K.[12] Consequently, the Fe^{2+} (or FeO) content of ferromagnesian silicates such as olivine, $(Mg,Fe)_2SiO_4$, and pyroxene, $(Mg,Fe)SiO_3$, is insignificant until temperatures below about 600 K where the reactions:

$$2MgSiO_3 \text{ (enstatite)} + 2Fe \text{ (metal)} + 2H_2O \text{ (gas)}$$
$$= Mg_2SiO_4 \text{ (forsterite)} + Fe_2SiO_4 \text{ (fayalite)} + 2H_2 \text{ (gas)} \quad (3.7)$$

$$Fe_2SiO_4 \text{ (fayalite)} + 2MgSiO_3 \text{ (enstatite)}$$
$$= Mg_2SiO_4 \text{ (forsterite)} + 2FeSiO_3 \text{ (ferrosilite)} \quad (3.8)$$

become thermodynamically favorable.[16,20] At these temperatures olivine and pyroxene solid solutions containing several tens of mol% of fayalite and ferrosilite are expected to form. Any unreacted Fe-metal is predicted to oxidize to magnetite at a pressure independent temperature of about 370 K *via* the reaction:

$$3Fe \text{ (metal)} + 4H_2O \text{ (gas)} = Fe_3O_4 \text{ (magnetite)} + 4H_2 \text{ (gas)} \quad (3.9)$$

However, the predicted gas–solid and solid–solid reactions are probably quite slow at the low temperatures where equilibrium incorporation of FeO into silicates is predicted. Calculations of gas–solid reaction rates indicate that the reactions forming magnetite and FeO-rich silicates are unlikely to proceed to equilibrium over the estimated 1–10 Ma lifetime of the solar nebula because of slow solid-state diffusion and the high activation energy of the gas–solid reactions.[36,37] The observed textural features and chemistry of FeO-rich olivines in meteorites may thus require high-temperature condensation under more oxidizing conditions in the solar nebula. Such conditions may be realized in dust-rich regions, such as the nebular midplane, where heating of the dust releases the oxygen in rock into the gas and increases the local O/H elemental ratio above the solar value. However, although many workers have now discarded the notion that FeO-rich silicates formed at low temperatures in the solar nebula, an origin by metamorphic reactions on the meteorite parent bodies is also still debated as an alternative to an origin under oxidizing conditions in the solar nebula.[12]

3.4.1.3 Moderately Volatile Elements. By convention, the moderately volatile elements are those with condensation temperatures intermediate between those of the major elements Fe, Mg and Si and the condensation temperature of troilite FeS. The geochemically diverse elements in this group include Na, K, Rb, Cr, Mn, Cu, Ag, Au, Zn, B, Ga, P, As, Sb, Bi, S, Se, Te, F and Cl (Figure 3.7). The condensation chemistry of many of these elements is uncertain because the thermodynamic properties of the trace element condensates and their solid solution properties with major host phases such as Fe metal, FeS and apatite minerals are not well known. For example, Ga can condense in solid solution in Fe metal, but Ga condensation in silicates such as feldspar where Ga may substitute for Al is also possible, and both host phases must be considered. Zinc can condense into FeS as ZnS, but it can also condense into Mg-silicates, *e.g.*, as Zn_2SiO_4 in olivine and $ZnSiO_3$ in pyroxene. Periodic behavior suggests that Rb, Cs, Br and I are moderately volatile elements but very little thermodynamic data are available for any plausible condensates of these elements. However, calculated condensation temperatures for Rb and Cs generally agree with expectations based on observed elemental abundance trends in chondritic meteorites.[22]

3.4.1.4 Highly Volatile Elements. The highly volatile elements have condensation temperatures below that of troilite (704 K). Only a few elements fall into this category: Cd, In, Tl, Hg, Br and I and the elements Se, Pb, Te and Bi are borderline cases.

Apatite minerals [$Ca_5(PO_4)_3X$, with X = Cl, Br, I, or OH] are likely host phases for the condensation of Br and I. However, the thermodynamic properties and condensation temperatures for Br and I apatites remain uncertain.

The meteoritic host phases of several highly volatile elements remain somewhat elusive, but metal and sulfide are likely hosts. Thallium, Pb and Bi are calculated to condense in solid solution in Fe metal while Cd and In are calculated to condense mainly in solid solution in FeS.[17,30] The classification of an element as moderately or highly volatile depends somewhat on the total pressure for which its 50% condensation temperature is calculated because many volatile elements have siderophile as well as chalcophile affinities. If half of the elemental inventory is condensed in metal before FeS forms, the element's 50% condensation temperature is above that of troilite, making it a moderately volatile element. For example, at 10^{-4} bar total pressure, half of Te and Pb condense into metal prior to FeS formation, but at 10^{-6} bar total pressure troilite becomes their major host phase and their 50% condensation temperatures are 694 K (Te) and 634 K (Pb).[30]

Mercury is a highly volatile element, but details of its condensation chemistry are unknown. Elemental Hg is calculated to condense below 200 K.[16] However, it is unlikely that Hg is as volatile as water ice. Condensation of HgS, HgSe and HgTe into troilite may occur at higher temperatures and leads to a 50% condensation temperature of about 250 K for Hg.

3.4.1.5 Chemically Reactive Atmophile Elements

Hydrogen. As the most abundant element, H_2 is the most abundant gas in solar composition material. Dissociation to atomic H occurs at high temperatures, *e.g.*, H_2 and H have equal abundances at 1880 K (10^{-6} bar total pressure) and 2230 K (10^{-4} bar total pressure). The atomic H abundance increases with increasing temperature until its thermal ionization becomes important. Monatomic H and H^+ have equal abundances at 7100 K (10^{-6} bar) and 8700 K (10^{-4} bar). The prevailing electron pressure is that due to the ionization of all elements in a solar composition gas – not only the electron pressure due to H ionization. In any case, the temperatures at which hydrogen is significantly ionized are much higher than those in the solar nebula.

Conversely, most H_2 remains in the gas with decreasing temperature because only a tiny fraction ($\sim 0.1\%$) of total hydrogen is removed by water ice condensation. Solid H_2 condenses at about 5 K, depending on the total pressure,[19] but temperatures this low are not reached in the solar nebula.

As noted above, about 0.1% of all hydrogen condenses out as water ice at temperatures of 150–250 K depending on the total pressure

(Figure 3.8). The condensation curve of water ice (or liquid water) is given by Equation (3.10):

$$\frac{10\,000}{T_C(H_2O)} = 38.84 - 3.83 \log_{10} P_T \qquad (3.10)$$

For example, water ice condenses at 185 K at 10^{-4} bar total pressure in solar composition material. Fifty percent of all water is condensed by 180 K. Liquid water condenses instead of water ice at total pressures of ~ 3.8 bar and above. Such high pressures probably did not occur in the solar nebula, but they may exist in protoplanetary disks around more massive proto-stars. The condensation of liquid water has important consequences such as the formation of aqueous ammonia solutions. Aqueous NH_3 solutions are stable down to 173 K at 1 bar total pressure, which is the eutectic point, *i.e.*, the lowest melting point, in the NH_3–H_2O phase diagram. The water ice (or liquid) condensation curve is an important boundary that separates the condensation of rocky material in the inner solar nebula from the condensation of icy material in the outer solar nebula.

If complete chemical equilibrium is maintained, hydrated silicates such as serpentine and talc are also expected to form by reactions such as (3.11) and (3.12) below ~ 450 K at 10^{-4} bar total pressure (see Figure 3.8 for talc):

$$2(Mg, Fe)_2SiO_4(\text{olivine}) + 3H_2O \text{ (gas)}$$
$$= (Mg, Fe)_3Si_2O_5(OH)_4 \text{ (serpentine)} + (Mg, Fe)(OH)_2 \text{ (ferrobrucite)} \qquad (3.11)$$

$$4(Mg, Fe)SiO_3 \text{ (pyroxene)} + 2H_2O \text{ (gas)}$$
$$= (Mg, Fe)_3Si_4O_{10}(OH) \text{ (talc)} + (Mg, Fe)(OH)_2 \text{ (ferrobrucite)} \qquad (3.12)$$

However, although they are thermodynamically favorable, these reactions probably did not occur in the solar nebula. Theoretical models predict that vapor phase hydration of rock in a near vacuum is a very slow process, taking much longer than the lifetime of the solar nebula.[36–38] The hydrated minerals that are observed in certain meteorite groups probably formed on the meteorite parent bodies, as indicated by theoretical models of hydration kinetics in the solar nebula and petrographic studies of these meteorites. Thus, water ice is probably the first H-bearing condensate to form.

Carbon. Carbon monoxide and methane are the two major C-bearing gases over the range of temperatures and pressures expected in the solar nebula (Figure 3.11). Carbon monoxide is the dominant carbon gas at high temperatures and low total pressures. Conversely, CH_4 is the dominant carbon gas at low temperatures and high pressures.[40,41] The two gases are converted by the net thermochemical reaction:

$$CO \ (gas) + 3H_2 \ (gas) = CH_4 \ (gas) + H_2O \ (gas) \tag{3.13}$$

The equilibrium constant K_{eq} for this reaction from 298–2500 K is given by:

$$\log K_{eq} = 11\,069.94/T - 1.17969 \log T - 8.96596 \tag{3.14}$$

Thermodynamic calculations using this equilibrium constant, and solar elemental abundances, give the distribution of carbon between CO

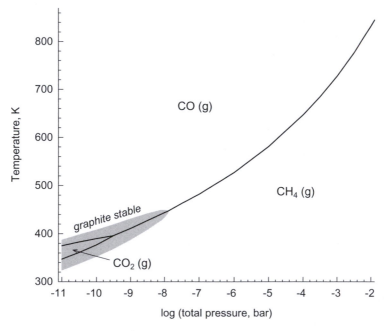

Figure 3.11 Distribution of carbon in a gas of solar composition as a function of temperature and total pressure. The longest curve shows where CO and CH_4 are equimolar in abundance, while the shorter curves show where abundances of CO equal that of CO_2, and CH_4 equals that of CO_2. The dominant C-bearing gases in the regions defined by these curves are indicated. The grey shaded area at low total pressures indicates where graphite becomes thermodynamically stable.

and CH_4 as a function of temperature and pressure (see curve in Figure 3.8, and carbon distribution diagram in Figure 3.11). At 10^{-4} bar total pressure CO is the dominant carbon gas at temperatures above 625 K, CH_4 is the major carbon gas at temperatures below 625 K, and the two gases have equal abundances at 625 K. The equal abundance curve shifts to higher temperatures as the total pressure increases. Although CO is the major carbon gas above the line and CH_4 is the major carbon gas below the line, CO is still present in the CH_4-rich region and CH_4 is still present in the CO-rich region.

In principle, 50% of total carbon is in CO and CH_4 along the CO/CH_4 equal abundance line. However, in practice CO and CH_4 constitute less than 100% of total carbon at the low pressures expected in the solar nebula because small amounts of CO_2 are also present. For example, 0.5% of total carbon is in CO_2 at 625 K and 10^{-4} bar on the CO/CH_4 equal abundance line. Carbon dioxide becomes increasingly important as temperature and pressure decrease because reactions such as CO disproportionation and CH_4 oxidation by water vapor produce CO_2:

$$2CO \text{ (gas)} = CO_2 \text{ (gas)} + C \text{ (gr)} \tag{3.15}$$

$$CH_4 \text{ (gas)} + 2H_2O \text{ (gas)} = CO_2 \text{ (gas)} + 4H_2 \text{ (gas)} \tag{3.16}$$

Carbon dioxide can become the major carbon gas in a temperature and pressure range below 408 K and 10^{-9} bar.[42]

Pure graphite is unstable over most of the temperatures and pressures shown in Figure 3.8, and only becomes stable at the lower pressures, as shown in Figure 3.11. However, elemental carbon can dissolve in Fe metal alloy. Considering this, one obtains a maximum dissolved carbon concentration in Fe metal alloy of $50 \, \mu g \, g^{-1}$ at 750 K in the inner solar nebula.[41] This corresponds to $15 \, \mu g \, g^{-1}$ carbon in the condensed solid, which is metal plus silicate. In contrast, carbon concentrations in ordinary chondrites are $1200–2350 \, \mu g \, g^{-1}$, about 100 times larger than produced by carbon dissolution in metal alloy. The large difference indicates the importance of C-bearing tar-like compounds in chondritic material.

However, graphite is thermodynamically stable and is a major repository for carbon at low temperatures and pressures similar to those where CO_2 is a major carbon gas. As noted above, the disproportionation of CO forms CO_2 and graphite. Graphite also forms *via* the reaction of CO and CH_4:

$$CO \text{ (gas)} + CH_4 \text{ (gas)} = H_2O \text{ (gas)} + H_2 \text{ (gas)} + 2C \text{ (gr)} \tag{3.17}$$

However, because of the low temperatures and pressures required for graphite stability, it is unclear whether the reactions forming it are fast enough to occur appreciably within the lifetime of the solar nebula.

Carbon occurs as carbon dissolved in Fe alloy, carbides, carbonates, diamond, graphite, organic matter and poorly graphitized carbon in chondritic meteorites. Most of the carbon in the ordinary (H, L, LL) chondrites occurs as C dissolved in metal, poorly graphitized carbon, $(Fe,Ni)_3C$ (known as cohenite or cementite), and an aromatic organic polymer. Most of the carbon in ordinary chondrites is apparently in the form of aromatic organic polymer, with C dissolved in metal, cohenite and poorly graphitized carbon being less abundant. The less metamorphosed unequilibrated ordinary chondrites also contain presolar diamonds and presolar silicon carbide. However, most of their carbon is probably also in polymeric organic material. Much of the carbon in CI and CM carbonaceous chondrites is in an aromatic organic polymer. Carbonates such as breunnerite, $(Mg,Fe)CO_3$, calcite $(CaCO_3)$ and dolomite, $CaMg(CO_3)_2$, are probably the second most abundant reservoir of carbon in these meteorites. Presolar grains (diamonds, SiC and graphite in decreasing order of abundance) are the next most important carbon host phases in CI and CM chondrites. Finally, most of the carbon in enstatite chondrites occurs as graphite, C dissolved in Fe alloy or cohenite, $(Fe,Ni)_3C$.

None of these C-bearing compounds is thermodynamically stable in solar composition material. Carbonates probably formed during aqueous alteration and metamorphism on meteorite parent bodies. Carbides such as silicon carbide SiC and titanium carbide TiC are relict presolar grains that formed in the circumstellar environments of carbon stars prior to the formation of the solar system. Their survival in meteorites shows that the presolar carbides were never heated to high temperatures in the solar nebula; otherwise, they would have been oxidized by reaction with nebular water vapor. The aromatic organic polymer probably results from disequilibrium processes (*e.g.*, ion–molecule reactions, photochemistry and grain catalyzed reactions) in the solar nebula and its parent molecular cloud.

Additional carbon gases to CO, CO_2 and CH_4 appear at higher temperatures as thermal dissociation of these gases becomes increasingly important. For example, CO thermally dissociates to its constituent elements at temperatures of several thousand kelvin and monatomic C becomes the major C-bearing gas. Likewise, CH_4 thermally dissociates to H atoms and CH_3 (methyl) radicals at high temperatures. In turn the CH_3 radicals dissociate to CH_2 (methylene) and CH (methylidene) radicals and H atoms. However, the temperatures under which thermal dissociation is important are much higher than those in the solar nebula and are not important for chemistry of planet-forming materials.

Urey (1953)[40] realized that the rate of the CO into CH_4 conversion may be too slow under the P-T conditions expected for the solar nebula. He suggested that CO was converted into tar-type compounds instead of being reduced to methane. Anders *et al.* (1973)[43] explored Urey's suggestion that reduction of CO formed tar-type compounds, and experimented with Fe-metal-catalyzed production of hydrocarbons from CO and H_2 by Fischer–Tropsch reactions. Theoretical models indicate that Fischer–Tropsch reactions may convert about 10% of all CO into hydrocarbons within the lifetime of the solar nebula.[36,37,39] This process may be the source of at least some of the organic material on primitive bodies (asteroids, comets, KBOs, satellites) in the outer solar system.

Lewis and Prinn (1980)[44] quantified Urey's thoughts on the slow conversion of CO into CH_4 by modeling the gas phase kinetics of this reaction. The gas phase reduction of CO turns out to be too slow to occur within the lifetime of the solar nebula. As a consequence, CO remained the major carbon gas in the solar nebula and the mole fraction of water vapor was about one half of what it would be if CO were converted quantitatively into methane.

However, the reduction of CO to CH_4 is expected to proceed in the higher density protoplanetary subnebulae present around the gas giant planets during their formation. The higher pressures in the subnebulae shift the equilibrium toward CH_4 and increase the rate of CO reduction. The distribution of carbon between CO and CH_4 in the subnebulae is that expected by chemical equilibrium and H_2O is more abundant than in the CO-rich solar nebula. Hence the water ice/rock ratios in "icy" satellites formed in the giant protoplanetary subnebulae, where CH_4 is the major carbon gas, are predicted to be higher than the water ice/rock ratios of "icy" bodies formed in the solar nebula, where CO was the major carbon gas.[38,39,45] To a first approximation, this distinction is observed.

At low temperatures in the outer solar nebula and in the giant protoplanetary subnebulae, CO and CH_4 may react with water ice to form the solid clathrate hydrates $CO \cdot 7H_2O$ and $CH_4 \cdot 7H_2O$:

$$CO \text{ (gas)} + 7H_2O \text{ (ice)} = CO \cdot 7H_2O \text{ (clathrate hydrate)} \qquad (3.18)$$

$$CH_4 \text{ (gas)} + 7H_2O \text{ (ice)} = CH_4 \cdot 7H_2O \text{ (clathrate hydrate)} \qquad (3.19)$$

Clathrate hydrates are cage compounds in which a gas molecule is trapped inside the water ice crystal lattice. One gas molecule is trapped for every seven water molecules. The condensation curves for $CH_4 \cdot 7H_2O$ and $CO \cdot 7H_2O$ are at temperatures below the water ice condensation curve in Figure 3.8. The formation of these clathrate hydrates requires sufficiently rapid diffusion of CO or CH_4 through the water ice crystal lattice. Theoretical models, which use experimentally determined activation energies for clathrate formation, indicate that CH_4 clathrate hydrate can form in the giant protoplanetary subnebulae but that CO clathrate hydrate cannot form in the much lower density environment of the outer solar nebula.[36,39] However, other workers[46] suggested that clathrate hydrate formation can still occur under special circumstances.

In any case, there is not enough water ice to condense all CO or all CH_4 as clathrate hydrates because the CO (or CH_4) to H_2O ratio is 1 : 7 in the clathrate hydrate while the solar C/O atomic ratio is 1 : 2. Thus, the residual CO or CH_4 condenses as CO or CH_4 ice at lower temperatures.

Nitrogen. Molecular nitrogen and ammonia are the two major N-bearing gases over the range of temperatures and pressures expected in the solar nebula. Dinitrogen is the dominant nitrogen gas at high temperatures and low total pressures whereas NH_3 is the dominant nitrogen gas at low temperatures and high pressures. The two gases are converted by the net thermochemical reaction:

$$N_2(gas) + 3H_2(gas) = 2NH_3(gas) \tag{3.20}$$

The equilibrium constant K_{eq} for this reaction from 298 to 2500 K is given by:

$$\log K_{eq} = 6051.59/T - 1.21176 \log T - 7.89739 \tag{3.21}$$

Chemical equilibrium calculations using this data and the solar elemental abundances show that N_2 is the dominant N-bearing gas at high temperatures and low pressures and that NH_3 is the major N-bearing gas at low temperatures and high pressures. Figure 3.8 shows the line along which N_2 and NH_3 have equal abundances. The N_2/NH_3 equal abundance line is slightly different than the CO/CH_4 equal abundance line because equal abundances of N_2 and NH_3 do not correspond to 50% of total nitrogen in each gas due to the two N atoms in N_2. For example, at 10^{-4} bar total pressure the N_2/NH_3 equal abundance line is at 345 K while 50% of total nitrogen is in each gas at 320 K.

However, the gas phase reduction of N_2 to NH_3, like the gas phase reduction of CO to CH_4, is predicted to be kinetically inhibited in the solar nebula. This should not be surprising because the industrial fixation of N_2 into NH_3 *via* the Bosch–Haber process requires high pressures, high temperatures and specially prepared Fe-based catalysts. Iron alloy grains are present in the solar nebula. However, even when the catalytic effects of Fe metal grains are taken into account very little NH_3 forms because it is not thermodynamically favored until low pressures and low temperatures where reaction rates are very slow (recall N_2/NH_3 is 1 : 1 at 345 K at 10^{-4} bar total pressure). In contrast, N_2 reduction to NH_3 is both thermodynamically favored and kinetically facile in the giant protoplanetary subnebulae.[38,39,44,45] Thus, N_2 is expected to be the dominant nitrogen gas throughout the solar nebula and NH_3 is expected to be dominant throughout the giant protoplanetary subnebulae.

The condensation chemistry of N_2 is fairly simple. At high temperatures only small amounts of N dissolve in Fe metal alloy. Most nitrogen condenses at low temperatures in the outer solar nebula when N_2 gas reacts with preexisting water ice to nitrogen clathrate hydrate $N_2 \cdot 7H_2O(s)$:

$$N_2 \text{ (gas)} + 7H_2O \text{ (ice)} = N_2 \cdot 7H_2O \text{ (clathrate hydrate)} \qquad (3.22)$$

However, formation of $N_2 \cdot 7H_2O(s)$ is probably inhibited by two factors. One is the limited availability of water ice, which may already be totally consumed by reactions at higher temperatures to form other hydrates and clathrates. The other is the expected kinetic inhibition of clathrate hydrate formation in the outer solar nebula.[38] In this case, N_2, like CO, will not condense until temperatures of about 20 K (at 10^{-4} bar pressure) where the solid ices form.

Ammonia starts condensing as ammonium bicarbonate (NH_4HCO_3) or as ammonium carbamate (NH_4COONH_2). The condensation temperatures and amounts of the solids that condense depend on the NH_3 and CO_2 abundances. In turn, the gas abundances depend on chemical reaction rates and dynamical mixing rates in the solar nebula and protoplanetary subnebulae. For example, either NH_4HCO_3 or NH_4COONH_2 condenses at ~ 170 K in the Jovian protoplanetary subnebula models of Prinn and Fegley.[45] However, most NH_3 condenses as ammonia monohydrate ($NH_3 \cdot H_2O$), also known as ammonium hydroxide (NH_4OH), *via* the reaction:

$$NH_3 \text{ (gas)} + H_2O \text{ (ice)} = NH_3 \cdot H_2O \text{ (ice)} \qquad (3.23)$$

Ammonia monohydrate is a distinct compound and is not a clathrate hydrate. It forms at a higher temperature than either NH_3 ice or liquid ammonia. For example, if all nitrogen is present as ammonia, $NH_3 \cdot H_2O$ condenses at 131 K at 10^{-4} bar total pressure. The N/O atomic ratio is 0.14 in solar composition material so condensation of ammonia monohydrate removes all NH_3 from the gas and leaves unreacted water ice for clathrate hydrate formation at lower temperatures. However, as mentioned above, quantitative condensation of CH_4, CO or N_2 as clathrate hydrates is impossible because not enough water ice is available.

Nitrogen in meteorites occurs as N dissolved in metal alloy, N dissolved in silicates, nitride minerals, organic matter and ammonium salts (in CI chondrites). Mass balance calculations show that nitrogen dissolved in silicates and/or trace amounts of (unobserved) nitride minerals are apparently the major N-bearing reservoirs in ordinary chondrites. Most of the nitrogen in CI and CM carbonaceous chondrites resides in the organic polymer. The nitride minerals sinoite (Si_2N_2O), nierite (α-Si_3N_4) and osbornite (TiN) are the major nitrogen reservoirs in enstatite chondrites. Nitrogen dissolution in metal and silicate probably took place during parent body metamorphism because the amounts of dissolved nitrogen are orders of magnitude greater than predicted for chemical equilibrium in the solar nebula. Likewise, at least some of the nitride minerals in enstatite chondrites probably formed during parent body metamorphism because N_2 partial pressures much higher than expected in the solar nebula are needed for the relevant equilibria. Nitrogen-bearing organic matter made by disequilibrium processes was probably the initial source of nitrogen on meteorite parent bodies.

3.4.1.6 Noble Gases.   The chemistry of the noble gases He, Ne, Ar, Kr and Xe in solar composition material is simple. All are present in the gas as the monatomic elements and Ar, Kr and Xe undergo condensation to either ices or clathrate hydrates at sufficiently low temperatures. Condensation of the pure ices will occur at slightly lower temperatures than condensation of the clathrate hydrates. The 50% condensation temperatures for the noble gas clathrate hydrates are indicated in Figure 3.7. However, the formation of these species, like the clathrates of CO and N_2, may be kinetically inhibited. Temperatures of about 20 K (at 10^{-4} bar pressure) are required for quantitative condensation of Ar, Kr and Xe as pure ices. Neither He nor Ne will condense out of the gas because temperatures of 5 K or below are required for this to happen.

3.5 SOLAR SYSTEM TIMESCALES

The timeline of events in the early solar system is traceable from the abundances of radioactive nuclides and/or their decay products. Several review articles cover this topic[47-50] and we only give a short description here. Like stable nuclides, radioactive nuclides are products of stellar nucleosynthesis, and some are spallation products from cosmic ray bombardment of interstellar materials. Some radioactive nuclides that were present in the early solar system may even have been produced within the solar nebula itself by particle radiation from the young Sun. Certainly, radionuclides from continuous nucleosynthesis in the Galaxy were present in the solar molecular cloud, and, depending on their production rates, half-lives and abundances, they were present in the solar nebula when the meteorite parent bodies and the planets formed and differentiated.

In principle, all radionuclides and their decay products can be used for radiometric dating. There is a common distinction between "long-lived" and "short-lived" radioactive nuclides with respect to their half-lives against decay and the age of the solar system. Long-lived means that a radioactive nuclide still naturally occurs in measurable quantities, although in reduced amounts since the time of solar system formation. Radioactive nuclides with half-lives above ~ 0.6 Ga usually qualify for this (Table 3.4).

The detectability of a "live" radionuclide today is limited by its half-life and initial abundance. It takes seven half-lives to decrease a radionuclide's abundance to below one percent of its original abundance, and 14 half-lives to reduce it below 0.01%. The shortest-lived isotope in solar system materials that is still alive today and was present during solar system formation is ^{235}U (half-life of 7.04×10^8 years). The presence of ^{235}U but the non-detection of ^{146}Sm (1.08×10^8 years) and ^{244}Pu (0.83×10^8 years) and other shorter-lived radionuclides in solar system rocks suggests that the age of the solar system must be at least several times that of the half-life of ^{235}U, assuming ^{235}U initially had a comparable abundance to ^{238}U.

With a solar system age of ~ 4.6 Ga, a nuclide with a half-life smaller than ~ 0.6 Ga is unlikely to remain in any detectable quantities, because the initial abundances of the now-extinct radionuclides were quite low to begin with (Table 3.5). Nuclides with shorter half-lives than ^{235}U, such as ^{146}Sm and ^{244}Pu, were likely present at the beginning but decayed below detection limits within the first several hundred million years of the solar system's history. These nuclides can only be traced through abundances of their stable decay nuclides. Because the short-lived

Table 3.4   Naturally occurring long-lived radionuclides in the solar system (in order of increasing half-life).

Nuclide	Half-life (years)	Decay mechanism	Stable decay product(s)
^{235}U	7.04×10^8	α	^{207}Pb
^{40}K	1.27×10^9	β^-, β^+	^{40}Ca, ^{40}Ar
^{238}U	4.47×10^9	α	^{206}Pb
^{232}Th	1.40×10^{10}	α	^{208}Pb
^{176}Lu	3.78×10^{10}	β^-	^{176}Hf
^{187}Re	4.22×10^{10}	β^-	^{187}Os
^{87}Rb	4.88×10^{10}	β^-	^{87}Sr
^{138}La	1.05×10^{11}	ec, β^-	^{138}Ba, ^{138}Ce
^{147}Sm	1.06×10^{11}	α	^{143}Nd
^{190}Pt	4.50×10^{11}	α	^{186}Os
^{123}Te	1.24×10^{13}	ec	^{123}Sb
^{152}Gd	1.1×10^{14}	α	^{148}Sm
^{115}In	4.4×10^{14}	β^-	^{115}Sn
^{186}Os	2.0×10^{15}	α	^{182}W
^{180}Ta	$> 1.2 \times 10^{15}$	ec, β^+	^{180}Hf
^{174}Hf	2.0×10^{15}	α	^{170}Yb
^{144}Nd	2.1×10^{15}	α	^{140}Ce
^{148}Sm	7×10^{15}	α	^{144}Nd
^{113}Cd	9×10^{15}	β^-	^{113}In
^{50}V	1.4×10^{17}	ec, β^-	^{50}Ti, ^{50}Cr

Decay mechanisms: α = alpha particle (He^{2+}) emission; β^- = electron emission; β^+ = positron emission; ec = electron capture.

radionuclides are now extinct in meteoritic and planetary materials, they are also called "extinct natural radionuclides" or "extinct natural radioactivities," a terminology introduced by Kohman (1956).[51] Note, however, that short-lived radionuclides, including several of those in Table 3.5, are still produced in various environments exposed to energetic particle radiation such as cosmic rays. Cosmogenic nuclides are found alive in small quantities in meteorites, and some are present in the terrestrial atmosphere. However, the relative quantities of short-lived radionuclides are much smaller than those that had been contributed by other nucleosynthesis sources to the molecular cloud by the time of solar system formation.

3.5.1   Chronometers with Long-lived Radioactivities

Chronometric systems with long-lived radioactive parent nuclides (Table 3.4) are the key for determining absolute ages of rocks and their components. This requires that the nuclide half-lives are well known, which, for example, limits the precision of the ^{87}Rb-^{87}Sr and ^{176}Lu-^{176}Hf chronometers. Ideally, the time interval to be dated is measured with an

Table 3.5 Known and potential short-lived radioactive nuclides in the early solar system.

Radionuclide	Decay mode	Decay nuclide	Half-life, t_H (Ma)	Approximate initial solar system abundance	Detected in
^{7}Be	ec	^{7}Li	53.3 days	Suspected, but unconfirmed	CAIs
^{41}Ca	ec	^{41}K	0.103	$1.4\times10^{-8}\times^{40}$Ca	CAIs
^{26}Al	β^+ (ec)	^{26}Mg	0.716	$5\times10^{-5}\times^{27}$Al	CAIs, chondrules, achondrites
^{10}Be	β^-	^{10}B	1.5	$7\times10^{-4}\times^9$Be	CAIs
^{53}Mn	ec	^{53}Cr	3.74	$2\text{--}4\times10^{-5}\times^{55}$Mn	CAIs, chondrules, achondrites
^{60}Fe	β^-	^{60}Ni	$\sim$1.49	$3\times10^{-7}\times^{56}$Fe	Chondrites, achondrites
^{107}Pd	β^-	^{107}Ag	6.5	$5\times10^{-5}\times^{108}$Pd	Iron meteorites, pallasites
^{182}Hf	β^-	^{182}W	9	$10^{-4}\times^{182}$Hf	CAIs, chondrites, achondrites
^{129}I	β^-	^{129}Xe	15.7	$1.1\times10^{-4}\times^{127}$I	Chondrules
^{92}Nb	ec, β^+	^{92}Zr	34.7	$0.1\text{--}1\times10^{-4}\times^{93}$Nb	CAIs, chondrites, mesosiderites
^{93}Zr		^{93}Nb		Decay to mono-isotopic Nb is difficult to infer	
^{244}Pu	α	Several fission products	82	$10\times10^{-3}\times^{238}$U	CAIs, chondrites, achondrites
^{146}Sm	α	^{142}Nd	103	$8\times10^{-3}\times^{144}$Sm	CAIs, chondrites, achondrites
^{247}Cm	α		15.6	$\leq2.9\times10^{-5}\times^{238}$U; uncert.	Angrites
^{135}Cs	β^-	^{135}Ba	2.3	$1\text{--}5\times10^{-4}\times^{133}$Cs	CAI, chondrites
^{36}Cl	β^- (ec)	^{36}Ar (^{36}S)	0.301	$\sim7\times10^{-6}\times^{35}$Cl	CAI
^{99}Tc			0.2		
^{205}Pb	ec	^{205}Tl	15.3	$\sim1\times10^{-4}\times^{204}$Pb	Irons
^{248}Cm	α		0.348	$\leq3.6\times10^{-7}\times^{238}$U	OC phosphate

isotopic system in which the radioactive nuclide's half-life is of similar order of magnitude. The long-lived radionuclides with half-lives similar to ^{235}U and above and their stable decay products (Table 3.4) are the preferred systems for dating "billions of years" events such as the ages of terrestrial, lunar and meteoritic rocks, whereas short-lived isotope systems (Table 3.5) provide information for relative timescales of events in the first few million to hundreds of millions years in the solar system. There are also some very long-lived radionuclides with half-lives above 10^{13} years listed in Table 3.4, but these have not much decayed since the birth of the solar system and any excesses in their daughter isotopes over normal isotopic compositions would be hard to detect. For all practical purposes, these nuclides can be treated as "stable." The radiometric system with the longest half-life applied to problems in meteoritics and planetary sciences is the ^{190}Pt-^{186}Os system, although this system gets much less attention than the better-established systems involving ^{40}K, ^{87}Rb, ^{147}Sm, ^{232}Th, ^{235}U or ^{238}U. Radiometric dating with these chronometers shows that most meteorites and their components formed and assembled into meteorite parent bodies around 4.5–4.6 billion years ago. The oldest known components are the calcium-aluminum-rich inclusions (CAIs); for example, CAIs in the Efremovka CV chondrite crystallized 4.567 billion years ago.[52]

The absolute and relative ages determined from radio-chronometers are always minimum ages of a system ("closure ages") and refer to the time (or time difference) when a radionuclide was captured in a mineral phase and no longer participated in diffusive exchange reactions with other phases. One caveat in measuring the true antiquity of rocks is that later disturbances in a system – be it impacts on meteorite parent bodies or planets, or large-scale melting and differentiation processes – can reset the radiometric isotope clocks through geologic time.

Radiometric dating makes use of the fact that the decay of a radionuclide produces a measurable increase in the amount of its daughter nuclide over time. However, the general problem is that the daughter nuclide is usually already present in the sample when the radioactive clock starts. Thus, the measurable amount of a daughter nuclide is at least a dual mixture of some initial amount plus the amount produced from decay since system closure. However, radiometric dating with long-lived nuclides can provide absolute ages of rocks because the initial concentration and the present-day concentration of the radionuclide and its daughter product can be inferred from isochron plots. Assume an originally isotopically homogeneous system containing a radionuclide. During chemical fractionations, an element distributes among different phases according to its cosmochemical and geochemical affinities, and all isotopes of this element will do so irrespective of whether they are radioactive or not. Different mineral phases retain different amounts of the element (and the relative proportions of its isotopes) according to the element's compatibility in the mineral lattice. Once a radionuclide is distributed into different phases and elemental exchanges between phases proceed no further, decay of the parent radionuclide increases the amount of the stable daughter nuclide, which is usually an isotope of a different element. The decay rate of the radionuclide and the (opposite) accumulation rate of the daughter nuclide are proportional to the original amount of the radionuclide enclosed in each phase. The proportionality (or decay) constant for this is related to the half-life of the radionuclide. Solving the rate equation leads to the well-known exponential radioactive decay equation (see also example below). If decay and accumulation started at the same time in all phases involved, each phase experiences a decrease in parent nuclide and an increase in daughter nuclide controlled by the same rate. However, because the absolute amounts of the radio-nuclide are different in each phase, one obtains a suite of phases that have correlating amounts of parent and daughter nuclides.

The ^{87}Rb-^{87}Sr system is useful for illustration. Beta-decay of radioactive ^{87}Rb increases stable ^{87}Sr over time. The measurable ^{87}Sr content today equals some initial amount of ^{87}Sr$_{\text{ini}}$ plus the amount of ^{87}Sr$_{\#}$ produced by exponential ^{87}Rb decay over time (where the present time is t). This also depends on the decay constant (λ_{87}) of ^{87}Rb:

$$^{87}\text{Sr}_t = {}^{87}\text{Sr}_{\text{ini}} + {}^{87}\text{Sr}_{\#} = {}^{87}\text{Sr}_{\text{ini}} + {}^{87}\text{Rb}_t(e^{\lambda_{87}t} - 1) \tag{3.24}$$

The decay constant is related to the half-life, t_{H}, as $\lambda = \ln(2)/t_{\text{H}} = 1.42 \pm 0.01 \times 10^{-11}\,\text{yr}^{-1}$ (note that in the older literature a decay constant of $1.39 \times 10^{-11}\,\text{yr}^{-1}$ was used; where necessary, quoted values are recalculated with the newer value; however, the decay constant of ^{87}Rb remains uncertain). Normalizing this relation to a stable isotope of Sr such as ^{86}Sr whose concentration has not changed during the decay interval gives:

$$^{87}\text{Sr}_t/^{86}\text{Sr} = {}^{87}\text{Sr}_{\text{ini}}/^{86}\text{Sr} + {}^{87}\text{Rb}_t/^{86}\text{Sr}(e^{\lambda_{87}t} - 1) \tag{3.25}$$

The normalization to a stable Sr isotope provides a proxy for the total amount of Sr that was originally incorporated into various phases, and is practical when isotopic ratios instead of absolute concentrations are measured. The time elapsed since closure of the system is calculated from:

$$t = 1/\lambda_{87} \ln\left\{1 + \left[{}^{87}\text{Sr}_t/^{86}\text{Sr} - {}^{87}\text{Sr}_{\text{ini}}/^{86}\text{Sr}\right]/\left({}^{87}\text{Rb}_t/^{86}\text{Sr}\right)\right\} \tag{3.26}$$

If several phases or components in a given system can be measured for their current $^{87}Sr/^{86}Sr$ and $^{87}Rb/^{86}Sr$ ratios, plotting the $^{87}Sr/^{86}Sr$ ratios *versus* $^{87}Rb/^{86}Sr$ ideally defines a linear relation in which the initial $^{87}Sr_{ini}/^{86}Sr$ follows from the intercept and the term $(e^{\lambda_{87}t} - 1)$ equals the slope. The initial $^{87}Sr_{ini}/^{86}Sr$ ratio and the closure age correspond to the time where the $^{87}Sr/^{86}Sr$ was first fixed in all phases involved.

Figure 3.12 shows Rb-Sr data for whole rock analyses of LL chondrites as an example.[53] The good correlation indicates that these chondrites formed contemporaneously and from a single isotopic reservoir, presumably the solar nebula. The isochron for the LL chondrites is well defined and leads to an $^{87}Sr_{ini}/^{86}Sr = 0.69882 \pm 0.00008$ [the uncertainties, usually 2 sigma, are often written abbreviated corresponding to the last digit(s) of the value, *e.g.*, 0.69882(8) for the example here]. The slope of $(e^{\lambda_{87}t} - 1) = 0.06588$ leads to a LL chondrite whole-rock isochron age of 4.493 ± 0.018 Ga when the decay constant $\lambda = 1.42 \times 10^{-11}\, yr^{-1}$ is used; with the older $\lambda = 1.39 \times 10^{-11}\, yr^{-1}$ one obtains 4.590 ± 0.018 Ga. Similar values are found from isochrons of H and E chondrites. Within uncertainties, all H, LL and E chondrite data combined[54] also define an isochron from which the age of chondrites is 4.498 ± 0.015 Ga (using the

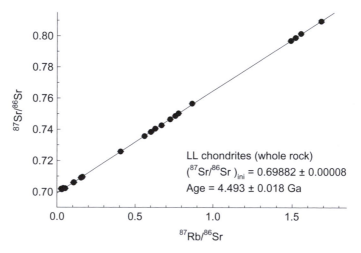

Figure 3.12 Rb-Sr systematics in LL chondrites (data from Minster and Allegre[53]). The measured $^{87}Sr/^{86}Sr$ ratios are plotted *versus* the $^{87}Rb/^{86}Sr$ ratios. Whole rock chondrites with higher Rb abundances (a proxy for the ^{87}Rb content if the present-day $^{87}Rb/^{85}Rb$ ratio is the same in all samples) also show higher ^{87}Sr abundances and the good correlation of the different LL chondrites indicates that they formed contemporaneously. The initial $^{87}Sr/^{86}Sr$ at the time of system closure follows from the intercept. The slope is related to the formation age of the LL chondrites and their age is computed using the decay constant of $1.42 \times 10^{-11}\, year^{-1}$.

more recent decay constant) and the representative chondrite initial $^{87}Sr_{ini}/^{86}Sr = 0.69885(10)$.

Individual chondrites may sample different parent bodies that may have formed over a small time interval. If each chondrite has an individual $^{87}Sr_{ini}/^{86}Sr$ and closure age, more scattering in the data set occurs. To obtain the age and initial isotopic composition of an individual chondrite, a mineral isochron must be constructed from measurements of different mineral separates of the meteorite. However, such well-defined "internal isochrons" are often difficult to obtain because disturbances from impacts, reheating and aqueous alterations can affect different minerals in a different fashion, so that not all minerals closed from isotopic exchange at the same time and thus deviate from the isochron. This is a problem for heavy aqueously altered CI carbonaceous chondrites where aqueous alteration and mobilization of Rb and Sr makes it impossible to obtain reliable Rb-Sr isochrons. The CM chondrites could be affected by such alterations as well.

In a large open reservoir such as the solar nebula, the $^{87}Sr/^{86}Sr$ ratio naturally increases over time. If different objects were isolated from the elemental and isotopic exchange reactions within the solar nebula at different times, objects that were removed later will start out with higher initial $^{87}Sr/^{86}Sr$. In that case, the different $^{87}Sr/^{86}Sr$ initials can be used to estimate the difference in formation times between different objects. Furthermore, if there were a known representative initial value for the solar system (or any other object of interest), it would be possible to derive some absolute "model" ages for objects that were derived from this system, even if no full isochrons can be established. In Equation (3.26) for the age, the $^{87}Sr_{ini}/^{86}Sr$ is the only unknown on the right-hand side of the equation since $^{87}Sr_t/^{86}Sr$ and $^{87}Rb_t/^{86}Sr$ are measured, and λ is a constant. Therefore, it is desirable to find samples with the lowest initial $^{87}Sr/^{86}Sr$ ratios that can be taken as representative for the time of solar system formation. Much research has been dedicated to hunting down the lowest initial values; not only for the Rb-Sr system but for other chronometric systems as well. For the Rb-Sr system, such searches are facilitated by measuring objects that experienced a very early fractionation of the volatile and incompatible Rb from the more refractory and compatible Sr. The initial $^{87}Sr/^{86}Sr$ ratios in Rb-poor objects were less "contaminated" by radiogenic ^{87}Sr during later decay. Then the measured $^{87}Sr/^{86}Sr$ ratios should be very close to the initial $^{87}Sr/^{86}Sr$ ratios so that only small corrections for radiogenic ^{87}Sr are necessary in Rb-poor samples. Among meteorites and their components, the refractory CAIs and basaltic achondrites (eucrites) show the lowest measured $^{87}Sr/^{86}Sr$ ratios because in both types of objects the more volatile Rb was either lost quite early or was never condensed.

One of the lowest measured $^{87}Sr_t/^{86}Sr$ ratios in CAIs from the Allende CV chondrite is $^{87}Sr/^{86}Sr = 0.69877$,[55-57] which was measured in very Rb poor mineral separated from these inclusions. This measured, present-day value is already lower than the initial value for chondrites (0.69885) from $\sim 4.5\,Ga$ ago.[54] Ideally, one would like to obtain the initial value for CAIs from mineral isochrons for comparison. However, well-defined isochrons cannot be constructed for CAIs because the Rb-Sr systematics of the minerals in the refractory inclusions was disturbed by later thermal and/or aqueous alterations. One can check the initial $^{87}Sr_{ini}/^{86}Sr$ ratio by back-calculating the measured values of $^{87}Sr/^{86}Sr$ in phases with low $^{87}Rb/^{86}Sr$ and assuming a certain age:

$$^{87}Sr_{ini}/^{86}Sr = {}^{87}Sr_t/^{86}Sr - {}^{87}Rb_t/^{86}Sr(e^{\lambda t} - 1) \qquad (3.27)$$

In phases with measured $^{87}Rb_t/^{86}Sr \approx 0.0002$ or less, adopting ages between 4.4 and 4.6 Ga only slightly lowers the measured $^{87}Sr_t/^{86}Sr = 0.69877(2)$ to a value of $^{87}Sr_{ini}/^{86}Sr = 0.69876(2)$, which are the same within uncertainties. This shows that the $^{87}Sr/^{86}Sr$ ratio in such Rb-poor phases is not increased much by radiogenic ^{87}Sr and that the initial $^{87}Sr/^{86}Sr$ is insensitive to adopted ages up to chondrite ages. Therefore, the measured $^{87}Sr/^{86}Sr$ value was used "as is" to define the initial $^{87}Sr_{ini}/^{86}Sr$ for refractory inclusions from the Allende meteorite. This CAI initial value was christened "ALL,"[55] but whether this value can be assumed as the initial $^{87}Sr/^{86}Sr$ for *all* solar system materials, and not just CAIs, is not really proven. The low initial values from CAIs may be representative for the early solar system and may imply that these inclusions are indeed the oldest preserved objects that solidified in the solar system, which can only be concluded from full isochron results. However, the Rb-Sr system for CAIs is disturbed and absolute ages are preferentially obtained with another chronometer such as U-Pb.

Some other caution is also advised. The validity of the chronometers involving radioactive nuclides and their decay products depends on the underlying assumption that compositional changes only occur in the parent and daughter isotopes, while all other stable isotopes of the elements involved remain in their same relative proportions over time. However, mass-dependent fractionations of isotopes, for example during condensation and evaporation processes, can change isotopic ratios as well. This is well known especially for lighter elements such as hydrogen and oxygen (Chapter 1).

Some Allende CAIs show mass-dependent anomalies in the light Sr isotopes such as ^{84}Sr and ^{86}Sr relative to the heavier ^{88}Sr and similar anomalies were reported for chondrules in the Allende meteorite.[58,59] If ^{86}Sr is enriched in some CAIs relative to the normal composition as found in chondritic and other solar system materials, the initial ^{87}Sr/^{86}Sr ratio of some CAIs may only appear smaller because the amount of ^{86}Sr was higher when CAIs formed, and not because the initial amount of ^{87}Sr was indeed lower. In addition, if some processes cause non-radiogenic abundance anomalies in the stable isotopes that are used for normalization, radiogenic isotopes (such as the ^{87}Sr) may be affected by such processes as well. In that case, the use of the initial values of CAIs for calculating ages without corrections for such isotope fractionation processes is not justified because the underlying assumption that all systems (CAIs, meteorites, planets) stem from the same unfractionated and isotopically homogeneous reservoir does not apply. In general, such problems with isotopic anomalies in stable isotopes may also be relevant to other chronometric systems.

Differentiated meteorites poor in volatile Rb also give low initial ^{87}Sr/^{86}Sr ratios, suggesting very early differentiation of small planetesimals. Well-defined isochrons from basaltic achondrites (eucrites) give an age of 4.30 ± 0.25 Ga and an ^{87}Sr/^{86}Sr initial of 0.69899(5), which has been nicknamed "BABI" for "basaltic achondrite best initial."[60–62] The Rb-Sr systematics of the volatile-poor and refractory element-rich angrites are more complicated, and, as for the refractory inclusions, isochrons are difficult to construct for them. The calculated angrites ^{87}Sr$_{ini}$/^{86}Sr initials (named "ADOR" for the Angra dos Reis meteorite) for 4.55 Ga ago are 0.69883(2),[63] and 0.69897(2).[64,65] The reasons why one out of the three research groups found a different value for the angrites remain unclear.

What is said for the Rb-Sr system applies to other radiometric chronometers. Among the other more important systems are the U-Pb and Th-Pb chronometers, which yield the most precise absolute ages. The U-Pb system has the added benefit that two similar parent–daughter pairs are available.

Both U isotopes and ^{232}Th decay to Pb isotopes. In an undisturbed sample, the amount of ^{206}Pb increases over time from the decay of ^{238}U. The number of ^{238}U$_\#$ decayed to ^{206}Pb$_\#$ equals the initial number of ^{238}U$_{ini}$ minus the number of ^{238}U$_t$ present after time t: ^{238}U$_\# = {}^{238}$U$_{ini} - {}^{238}$U$_t$. The ^{238}U$_{ini}$ and ^{238}U$_t$ are related through the decay probability law as a function of time: ^{238}U$_{ini} = {}^{238}$U$_t$exp $(\lambda_{238}t)$, where $\lambda_{238} = \ln(2)/t_{H238}$ is the decay constant of ^{238}U and t_{H238} is its half-life. The number of ^{206}Pb nuclei produced during the time t equals that of ^{238}U decayed, and the amount of radiogenic $^{206}Pb_\#$ is:

$$^{206}Pb_\# = {}^{238}U_\# = {}^{238}U_t[\exp(\lambda_{238}t) - 1] \tag{3.28}$$

If some initial ^{206}Pb was present ($^{206}Pb_{ini}$), the total amount of $^{206}Pb_t$ after the time t is:

$$^{206}Pb_t = {}^{206}Pb_{ini} + {}^{206}Pb_\# = {}^{206}Pb_{ini} + {}^{238}U_t[\exp(\lambda_{238}t) - 1] \tag{3.29}$$

Similarly, for ^{207}Pb from ^{235}U decay:

$$^{207}Pb_t = {}^{207}Pb_{ini} + {}^{207}Pb_\# = {}^{207}Pb_{ini} + {}^{235}U_t[\exp(\lambda_{235}t) - 1] \tag{3.30}$$

and ^{208}Pb from ^{232}Th decay:

$$^{208}Pb_t = {}^{208}Pb_{ini} + {}^{208}Pb_\# = {}^{208}Pb_{ini} + {}^{232}Th_t[\exp(\lambda_{232}t) - 1] \tag{3.31}$$

For practical purposes, the expressions above are normalized to ^{204}Pb. There are no contributions from long-lived isotopes to ^{204}Pb, and although ^{204}Pb is radioactive it can be treated like a stable nuclide as its half-life is $\geq 1.4 \times 10^{17}$ yr. Applying the ^{204}Pb normalization and skipping the subscript "t" for present-day values leads to expressions similar to that already described for the Rb-Sr system above:

$$^{206}Pb/^{204}Pb = {}^{206}Pb_{ini}/^{204}Pb + {}^{238}U/^{204}Pb \exp(\lambda_{238}t - 1) \tag{3.32}$$

$$^{207}Pb/^{204}Pb = {}^{207}Pb_{ini}/^{204}Pb + {}^{235}U/^{204}Pb \exp(\lambda_{235}t - 1) \tag{3.33}$$

$$^{208}Pb/^{204}Pb = {}^{208}Pb_{ini}/^{204}Pb + {}^{232}Th/^{204}Pb \exp(\lambda_{232}t - 1) \tag{3.34}$$

In the ideal case, these three independent isotope systems are concordant, meaning that they yield the same age. In reality, these systems can be disturbed by fractionations and loss of U, Th, Pb or the intermediate decay chain nuclides. One should expect that especially the two U-Pb systems are concordant in a closed system because the same elements are involved and any chemical fractionations should cancel out. For example, one potential problem with the U-Pb systems is radon loss over longer time, which could be important for smaller planetary and asteroidal objects with large porosities. In the ^{235}U decay chain, ^{219}Rn with a half-life of 3.96 s is one of the intermediate decay products. This short half-life should not allow much diffusive loss before ^{219}Rn alpha-decays to ^{215}Po. On the other hand, in the ^{238}U decay chain, ^{222}Rn

with a half-life of 3.82 days occurs. The longer stability of ^{222}Rn allows more diffusive redistribution to other phases or even loss from the sample. Loss of Rn over a long time period from the ^{238}U decay chain would result in smaller amounts of radiogenic ^{206}Pb, and sample ages derived from ^{238}U–^{206}Pb system would appear younger than those from the ^{235}U–^{207}Pb system.

An often applied method to derive ages of samples is "Pb-Pb" dating, which exploits the fact that there are two U-Pb systems. It requires the measurements of ^{207}Pb/^{204}Pb, ^{208}Pb/^{204}Pb and ^{235}U/^{238}U ratios in the samples. Rearranging for the amounts of radiogenic ^{207}Pb$_{\#}$ and ^{208}Pb$_{\#}$ and forming the quotient of the equations gives:

$$
\begin{aligned}
^{207}\text{Pb}_{\#}/^{206}\text{Pb}_{\#} \\
= [^{207}\text{Pb}/^{204}\text{Pb} - {}^{207}\text{Pb}_{\text{ini}}/^{204}\text{Pb}]/[^{206}\text{Pb}/^{204}\text{Pb} - {}^{206}\text{Pb}_{\text{ini}}/^{204}\text{Pb}] \ldots \\
= {}^{235}\text{U}/^{238}\text{U}\{[\exp(\lambda_{235}t) - 1]/[\exp(\lambda_{238}t) - 1]\}
\end{aligned}
$$

$$(3.35)$$

This equation has the initial ^{207}Pb$_{\text{ini}}$/^{204}Pb and ^{206}Pb$_{\text{ini}}$/^{204}Pb ratios and time as unknown variables and can be solved iteratively. The present day ^{235}U/^{238}U ratio is approximately constant (1/137.88)[66] in essentially all known analyzed samples. However, there might be hints of a higher ratio from decay of short-lived ^{247}Cm in some samples and, recently, highly resolved measurements of ^{238}U/^{235}U in CAIs indicate that there are real, yet to be understood, variations.[67] The initial ^{206}Pb/^{204}Pb and ^{207}Pb/^{204}Pb ratios must also be known for calculating model ages, which is a similar problem to finding the initial ^{87}Sr/^{86}Sr in the Rb-Sr system. Low ratios for radiogenic Pb isotopes are found in iron meteorites, because Th and U are incompatible and essentially absent in FeNi-metal. In that case, the measured values closely approximate the "true" initials. On the other hand, contamination of meteorites with terrestrial lead, especially from widespread use of Pb-tetraethyl in gasoline in the last century, is a common problem.

Many studies use the Pb ratios measured in the sulfide from the Canyon Diablo iron meteorite[68] as the "initial," "primitive" or "primordial" Pb isotopic composition, also often referred to as "PAT." Metal of this meteorite and of other iron meteorites has similarly low Pb isotopic ratios.[69]

As in the Rb-Sr system, the refractory inclusions make a special case. According to Chen and Wasserburg,[70] the lead isotopes in CAIs are consistent with an isochron age of 4.559 ± 0.015 Ga and the initials found for the Canyon Diablo meteorite. They assume that there are two groups of CAIs, one formed at 4.559 Ga and another with 4.468 Ga. However, according to Tera and Carlson[71] all CAI data from Chen and Wasserburg fit onto one single line if one makes no assumption about the initial Pb. Tera and Carlson found a CAI age of 4.566 ± 0.008 Ga and simultaneously derived Pb initials for CAIs from their isochron fits. The correlation misses the standard "primordial" Pb from the Canyon Diablo meteorite and plots below it. Table 3.6 summarizes some data for Pb-initial values.

The age of 4566 ± 8 Ma for refractory inclusions in the Allende chondrite derived by Tera and Carlson agrees with that of CAIs in the Efremovka carbonaceous chondrites, for which a very precise age of 4567.2 ± 0.7 Ma was determined by Amelin *et al.* in 2002.[52] These are the oldest and best determined ages of any solar system solids. If the CAIs are indeed the oldest solidified materials, and their initials are lower than in any other materials, defining the Pb isotopic composition of the Canyon Diablo iron meteorite as the "primordial" Pb composition and using its initials to calculate model Pb-Pb ages may not appear to be a wise choice, but seems to have become a traditional choice by now.

Aside from refractory inclusions, well-defined ages for chondrites and differentiated meteorites have been measured with the U-Pb and Pb-Pb systems. Chondrules formed around 4566.2 ± 2.5 Ma[72] and their isochrons are consistent with the Pb initial values found for Canyon Diablo. Entire chondrites assembled in about the same time interval.

Phosphates in chondrites are another preferred target for Th-U-Pb dating because they readily incorporate Th and U, but not Pb. Their Pb isotopic composition is largely determined by radiogenic Pb. Phosphates are preferentially observed in chondrites of higher petrologic type (Chapter 2) because they are secondary phases that form during thermal metamorphism. The "phosphate ages" measure the time when the parent body metamorphism ended exchange reactions with phosphates. The limits of metamorphism ages coincide with whole rock chondrite ages but chondrite parent body metamorphism occurred at least from 4.563 to 4.502 Ga ago.[73]

Typical crystallization ages for angrites are between 4.551 and 4.558 Ga[68,70,71,74] and between 4.13 and 4.56 Ga for eucrites.[68,75] The "younger" range of the eucrite crystallization ages indicates that some regions of the eucrite parent body experienced late disturbances and shock-resetting of the Rb-Sr systematics. Age determinations with the Sm-Nd chronometer yield similar results[75]. Both the Rb-Sr and the U-Pb dating systems consist of element pairs in which one element is more volatile than the other. In the Rb-Sr system, the parent nuclide (^{87}Rb) is more volatile, while in the U-Pb system the daughter nuclides

Table 3.6 Lead isotope initials.

Sample	$^{206}Pb_i/^{204}Pb$	$^{207}Pb_i/^{204}Pb$	$^{208}Pb_i/^{204}Pb$	Source
Canyon Diablo (IA) troilite	9.307 ± 0.006	10.294 ± 0.006	29.476 ± 0.018	Tatsumoto et al.[68]
Canyon Diablo (IA) metal	9.3028 ± 0.0074	10.291 ± 0.010	29.450 ± 0.037	Göpel et al.[69]
Average metal in iron meteorites	9.306 ± 0.008	10.298 ± 0.009	29.476 ± 0.031	Göpel et al.[69]
CAIs, isochron results	9.068	9.822	Not reported	Tera and Carlson[71]

(Pb isotopes) are more volatile. Redistribution of the more volatile Pb through secondary heating (metamorphism, impact heating) is a problem for dating eucrites, as well as lunar samples because Pb isotopes become more homogenized though such processes, and the decay record is easily disturbed. Consequently, age determinations are subject to (subjective) debates, and the interpretations of ages and initials inferred from easily disturbed chronometers are not necessarily unique. The interested reader may find the original literature enlightening in this respect.

3.5.2 Short-lived Radioactivities

The potential importance of short-lived nuclides for radiometric dating was first discussed by Brown in 1947.[76] In 1955, Urey[77] considered ^{26}Al, ^{36}Cl, ^{93}Zr, ^{107}Pd, ^{129}I, ^{135}Cs, ^{236}U and ^{237}Np. Urey did not include ^{60}Fe in his investigation because the recognition of ^{60}Fe as a short-lived nuclide only came in the late-1950s. These early predictions about the possible presence of short-lived nuclides in the protoplanetary disk and the best possibility of detecting them in meteorites were fulfilled when refined analytical instrumentation led to the discovery of relative excesses in the nuclides from decay of ^{129}I by Reynolds (1960)[78] and ^{244}Pu by Rowe and Kuroda (1965)[79] in several meteorites. Evidence for the former presence of ^{146}Sm, ^{53}Mn and ^{60}Fe followed later.[80–82] These and other short-lived nuclides and their decay products are used to determine early timescales within the solar system. One of the most important short-lived chronometers is the ^{26}Al-^{26}Mg system. The former presence of ^{26}Al was shown for minerals of Ca-Al-rich inclusions (CAIs) by Gray and Compston (1974).[83] These refractory meteorite components were long recognized as the oldest preserved phases from the early solar system – hence they were the best bets for detecting excesses in daughter nuclides from short-lived radioactivities.

The former existence of a short-lived radionuclide trapped in a mineral phase becomes apparent through excesses in its daughter nuclide when compared to the normal isotopic abundance of the daughter element. For example, short-lived ^{26}Al decays into ^{26}Mg, and if ^{26}Al was present in a mineral the ^{26}Mg/^{24}Mg isotopic ratio was enhanced through ^{26}Al decay over the ^{26}Mg/^{24}Mg ratio found in normal (terrestrial) rocks. Since there are mass-dependent chemical and physical processes that can alter the ^{26}Mg/^{24}Mg ratio, one can compare the ^{26}Mg/^{24}Mg and ^{25}Mg/^{24}Mg ratios. ^{25}Mg usually has no radiogenic contributions and the ^{25}Mg/^{24}Mg ratio can provide a measure for the extent of any mass-dependent isotopic fractionation processes relative to a standard value. If there are fractionations of the ^{25}Mg/^{24}Mg ratios from the standard, the $^{26}Mg/^{24}Mg$ ratios must also be corrected for mass fractionation effects. For reference, the $^{26}Mg/^{24}Mg$ measurements are usually corrected for mass fractionation effects using:

$$\left(^{26}Mg/^{24}Mg\right)_{corrected}$$
$$= \left(^{26}Mg/^{24}Mg\right)_{measured} / \left[\left(^{25}Mg/^{24}Mg\right)_{measured} / \left(^{25}Mg/^{24}Mg\right)_{standard} \right]^2$$

$$(3.36)$$

Often it is found that the $^{25}Mg/^{24}Mg$ ratios are normal in meteoritic phases that clearly have $^{26}Mg/^{24}Mg$ ratios above normal, which indicates that the excess of ^{26}Mg must be due to radioactive decay of ^{26}Al at an early time in the solar system.

As an example, Figure 3.13 shows the $^{25}Mg/^{24}Mg$ and $^{26}Mg/^{24}Mg$ ratios plotted against the $^{27}Al/^{24}Mg$ for hibonites from the Dhajala H3 chondrite.[84] The $^{26}Mg/^{24}Mg$ clearly correlate with the $^{27}Al/^{24}Mg$ ratios and define a "fossil" isochron whereas the $^{25}Mg/^{24}Mg$ are not significantly different from the standard Mg-isotopic ratio.

The correlation of the $^{26}Mg/^{24}Mg$ and $^{27}Al/^{24}Mg$ ratios is explained as follows. The amount of ^{26}Mg now present in a sample depends on the original amount of ^{26}Mg present (subscript ini) plus the ^{26}Mg produced through decay of ^{26}Al that was incorporated into the sample when it initially formed, so $^{26}Mg = {}^{26}Mg_{ini} + {}^{26}Al_{ini}$. It is convenient to normalize to the major Mg isotope ^{24}Mg. Expanding the last quotient gives the "fossil" isochron equation for extinct radioactivities applied to the ^{26}Al-^{26}Mg system from which the initial $^{26}Al_{ini}/^{27}Al$ ratio can be derived:

$$^{26}Mg/^{24}Mg = {}^{26}Mg_{ini}/^{24}Mg + {}^{26}Al_{ini}/^{24}Mg$$
$$= {}^{26}Mg_{ini}/^{24}Mg + \left(^{26}Al_{ini}/^{27}Al\right)\left(^{27}Al/^{24}Mg\right)$$

$$(3.37)$$

If the measurements cover several different phases that crystallized or formed from the same reservoir at the same time but have different Al and Mg concentrations, the $^{26}Mg/^{24}Mg$ and $^{27}Al/^{24}Mg$ ratios will correlate according to the isochron equation for the extinct radioactivity. From a linear fit to the data, the initial $^{26}Al_{ini}/^{27}Al$ ratio follows from the slope, and the intercept gives the initial $^{26}Mg_{ini}/^{24}Mg$.

The hibonite in the Dhajala H3 chondrite defines an initial of $^{26}Al/^{27}Al = 8.4 \pm 0.5 \times 10^{-6}$. This value is lower that the canonical value of $^{26}Al/^{27}Al \approx 5 \times 10^{-5}$ from CAIs in carbonaceous chondrites, which is the initial $^{26}Al/^{27}Al$ ratio adopted for the solar system.[85,86] This difference in initial ratios may indicate a relative age difference, if ^{26}Al was homogeneously distributed within the solar nebula. It is important

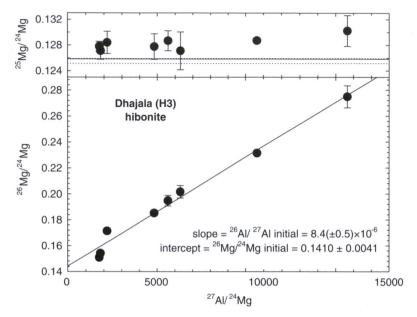

Figure 3.13   A "fossil" isochron diagram for deriving the initial abundance ratio of a now extinct isotope relative to a stable isotope of the same element, illustrated here for the ^{26}Al-^{26}Mg systematics in hibonite from the Dhajala H3 chondrite. The top diagram shows the uncorrected ^{25}Mg/^{24}Mg ratios plotted against the ^{27}Al/^{24}Mg ratios. There is little variation of the ^{25}Mg/^{24}Mg data from the standard value (shown by the horizontal line. The dotted lines show the standard deviation of the ^{25}Mg/^{24}Mg standard value). The ^{25}Mg/^{24}Mg data are used to correct for the mass-dependent isotope fractionation of the ^{26}Mg/^{24}Mg data. The corrected ^{26}Mg/^{24}Mg data plotted against the ^{27}Al/^{24}Mg ratios show a good correlation that indicates *in situ* decay of ^{26}Al that was incorporated into the hibonites. The slope corresponds to a ^{26}Al/^{27}Al initial of $8.4 \pm 0.5 \times 10^{-6}$. Within uncertainties, the intercept of ^{26}Mg/^{24}Mg $= 0.1410 \pm 0.0041$ equals the measured standard value of 0.13778 ± 0.00151. Data from Hinton and Bischoff.[84]

to realize that this is a regularly made but unproven assumption for all relative age determinations with the ^{26}Al-^{26}Mg system. Similar assumptions are made about the initial distributions for the other short-lived nuclides because an initial homogeneous distribution is the premise for using chronometers involving radionuclides.

But let us play the age game as everyone does. The time difference (Δt) implied by different initial ^{26}Al/^{27}Al ratios for two objects (subscript 1 and 2, and 2 has the larger ratio and is presumably older) is:

$$
\begin{aligned}
\Delta t &= 1/\lambda_{26} \ln\left[\left(^{26}\text{Al}/^{27}\text{Al}\right)_2 / \left(^{26}\text{Al}/^{27}\text{Al}\right)_1\right] \\
&= t_{\text{H}} / \ln(2) \times \ln\left[\left(^{26}\text{Al}/^{27}\text{Al}\right)_2 / \left(^{26}\text{Al}/^{27}\text{Al}\right)_1\right]
\end{aligned}
\tag{3.38}
$$

Inserting the values for the hibonite found in Dhajala and the canonical ratio in CAIs leads to an age difference of 1.9 Ma between these two systems. When measureable, chondrules also have lower $^{26}Al/^{27}Al$ ratios, although some chondrules show ratios approaching the canonical value of CAIs. The average from these data is $\sim 8 \times 10^{-6}$, similar to the hibonite data in the Dhajala H3 chondrite. The glass portion in chondrules of the Semarkona LL3 chondrite give initial $^{26}Al/^{27}Al$ ranging from 6×10^{-6} to 9×10^{-6} (ref. 87). This implies that these chondrules formed ~ 1.8 to ~ 2.3 Ma after CAIs.

A related question to the homogeneous distribution of ^{26}Al and other radionuclides is whether ^{26}Al was indeed "alive" in the early solar system so that *in situ* decay led to the observed ^{26}Al excesses in various meteorite components.[88] One could suppose that these excesses were caused by "fossil" decay products such as seen in presolar grains.[89] Several presolar grain types show large $^{26}Mg/^{24}Mg$ ratios. These likely stem from decay of ^{26}Al that these grains had incorporated when they formed in stellar ejecta (see ref. 1). However, ^{26}Al most likely had decayed before these grains were incorporated into solar system materials. Since these presolar grains exist, the solar nebula obviously was not homogenized, and the assimilation of a small presolar grain with large anomalous isotopic compositions in $^{26}Mg/^{24}Mg$ into meteoritic components such as CAIs or chondrules could have caused the small $^{26}Mg/^{24}Mg$ excesses that are now measured. However, this cannot explain why the $^{26}Mg/^{24}Mg$ ratios *correlate* with Al/Mg ratios when different minerals of a given meteoritic component are measured. For example, if a presolar grain with high $^{26}Mg/^{24}Mg$ ratio but no live ^{26}Al was incorporated into condensing or crystallizing minerals, the ^{26}Mg from the grain would follow the ^{24}Mg because they are chemically the same. Since Mg and Al have different compatibilities in different minerals, Mg and Al become fractionated and no positive correlation of $^{26}Mg/^{24}Mg$ with Al/Mg is expected to develop. All these minerals will have essentially the same $^{26}Mg/^{24}Mg$ ratio. On the other hand, live ^{26}Al would fractionate from Mg like stable ^{27}Al does. Minerals that incorporated the radioactive ^{26}Al subsequently develop higher $^{26}Mg/^{24}Mg$ ratios that correlate with the Al content. Therefore, searches for evidence of ^{26}Al are best done in Al-rich and Mg-poor components. High amounts of Al make it more likely that larger quantities of ^{26}Al were incorporated, and low concentrations of Mg ensure that the ^{26}Mg from ^{26}Al decay is not lost against a sea of background ^{26}Mg already present in the sample. Lee and co-workers[88] demonstrated that there are correlated $^{26}Mg/^{24}Mg$ and $^{27}Al/^{24}Mg$ ratios in mineral phases from a single CAI. This clearly showed that ^{26}Al was alive when these minerals formed, and that the ^{26}Mg excess was not due to incorporation of ^{26}Mg from some other (pre-solar stellar) source. Thus, the Al-Mg system is a good choice to track high temperature volatility fractionations of refractory Al from Mg and high temperature igneous processes.

We only briefly mention some other short-lived nuclide chronometers here that are currently used in meteoritics; ref. 49 is a useful starting point for more information. In the Mn-Cr system, ^{53}Mn decays to ^{53}Cr by electron capture with a half-life of 3.74 Ma. This system traces volatility and igneous fractionation timescales as does the Al-Mg system, but Mn and Cr have higher volatilities and different mineral compatibilities. Therefore, the Al-Mg and Mn-Cr systems do not necessarily "date" the same events. The elements Al, Mg, Mn and Cr are mainly lithophile in most meteorites. In addition, Mn and Cr also have slightly siderophile tendencies and the Mn-Cr system can also be applied to iron meteorites and pallasites. The Fe-Ni and Pd-Ag systems are potentially useful for dating metal fractionation events. The ^{60}Fe undergoes beta decays to ^{60}Co, which then decays to ^{60}Ni. The best chances of detecting excesses in ^{60}Ni are Fe-rich and Ni poor phases. However, Fe-rich phases also often contain large amounts of Ni, and, thus, detecting a small excess of ^{60}Ni against a large initial ^{60}Ni in metal such as kamacite or taenite is fairly difficult. The former presence of ^{60}Fe has been established in relatively Ni-poor troilite in chondrites of petrologic type 3, in chondrules and some differentiated eucrites. The ^{107}Pd-^{107}Ag system is ideal for measuring the timing of metal alloy crystallization. The Pd and Ag fractionate during crystallization and if ^{107}Pd is alive during crystallization the conditions for developing "fossil" isochrons are met. Another widely used system involves ^{182}W decay to ^{182}Hf. In meteorites and planetary materials Hf behaves as a lithophile element whereas W is mainly siderophile. Both elements are also highly refractory and any fractionation of these elements is related to metal-silicate fractionations. This makes the ^{182}W-^{182}Hf system very attractive for dating metal core formation on the Earth and differentiated meteorite parent bodies, which most likely also include Mars (Chapter 4).

3.6   AN EARLY SOLAR SYSTEM TIMELINE

The derived initial values of short-lived radionuclides can be tied to the absolute timescale if an independent measurement of the absolute age can be determined in at least one system for which the initial abundance of the short-lived nuclide has also been measured. The absolute ages of CAIs, chondrites and eucrites are well determined from Pb-Pb dating and provide such anchor ages to place the estimates of the relative time scales from short-lived chronometers onto an absolute age scale. For example, for the relative ages derived from the $^{26}Al/^{27}Al$ ratios, the absolute age of CAIs [4567.2 Ma (ref. 52)] can be used because the initial $^{26}Al/^{27}Al$ in many CAIs is also well determined (5×10^{-5}). Similarly, the Pb-Pb age of angrites [4557.8 Ma (ref. 64)] is taken as a reference age to tie relative Mn-Cr ages to absolute ones.

The results from many measurements of the different long-lived and short-lived isotope chronometers place the formation of meteoritic components and meteorite parent bodies and subsequent metamorphic events or crystallization processes into a timeline of events. In Figure 3.14, the CAIs are placed as the oldest known solids at 4.567 Ga, which we take as the minimum age of the solar system.

The top bar in Figure 3.14 shows the range in estimates of the maximum lifetime of the solar nebula, with the longest duration of about 10 Ma. The CAIs formed within 2–4 Ma and underwent thermal and/or aqueous alterations over ~2 Ma afterwards, for example, the $^{26}Al/^{27}Al$ radiometric system indicates that aqueous alteration of the mineral assemblage of CAIs led to secondary phases such as sodalite within about 2 Ma. Iodine-Xe dating places formation of halite (not shown) in

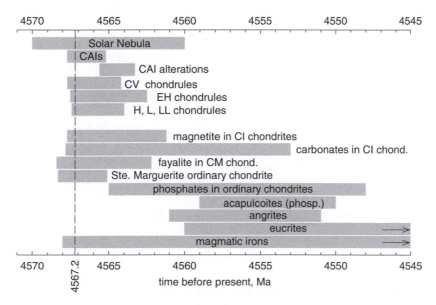

Figure 3.14 Timeline of events for the early solar system development assembled from absolute and relative age dating with long-lived and short-lived radioactive nuclide systems. The bar lengths reflect the range in measurements as well as their uncertainties.

ordinary chondrites within the first 2 Ma of solar system formation. Absolute (with Pb-Pb) and relative (with $^{26}Al/^{27}Al$) age dating show that chondrule formation started contemporaneously with CAIs and lasted for at least about 1–4 Ma. Chondrule formation requires the presence of nebula gas so chondrule ages extending to about 4 Ma after CAIs solidification implies that solar nebula gas did not dissipate earlier. This duration of solar nebula persistence is consistent with the 2–3 Ma lifetimes of classical T Tauri stars and most disks currently observable around other young stars. However, since we do not know how old the solar nebula was when the first CAIs solidified, lifetimes of up to ∼ 10 Ma as observed for some T Tauri stars cannot be ruled out.

Most magnetite in the CI chondrites is most likely of secondary nature but still is among the oldest phases, as suggested from I-Xe dating. Incorporation of Fe into silicates as indicated by fayalite formation on the CM chondrite parent body happened during similar time scales as magnetite formation. Aqueous alteration on the CI chondrite parent must have started quite early and extended to ∼ 20 Ma after CAI formation. This duration comes from dating the formation of carbonates in CI chondrites with the ^{53}Mn-^{53}Cr chronometer.

The Ste. Marguerite H4 chondrite is shown as an example of an ordinary chondrite formation age. Secondary phosphates formed during parent asteroid metamorphism over ∼ 15 Ma after formation of the chondrite parent bodies. Phosphates in the acapulcoites show similar duration of metamorphism on other meteorite parent bodies. The differentiated meteorites, angrites and eucrites have relatively high absolute Pb ages, indicating very early differentiation and metal-silicate separation in planetesimals. Although more uncertain, the ages of magmatic irons of up to 4.557 ± 0.012 Ga are compatible with this scenario. These ages indicate that planet formation must have been relatively fast and differentiation processes on planetary objects likely began as soon as larger planetesimals and rock planets had formed. The formation of the gas giant planets also must have been rapid (few Ma) so that they could capture solar nebula gas before it dissipated. Thus, it seems that the major characteristics of planets were imprinted in the first few million years of solar system development.

FURTHER READING AND RESOURCES

W. Benz, R. Kallenbach and G. W. Lugmair (eds.), *From Dust to Terrestrial Planets*, Springer, reprinted from Space Science Reviews, 2000, **92** (issues 1–2).

J. F. Kerridge and M. S. Matthews (eds.), *Meteorites and the Early Solar System*, University of Arizona Press, Tucson, 1988.

K. Lodders and B. Fegley, *The Planetary Scientist's Companion*, Oxford University Press, New York, 1998.

Y. J. Pendleton and A. G. G. M. Tielens (eds.), *From Stardust to Planetesimals*, Astronomical Society Pacific Conference Series, vol. 122, 1997.

Protostars and Planets, vols I–V, published in 1978, 1985, 1993, 2000, 2007 by University of Arizona Press, Tucson, AZ.

REFERENCES

1. Selected reviews on presolar grains: E. Anders and E. Zinner, *Meteoritics*, 1993, **28**, 490; K. Lodders and S. Amari, *Chem. Erde*, 2005, **65**, 93, 2005 (in English); P. Hoppe and E. Zinner, *J. Geophys. Res.*, 2000, **105**, 10371.
2. A. P. Jones, A. G. G. M. Tielens, D. J. Hollenbach and C. F. McKee, *Astrophys. J.*, 1994, **433**, 797; A. P. Jones, in *Astrophysics of Dust*, A. N. Witt, G. C. Clayton and B. T. Draine, (ed.), Astronomical Society Pacific Conference Series **vol. 309**, 2004, p. 347.
3. B. D. Savage and K. R. Sembach, *Annu. Rev. Astron. Astrophys.*, 1996, **34**, 279.
4. C. M. Hohenberg, *Science*, 1969, **166**, 212; A. G. W. Cameron and J. W. Truran, *Icarus*, 1977, **30**, 447.
5. J. P. Bradley, D. E. Brownlee and T. P. Snow, Astronomical Pacific Conference Series **vol. 122**, 1997, p. 217; J. P. Bradley, *Science*, 1999, **285**, 1716.
6. D. Saumon and T. Guillot, *Astrophys. J.*, 2004, **609**, 1170.
7. J. S. Lewis, *Physics and Chemistry of the Solar System*, Academic Press, 1995.
8. A. G. W. Cameron, *Meteoritics*, 1995, **30**, 133; J. S. Lewis, *Science*, 1974, **186**, 440; C. Terquem, J. C. B. Papaloizou and R. P. Nelson, *Space Sci. Rev.*, 2000, **92**, 323; K. Willacy, H. H. Klahr, T. J. Millar and Th. Henning, *Astron. Astrophys.*, 1998, **338**, 995.
9. F. H. Shu, J. R. Najita, H. Shang and Z. Y. Li, *Protostars and Planets IV*, University of Arizona Press, Tucson, AZ, 2000, p. 879.
10. J. S. Greaves, W. S. Holland and M. C. Wyatt, *et al.*, *Astrophys. J.*, 2005, **619**, L187; G. F. Benedict, B. E. McArthur and G. Gatewood, *et al.*, *Astron. J.*, 2006, **132**, 2206.
11. B. Fegley, *Space Sci. Rev.*, 1999, **90**, 239.

12. A. N. Krot, B. Fegley Jr., H. Palme and K. Lodders, in *Protostars and Planets IV*, University of Arizona Press, Tucson, AZ, 2000, p. 1019.
13. R. Wildt, *Z. Astrophys.*, 1933, **6**, 345.
14. H. N. Russell, *Astrophys. J.*, 1934, **79**, 317.
15. H. C. Lord, *Icarus*, 1965, **4**, 279.
16. J. W. Larimer, *Geochim. Cosmochim. Acta*, 1967, **31**, 1215.
17. J. W. Larimer, *Geochim. Cosmochim. Acta*, 1973, **37**, 1603; J. W. Larimer, in *Meteorites and the Early Solar System*, J. F. Kerridge and M. S. Matthews, (ed.), University of Arizona Press, Tucson, AZ, 1988, p. 375.
18. L. Grossman, *Geochim. Cosmochim. Acta*, 1972, **36**, 597.
19. J. S. Lewis, *Icarus*, 1972, **16**, 241.
20. L. Grossman and J. W. Larimer, *Rev. Geophys. Space Phys.*, 1974, **12**, 71.
21. W. V. Boynton, *Geochim. Cosmochim. Acta*, 1975, **39**, 569.
22. C. M. Wai and J. T. Wasson, *Earth Planet. Sci. Lett.*, 1977, **36**, 1; M. Wai and J. T. Wasson, *Nature*, 1979, **282**, 790.
23. D. W. Sears, *Earth Planet. Sci. Lett.*, 1978, **41**, 128.
24. B. Fegley and J. S. Lewis, *Icarus*, 1980, **41**, 439.
25. S. K. Saxena and G. Eriksson, *Earth Planet. Sci. Lett.*, 1983, **65**, 7.
26. B. Fegley and H. Palme, *Earth Planet. Sci. Lett.*, 1985, **72**, 311.
27. A. S. Kornacki and B. Fegley, *Earth Planet. Sci. Lett.*, 1986, **75**, 297.
28. H. Palme and B. Fegley, *Earth Planet. Sci. Lett.*, 1990, **101**, 180.
29. D. S. Ebel and L. Grossman, *Geochim. Cosmochim. Acta*, 2000, **64**, 339.
30. K. Lodders, *Astrophys. J.*, 2003, **591**, 1220.
31. T. R. Ireland and B. Fegley, *Int. Geol. Rev.*, 2000, **42**, 865.
32. A. El Goresy, K. Nagel and P. Ramdohr, *Proc. Lunar Planet. Sci. Conf.*, 1978, **9**, 1249.
33. J. D. Blum, G. J. Wasserburg, I. D. Hutcheon, J. R. Beckett and E. M. Stolper, *Geochim. Cosmochim. Acta*, 1988, **53**, 543.
34. A. Bischoff. and H. Palme, *Geochim. Cosmochim. Acta*, 1987, **51**, 2733.
35. B. Fegley and A. S. Kornacki, *Earth Planet. Sci. Lett.*, 1984, **68**, 181.
36. B. Fegley, in Workshop on the Origins of Solar Systems, J. A. Nuth and P. Sylvester, (ed.), LPI Technical Report No. 88-04, Houston, 1988, pp. 51–60.
37. B. Fegley, *Space Sci. Rev.*, 2000, **92**, 177.
38. B. Fegley and R. G. Prinn, in *The Formation and Evolution of Planetary Systems*, H. A. Weaver and L. Danly, (ed.), Cambridge University Press, Cambridge, 1989, pp. 171–211.

39. R. G. Prinn and B. Fegley, in *The Origin and Evolution of Planetary and Satellite Atmospheres*, S. K. Atreya, J. B. Pollack and M. S. Matthews, (ed.), University of Arizona Press, Tucson, AZ, 1989, p. 78.

40. H. C. Urey, Chemical evidence regarding the Earth's origin. in XIIIth International Congress Pure and Applied Chemistry and Plenary Lectures, Almqvist and Wiksells, Stockholm, 1953, p. 188.

41. J. S. Lewis, S. S. Barshay and B. Noyes, *Icarus*, 1979, **37**, 190.

42. K. Lodders and B. Fegley, *Icarus*, 2002, **155**, 393.

43. E. Anders, R. Hayatsu and M. H. Studier, *Science*, 1973, **182**, 781.

44. J. S. Lewis and R. G. Prinn, *Astrophys. J.*, 1980, **238**, 357.

45. R. G. Prinn and B. Fegley, *Astrophys. J.*, 1981, **249**, 308.

46. J. L. Lunine, in *The Formation and Evolution of Planetary Systems*, H. A. Weaver and L. Danly, (ed.), Cambridge University Press, Cambridge, 1989, p. 213.

47. T. P. Kohman, *J. Chem. Ed.*, 1961, **38**, 73.

48. T. D. Swindle, in *Protostars and Planets III*, E. H. Levy and J. I. Lunine, (ed.), University of Arizona Press, Tucson, AZ, 1993, p. 867.

49. R. W. Carlson and G. W. Lugmair, in *Origin of the Moon*, R. M. Canup and K. Righter, (ed.), University of Arizona Press, Tucson, AZ, 2000, p. 25.

50. J. N. Goswami, K. K. Marhas, M. Chaussidon, M. Gounelle and B. S. Meyer, in *Chondrules and the Protoplanetary Disk*, A. N. Krot, E. R. D. Scott and B. Reipurth, (ed.), ASP Conference Series **vol. 341**, 2005, p. 485.

51. T. P. Kohman, *Ann. N.Y. Acad. Sci.*, 1956, **62**, Art. 21, 503.

52. Y. Amelin, A. N. Krot, I. A. Hutcheon and A. A. Ulyanov, *Science*, 2002, **297**, 1678.

53. J. F. Minster and C. J. Allegre, *Earth Planet. Sci. Lett.*, 1981, **56**, 89.

54. J. F. Minster, J. L. Birck and C. L. Allegre, *Nature*, 1982, **300**, 414.

55. C. M. Gray, D. A. Papanastassiou and G. J. Wasserburg, *Icarus*, 1973, **20**, 213.

56. G. W. Wetherill, R. Mark and C. Lee-Hu, *Science*, 1973, **182**, 281.

57. F. A. Podosek, E. K. Zinner, G. J. MacPherson, L. L. Lundberg, J. C. Brannon and A. J. Fahey, *Geochim. Cosmochim. Acta*, 1991, **55**, 1083.

58. P. J. Patchett, *Nature*, 1980, **283**, 438.

59. P. J. Patchett, *Earth Planet. Sci. Lett.*, 1980, **50**, 181.

60. D. A. Papanastassiou and G. J. Wasserburg, *Earth Planet. Sci. Lett.*, 1969, **5**, 361.
61. C. J. Allegre, J. L. Birck, S. Fourcade and M. P. Semet, *Science*, 1975, **187**, 436.
62. M. I. Smoliar, *Meteoritics*, 1993, **28**, 105.
63. G. J. Wasserburg, F. Tera, D. A. Papanastassiou and J. C. Hueneke, *Earth Planet. Sci. Lett.*, 1977, **35**, 294.
64. G. W. Lugmair and S. J. G. Galer, *Geochim. Cosmochim. Acta*, 1992, **56**, 1673.
65. L. E. Nyquist, B. Bansal, H. Wiesmann and C. Y. Shih, *Meteoritics*, 1994, **29**, 872.
66. J. H. Chen and G. J. Wasserburg, *Geophys. Res. Lett.*, 1980, **7**, 275.
67. G. A. Brennecka, S. Weyer, M. Wadhwa, P. E. Janney and A. D. Anbar, *Lunar Planet Sci. Conf. 40*, 2009, #1061.
68. M. Tatsumoto, R. J. Knight and C. J. Allegre, *Science*, 1973, **180**, 1278.
69. C. Göpel, G. Manhes and C. J. Allegre, *Geochim. Cosmochim. Acta*, 1985, **49**, 1681.
70. J. H. Chen and G. J. Wasserburg, *Earth Planet. Sci. Lett.*, 1981, **52**, 1.
71. F. Tera and R. W. Carlson, *Geochim. Cosmochim. Acta*, 1999, **63**, 1877.
72. Y. Amelin and A. Krot, *Meteoritics Planet. Sci.*, 2007, **42**, 1321.
73. C. Göpel, G. Manhes and C. J. Allegre, *Earth Planet. Sci. Lett.*, 1994, **121**, 153.
74. G. W. Lugmair and S. J. Galer, *Geochim. Cosmochim. Acta*, 1992, **56**, 1673.
75. R. Tera, R. W. Carlson and N. Z. Boctor, *Geochim. Cosmochim. Acta*, 1997, **61**, 1713.
76. H. Brown, *Phys. Rev.*, 1947, **72**, 348.
77. H. C. Urey, *Proc. Natl. Acad. Sci. USA*, 1955, **41**, 127.
78. J. H. Reynolds, *Phys. Rev. Lett.*, 1960, **4**, 8.
79. M. W. Rowe and P. K. Kuroda, *J. Geophys. Res.*, 1965, **70**, 709.
80. G. W. Lugmair, N. B. Scheinin and K. Marti, *Earth Planet. Sci. Lett.*, 1975, **27**, 79.
81. J. L Birck and C. J. Allegre, *Geophys. Res. Lett.*, 1985, **12**, 745.
82. S. Tachibana and G. R. Huss, *Astrophys. J.*, 2003, **588**, L41.
83. C. M. Gray and W. Compston, *Nature*, 1974, **251**, 495.
84. R. W. Hinton and A. Bischoff, *Nature*, 1984, **308**, 169.
85. T. Lee, D. A. Papanastassiou and G. J. Wasserburg, *Geophys. Res. Lett.*, 1976, **3**, 41.

86. G. J. MacPherson, A. M. Davis and E. K. Zinner, *Meteoritics*, 1995, **30**, 365.
87. N. T. Kita, H. Nagahara, S. Togashi and Y. Morishita, *Geochim. Cosmochim. Acta*, 2000, **64**, 3913.
88. T. Lee, D. A. Papanastassiou and G. J. Wasserburg, *Astrophys. J.*, 1977, **211**, L107.
89. D. D. Clayton, *Astrophys. J.*, 1975, **199**, 765.

CHAPTER 4

The Bodies in the Inner Solar System

"Ay, for 't were absurd
To think that nature in the earth bred gold
Perfect i' the instant: something went before.
There must be remote matter."
Ben Jonson, *The Alchemist*, 1610

4.1 INTRODUCTION

In this chapter we visit the terrestrial planets Mercury, Venus, Earth and Mars as well as the Earth's Moon, and the differentiated asteroid Vesta. A few facts about other asteroids are also included in this chapter. The terrestrial planets – or Earth-like planets – and their smaller differentiated asteroid cousins consist mainly of silicate rocks, metal and sulfide in varying proportions and are quite poor in H and He when compared to the compositions of the Sun and the outer planets. Water and other volatile substances are relatively rare, which suggests that high temperature processes played a major role during the formation of the terrestrial planets.

The terrestrial planets experienced substantial melting of silicates, sulfides and metal during their formation, and the denser metal (and sulfide) separated into a metallic core that is surrounded by a silicate mantle. Such differentiation of metal from silicate took place on some asteroids from which the silicate-rich achondrites originated. Some of the iron meteorites may represent cores of such shattered differentiated asteroids (Chapter 2). The Moon's materials experienced substantial

Chemistry of the Solar System
By Katharina Lodders and Bruce Fegley, Jr.
© K. Lodders and B. Fegley, Jr. 2011
Published by the Royal Society of Chemistry, www.rsc.org

melting and crystallization, but the Moon notably distinguishes itself by having only a very small metallic core, if any at all. Igneous melting and crystallization processes further led to the early formation of silicate crusts on planetary surfaces. Igneous processing is still ongoing today on at least one planet – the Earth – but is no longer active on the Moon, Mercury or Mars. However, active volcanism may still occur on Venus now.

4.2 PHYSICAL AND CHEMICAL PROPERTIES

The Earth, Moon, Mars and Vesta are the only rocky planetary bodies for which sufficient geophysical and geochemical information is available to test accretion models. The bulk densities are known for all terrestrial planets, but the moment of inertia, which gives insight into the mass distribution within a planet, is only more reliably known for the Earth, Moon and Mars. The chemical composition of bulk planetary silicates (mantle and crust) is known for the Earth and to some extent for the Moon, Mars, Mercury, Venus and Vesta. Much information about the silicate composition of Mars and Vesta is based on the general notion that the shergottites, nakhlites and chassignites (SNC meteorites) are from Mars and that the eucrites, howardites and diogenites (EHD meteorites) are from Vesta. The discussion here focuses on the Earth–Moon system, Mars and Vesta because geophysical and geochemical data for Mercury and Venus are still insufficient.

To understand how the terrestrial planets, the Moon and the differentiated asteroids formed from the materials in the solar nebula we need to know their physical and chemical properties which can serve as constraints for the accretion models. Some of the major physical planetary properties to be explained are the sizes of the crust, mantle and core, the bulk density, the moment of inertia, the presence or absence of a magnetic field and, where applicable, the seismological observations. The chemical properties include the compositions of the silicate portion and core, the silicate mantle oxidation state, trace element chemistry in the mantle (siderophile and chalcophile element contents), and volatile element content (alkalis, halogens, carbon, water).

4.2.1 Physical Properties

Table 4.1 summarizes some physical properties of the terrestrial planets, the Moon and Vesta. The masses, sizes and bulk densities are known for all these objects. Their densities are indicative of their bulk chemical compositions. One problem in comparing the observed bulk densities is

Table 4.1 Some physical properties of the terrestrial planets, the Moon and the asteroid Vesta.

Property	Mercury	Venus	Earth	Mars	Moon	Vesta
Mass (10^{24} kg)	0.33022	4.8685	5.9736	0.64185	0.07349	$2.6{-}3.0\times10^{-4}$
Radius, mean (km)	2437.6	6051.84	6371	3389.92	1737.1	260
ρ Obs. density (g cm^{-3})	5.43	5.243	5.515	3.934	3.344	3.53–4.07
ρ Uncompressed at 10 kbar	5.3^a	$3.96{-}4.1^a$	4.05^a	3.74^a	3.4^a	3.4
C/MR^2	0.33	0.324–0.334	0.3307	0.3662	0.3932	0.33
J_2 ($\times10^{-6}$)	60	4.46	1082.636	1980.45	203.43	–
Core % of total mass	$\sim68.5^a$	$\sim30{-}35^a$	32.5	21^a	<5	$12{-}15^a$
r_{core}/R	$\sim0.78^a$	$\sim0.46{-}54^a$	0.546	0.48^a	<0.14	–
Mole fraction sulfur in core	?	?	0.05^a	0.17^a	0	0^a
Silicates (mol fraction); #Fe = FeO/(FeO + MgO)	<0.045	0.08	0.112	0.243	0.191	0.334
$\rho_{silicates}$ (g cm^{-3})	$\sim3.3^a$	$\sim4^a$	2.7–3 crust, 3.3–5.7 mantle	3.5^a	3.3	3.2
ρ_{core} (g cm^{-3})	9.5^a	$\sim12^a$	9.9–12.2 OC 12.8–13.1 IC	$\sim7.3^a$	–	7.8^a
Surface temperature (K)	Day 590–700; night 100	740	288; 180–330	214; 140–300	120–390	~200
Rotation period	58.646 d	243.02 d	23.9345 h	24.62 h	27.32 d	–
Dipole field moment (gauss km^{-3})	0.0033	None detected	0.61	None at present	None at present	≤0.002

^{a}Indicates a model-dependent quantity.

that higher interior densities due to compression must be considered for the more massive objects. Therefore, uncompressed densities are chosen for comparisons (densities for 298 K standard temperature and 1 bar or 10 kbar pressure are often used in the literature).

In contrast to the terrestrial planets, the mineralogy of chondritic meteorites suggests alterations at relatively low total pressures and chondrite parent bodies should be relatively small. This points to asteroids. The high-pressure minerals that are occasionally encountered in meteorites are plausibly explained by formation through shock. Diamond (cubic β-C), londsdaleite (hexagonal diamond) and chaoite (hexagonal C) in ureilites most likely formed through shock from collisions in space, and diamond in some iron meteorites, such as in Canyon Diablo from the Barringer meteorite crater in Arizona, formed by the shockwave as meteorites impacted on Earth. Formation of diamond requires total pressures above 14 kbar and the only reason that it is still observed in meteorites and terrestrial rocks is that the conversion of metastable diamond into graphite, the stable low pressure modification of carbon, is kinetically inhibited. Other minerals indicative of higher total pressures are the SiO_2 polymorphs tridymite and cristobalite. Monoclinic and triclinic tridymite is stable above 3 kbar and tetragonal cristobalite above > 5 kbar. However, these polymorphs can be produced as metastable phases in mineral break-down reactions at lower total pressures, so they are not unique indicators of high total pressures. Other high-pressure minerals like magnesiowüstite, $(Mg,Fe)O$, ringwoodite, $(Mg,Fe)_2SiO_4$ with spinel structure, or majorite, $Mg_3(MgSi)Si_3O_{12}$, a garnet, or pyrope $(Mg_3Al_2(SiO_4)_2)$ are common minerals in the earth's mantle but only rarely observed in meteorites. A few, relatively rare, metallic phases in meteorites that are likely products of shock transformation are martensite, α_2-FeNi, or suessite, $(Fe,Ni)_3Si$.

All these phases require higher pressures than the highest hydrostatic pressures that can be attained in the largest asteroids still present today. Ceres, the largest asteroid, or rocky "dwarf planet" has a diameter of ~ 960 km, followed by Vesta (~ 520–560 km) and Pallas (523 km). Among the more than 20 000 asteroids, only ~ 140 have diameters larger than 100 km. The hydrostatic pressure (P in bar) experienced by material in a uniform body with constant density ρ (in g cm^{-3}) is:

$$P(\text{bar}) = 1.4 \times 10^{-3} \rho^2 (R^2 - z^2) \qquad (4.1)$$

where R (in km) is the radius of the body and z (in km) is the distance of the sample from the center. Ceres, with a density of 2.14 g cm^{-3} has a central pressure of ~ 2 kbar.

This equation cannot be applied to the terrestrial planets, and without direct measurements of the distributions of temperature, pressure and composition in the planetary interiors, estimates for the uncompressed densities have larger uncertainties.

The moments of inertia describe the mass distribution within a planet and give information about material compression for the crust, mantle and core. These have only been determined for the Earth, Moon and Mars, and are not well-known for Mercury, Venus and Vesta. A perfect sphere of uniform composition has a moment of inertia $C/MR^2 = 0.4$ (C is the polar axis). For reference, a sphere with a total density ρ_{tot} and total radius R, made of a core with density ρ_C and radius r_C, and a surrounding silicate mantle layer with density ρ_M, has:

$$C/MR^2 = \tfrac{2}{5}\left[\rho_M/\rho_{tot} + (1 - \rho_M/\rho_{tot})(r_C/R)^2\right] \tag{4.2}$$

Second-order correction terms (such as the gravitational coefficients J_2 and C_{22}) are for the density distribution and the changes in the gravitational potential of a rotating planet with flattened poles, which is of considerable interest for keeping artificial satellites in orbit. In this context, the historical debate about the figure of the Earth given below may be interesting to the reader. The anomalies in mass distributions that can be caused by impacts or by melting of polar ice shields are also covered by these coefficients. Seismological studies constraining the interior structures are restricted to the Earth and Moon (only one seismic event was registered by the Mars seismometer on the lander during the Viking mission). We only have a fair idea about the interior structure of the Earth (Figure 4.1) and, possibly, the Moon.

Magnetic fields are detected on the Earth and Mercury, but not on Venus or Mars. Among the planets, the Earth has the strongest magnetic field, which is generated through electric currents created by convective flows in the liquid outer core. Although Mercury is the smallest and most slowly rotating planet, it possesses a weak magnetic field. A thin outer layer of the Mercurian core may be kept liquid through tidal heating, and the magnetic field is probably created within this layer. The Galilean moon Ganymede, of a similar size to Mercury, has a magnetic field that is probably driven by tidal heating of its silicate mantle.

Very early Mars (*i.e.*, before $\sim$4.2 Ga ago) may have had a magnetic field when its core was (partially) liquid, but the magnetic field was presumably lost during the heavy bombardment period (4.5 to 3.8 Ga ago). The Utopia basin in the Martian northern hemisphere with a diameter of $\sim$3000 km is a remaining impact structure from that period,

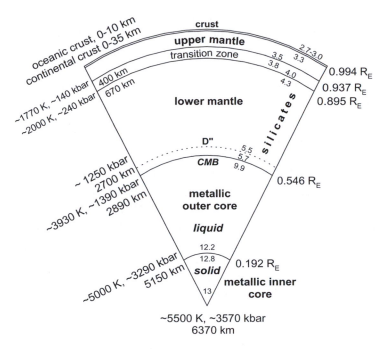

Figure 4.1 The Earth's interior. Numbers at the phase boundaries are approximate densities in g cm^{-3}. Three seismic discontinuities are located in the silicate mantle at about 400, 670 and 2700 km depth. The first two discontinuities come from changes in mineral phases. The nature of the discontinuity marked D″ is more uncertain but most likely corresponds to a change in chemical composition (CMB = core–mantle boundary).

and these ancient crustal rocks show no signs of residual magnetism, which should be there had these rocks crystallized in the presence of a magnetic field.

Venus, often taken as Earth's twin, does not possess a large detectable internally generated magnetic field. This is somewhat surprising if the Venusian core is similar to that of the Earth's. Only a small solar-wind induced magnetic field was detected by the Pioneer Venus Orbiter in 1980. The lack of a comparable magnetic field on Venus is a major difference from the Earth, as is the lack of plate tectonics on Venus.

Venus' rotation rate may be simply too slow to sustain strong convection in the core to create a dynamo-driven field. The absence of such a field suggests that the core of Venus is either completely liquid or completely solid. For a dynamo field, the configuration of a solid inner core and an outer liquid core is required. Without a crystallizing solid inner core, there is no heat source driving convective flows in the outer liquid core. Therefore, no electric currents to build a magnetic field result. The core may be entirely liquid because the planet has not cooled enough. The lack of plate tectonics hinders cooling of the planet, and a hotter interior may prevent a dynamo from forming. The presence of a liquid core is facilitated if larger amounts of light alloying elements in the core can suppress the melting point. If the core is entirely solid, no dynamo effect results either. A solid FeNi core would be favored if the amounts of light alloying elements are minimized.

The lack of a magnetic field can have serious consequences. On Venus, it may have facilitated loss of H from the atmosphere. Photochemical dissociation of water above the Venusian clouds liberates H (see Venus' atmospheric chemistry in Section 4.4). Without a magnetic field shielding the planet, solar wind ions can easily impact and interact with the atmosphere and pull H from it. Interestingly, the lack of plate tectonics may be due to the very dry Venusian mantle silicates. This implies that once a magnetic field dissipates atmospheric water loss is enhanced, which in turn dries out the planet over a long time. With a hot interior and lack of plate tectonics, degassing of water and other volatiles from the mantle silicates should have been expedited.

Given that there are not many well-defined constraints, the models of the interiors of Mercury, Venus and Mars are non-unique. Figure 4.2 displays some plausible models considering the physical properties mentioned above and results from geochemical modeling (see below).

Much more information is available about the Earth. The following sub-section describes some of the efforts to decipher the Earth's figure, density and the recognition and the nature of the Earth's core. After that, we continue with the chemical properties of the large inner solar system objects.

4.2.1.1 Figure, Density and Structure of the Earth.   Finding out what the Earth is made of has occupied philosophers of science since ancient times. The problem is that to infer something intelligent about the nature of the planet requires observational data, preferentially in quantitative form. We start our exploration of the Earth's interior in the Renaissance and note in advance that many erroneous deductions along the way could have been prevented had the different branches of science interacted more.

Some of the basic physical data of the Earth are its size, figure, mass and density (Table 4.1), but the efforts required to obtain the data are easily forgotten when one opens any handbook or textbook to look up the numbers. The phenomenon of magnetism can be added to the early observations, as it is directly related to the composition of the Earth's

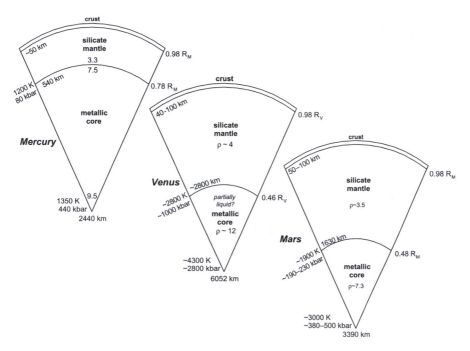

Figure 4.2   Model interiors of Mercury, Venus and Mars. Numbers at the phase boundaries are approximate densities in g cm^{-3}.

interior and the presence of a metallic core. Diagrams of physical models for the Earth that have been proposed over time are shown in Figure 4.3; Figure 4.1 gives a more detailed summary of the structural features of the Earth.

We begin with William Gilbert (1544–1603), who was first to point out that the Earth itself is a huge magnet and must have large amounts of iron in its interior. In his *De Magnete* (published in 1600), Gilbert states:

" . . . That the centre of the magnetic forces in the Earth is the centre of the Earth"

and

"Magnetic bodies are governed and regulated by the Earth and they are subject to the Earth in all their movement. All the movements of the loadstone are in accord with the geometry and form of the Earth and are strictly controlled thereby . . . "

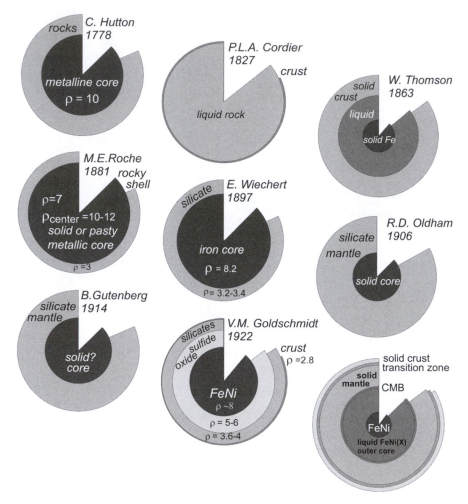

Figure 4.3 Historical development of interior structure models for the Earth.

Gilbert then asked where the substance causing the magnetism is located since it is not found among the surface materials:

"Yet we do not hold the whole interior of this our globe to be of rock or of iron, albeit the learned Franciscus Maurolycus' deems the Earth in its interior to consist throughout of rigid rock."

This brings us to the mathematician Franciscus Maurolycus (1494–1575) who was concerned with the figure of the Earth. In his *Cosmographia* from 1543 he devised a method to determine the polar radius of the Earth by measuring the arc of a meridian. This method was applied some 120 years later by Jean-Felix Picard (1620–1682). Picard used large-scale equipment to make fine scale resolution measurements of an arc of a meridian of the Earth between Paris and Amiens in France. Per one degree of latitude, he found 57060 toises or 111.21 km (before 1812, one toise = 1.949 m). This gave 111.21 km × 180°/π = 6371.9 km, which Picard reported in his *Mesure de la Terre* from 1671. The currently accepted value for the polar radius is 6356.753 km, so Picard was only off by ∼0.23%.

Several other surveys and measurements of the polar and equatorial radius of the Earth followed, and the question of the true figure of the Earth extended into the nineteenth century. There was consensus that the Earth is a spheroid, but the divided schools of thought had yet to agree on whether it is oblate or oblong. The figure is described by the lengths of the polar and equatorial axes and is of serious interest for mapping and navigation. By 1830, the English astronomer Sir George Airy (1801–1898) had carried out investigations of the axes lengths from measurements of surface arcs as a function of latitude and longitude. The German astronomer Friedrich Wilhelm Bessel (1784–1846) did the same and produced similar values in the 1840s. These values dominated the scene for the next 25 years, hindering progress as both data sets were found to be in error in 1866 by the British surveyor Alexander Clarke (1828–1914). Airy's and Bessel's values agreed within 150 feet but Clarke's data gave a 1/5 mile longer equatorial semi-axis and a 3/10 mile longer polar semi-axis. This may not sound like much, but since the radius enters into the arc of the distance of two surface points, it does matter on map projections.

Members of the Newtonian school like Clairaut, Huygens and Pierre Louis Maupertuis defended the view of an Earth flattened at the poles. The members of the Cassinian school insisted that the Earth is elongated. They based their reasoning on measurements of surface arcs from which the axes' lengths are easily derived, had these measurements been correct. To settle the issue, two parties of mostly mathematicians and astronomers were sent to measure a meridian arc at the Arctic circle and at the equator. During five months in 1736 the team with Pierre Louis Moreau de Maupertuis (1698–1759), Charles Camus (1699–1768), Anders Celsius (1701–1744), Alexis Claude de Clairaut (1713–1765), Pierre Charles Le Monnier (1715–1799) and Reginald Outhier (1694–1774) re-measured the Lapland meridional arc that was one of the pillars of the Cassinian hypothesis. With the new measurement of the radius at the arctic, the Cassianian theory crumbled. Voltaire applauded his friend Maupertuis on having "flattened the poles and the Cassinis" (quoted in Woodward 1889[1]).

The density determination of the Earth was subject to numerous investigations over the centuries, as were models of the internal state of the Earth. In 1687, Sir Isaac Newton (1642–1727) concluded in his Principia:[2]

"....the common matter of our Earth on the surface thereof is about twice as heavy as water, and a little lower, in mines, is found about three, or four, or even five times more heavy, it is probable that the quantity of the whole matter of the Earth may be five or six times greater than if it consisted all of water; especially since I have before shewed that the Earth is about four times more dense than Jupiter. If, therefore, Jupiter is a little more dense than water, ..."

However, the absolute value for the mean density of the Earth from the density ratio was at least as uncertain as the value for Jupiter; so Newton came up with a range of 5–6 g cm^{-3} for the Earth. The mean density of Jupiter is ~ 1.3 g cm^{-3}, so Newton's estimate for the terrestrial/Jovian density ratio of 4 leads to 5.2 g cm^{-3} for the density of the Earth. This is close to the modern mean density value of 5.515 g cm^{-3} for the Earth. However, Newton also found a Moon/Earth density ratio of 11/9, making the Moon much denser than it actually is.

Until 1743, models of the interior of the Earth widely assumed that it was a simple fluid. This began to change when Clairaut presented his results on the *Theory of the Figure of the Earth*. He added the concept of compressibility to the existing modeling of a fluid under the influence of gravitation. This led to a description of the Earth's gravity field and to consequences from fluid mechanics such as density gradients through hydrostatic pressure and ellipticity. It provided a tool to estimate the pressure conditions in the Earth's interior, which are not accessible by direct measurement. By 1784, Pierre-Simon Laplace (1749–1827) had expanded on Clairaut's results and introduced spherical harmonics in the solution to the equations. This described a "layered" density distribution in a spheroid that increases from surface to the center, arranged symmetrically about the center of the Earth. All this still assumed a fluid interior. The model of Laplace showed that the crushing strength of steel is reached a few km below the surface, and that the density varies from ~ 3 to ~ 11 g cm^{-3} towards the center; where a total pressure of ~ 3000 kbar is reached. Current estimates of the central pressure are 3570 kbar.

In 1778, the English mathematician Charles Hutton (1737–1823) set out to determine the density of the Earth from a survey and plumb deflection measurements by the English astronomer Nevil Maskelyne (1732–1811) at Schehallien, Scotland. This would give the density ratio of the Earth to that of the hill at Schehallien. Hutton alludes to the painstaking calculations and conveys a flair for them in his article, but he finally obtained an Earth to hill density ratio of $17804 : 9933 \approx 1343 : 800 \approx 9 : 5$. The next dilemma was to find the density of the hill, and Hutton assumed a density of $\sim 2.5\,\mathrm{g\,cm^{-3}}$ for common stone, leading to a mean density of the Earth of $\sim 4.5\,\mathrm{g\,cm^{-3}}$. So the entire Earth density was about twice the density of the surface rocks, and Hutton concluded that "there must be somewhere in the Earth, towards more central parts, greater quantities of metals." Adopting a density of $10\,\mathrm{g\,cm^{-3}}$ for metal, he finds that it:

"requires 16/27, or a little more than one-half of the matter in the whole Earth to be metal of this density . . . or 4/15 or between 1/3 and 1/4 of the whole magnitude will be metal; and consequently, 20/31 or nearly 2/3 of the diameter of the Earth, is the central or metalline part."[3]

His result is an overestimate of the core dimension and mass since it was based on an incorrect result for the total density of the Earth. However, it reminds us that the Earth is not homogeneous in chemical composition throughout, and that there is a need for a metallic core within the Earth.

Hutton's view on the metallic interior of the Earth was fully entertained by the father of meteorite studies, Ernst F. Chladni in 1794, who likened the Earth's composition to that of meteorites. A larger fraction of the Earth's mass in the form of Fe metal is consistent with the common occurrence of Fe metal in meteorites. Among other similarities, Chladni noted that Fe is common not only in the surface of the Earth, but also in plants and animals, a theme later picked up again by Eli de Beaumont (Chapter 1). Chladni reiterated Gilbert's results that the magnetic properties of the Earth suggest the presence of a considerable "stock" of metal in the Earth's interior.

In 1799, Henry Cavendish (1731–1810) reported the results of a density measurement of the Earth from his well-known torsion balance experiment as $5.48\,\mathrm{g\,cm^{-3}}$, which is only $\sim 0.6\%$ different from the accepted value of $5.515\,\mathrm{g\,cm^{-3}}$. Several other density determinations by pendulum measurements in mines lead to a wide spread of density values, ranging from ~ 4.7 to $6.5\,\mathrm{g\,cm^{-3}}$.

Over time, the temperature within the Earth caught the attention of several researchers. The French mathematician Jean Baptiste Joseph

Fourier (1768–1830) devised equations to address the temperature of the Earth:

> "The question of terrestrial temperatures has always appeared to us of the grandest objects of cosmological studies, and we have had it constantly in view in establishing the mathematical theory of heat" (quoted in Woodward, 1889[1]).

Although he could not solve the issue entirely, his work *Analytical Theory of Heat* from 1820 was a big step ahead. Fourier's theory on thermal conductivity led to the result that the Earth cooled exponentially. Fourier preferred a solution indicating that the cooling Earth had reached a uniform heat distribution through convection throughout its mass, and that the crust cooled first.

In 1827, Pierre Louis Antoine Cordier (1777–1861), who built up the geology and meteorite collection at the Musée Nationale d'Histoire Naturelle in Paris, was inspired by Fourier's work. He had noticed that the temperatures in mines increased by about 25 degrees per kilometer of increasing depth, which upon extrapolation, and not considering pressure effects, led him to conclude that rocks at greater depths (below ~ 50–60 km) should be entirely melted. He assumed that the entire interior of the planet was fluid. In 1839, the English mathematician William Hopkins (1793–1866) showed that once a hot liquid sphere cools, it may not form a solid crust first, as assumed by previous researchers. He argued that high pressures within the Earth lead to solidification long before a solid crust might form. Hopkins analyzed how motions of a completely fluid Earth affect the precession and nutation movements. Precession describes the oscillation of the Earth's spin axis relative to the pole of the ecliptic. It takes about 26 000 years to complete an oscillation about this inertial space. Nutation is a short superimposed motion. He found that a completely fluid Earth with a crust 20–30 miles thick cannot account for the observations and emphasized that pressure effects must be considered. He deduced a thickness of 800–1000 miles for the continents and suggested that a solid center is required. Also in 1839, the German mathematician Carl Friedrich Gauss (1777–1855) proved that the magnetic field must be internal to the Earth, as had been suspected by Gilbert. He deduced this from a spherical harmonic analysis of the magnetic field observations.

By 1848, Edouard Roche (1820–1883) had derived that the rate of Earth's density increase varies as the radius squared and he found a solution for the interior structure that satisfies the mean density, surface density, ellipticity and precession. His solution involved a solid Fe-core of 5/6th of the Earth's radius, and crust of 1/5th of the radius. In 1850, A. Boisse proposed that meteorites, sorted by their specific density, resemble the internal structure of the Earth.

In 1863 William Thomson (Lord Kelvin, 1824–1907) investigated the Earth's interior and rigidity in order to explain the observed amounts of nutation and precession, and effects on tides. At the time, Hopkin's earlier work had not made much impact and many geologists still assumed that the entire Earth was composed of a liquid except for a solid crust. However, Thomson confirmed that this structure was inconsistent with the nutation and precession movements of the Earth.

Another constraint on the Earth's structure is the deformation of the Earth caused by lunar and solar tidal actions. These are smaller than expected if the Earth were perfectly fluid. Thomson mentioned that an almost completely fluid planet would be at odds with the velocities of tidal waves and earthquake waves. At the time, it was known from earthquake waves that the Earth's rigidity and resistance to compression below the near surface is much less than that of iron, but that these properties are much greater several hundred miles below the surface. Thomson found that gravitational effects on rigidity are quite important for describing the Earth's shape, and, unless the average matter was more rigid than steel, tidal forces of the Moon and Sun would distort the Earth considerably.

Thomson came up with a model of a solid, sunken iron core of 2000 miles in diameter that was magnetic to within 100 miles of its surface. This was surrounded by a lighter liquid 3000 miles in diameter, which in turn was encased by the solid crust of 2500 miles thickness. Given the rigidity of rocks on the surface, a solid crust of at least 2500 miles thick was needed because otherwise, the crust's rigidity was too low and it would respond to the Moon's tidal attractions to:

" . . . simply carry the waters of the ocean up and down with it, and there would be no sensible tidal rise and fall of water relatively to land." (Thomson 1867).[4]

In a footnote, Thomson (1863)[5] remarks about the iron nucleus:

"An ancient cold iron meteorite which may have entered a nebula of smaller bodies and formed the nucleus of our present Earth, which under such circumstances could not but be built up and heated by attracting them to itself" which is an interesting precursor of later heterogeneous accretion theories of the Earth (see below).

In 1881, shortly before his death, Roche investigated the compressibility and the interior structure of the Earth again.[6] Like Thomson in 1863, Roche found that a completely fluid Earth with a thin crust is not consistent with the measurements of the flattening of the Earth and with its precession. Roche favored an Earth model "of a nucleus or solid mass very nearly homogenous, covered with a lighter shell whose density, from geological considerations, can be estimated as 3 with respect to water." The nucleus (what we now call the core) had an average density of $7 \, \text{g cm}^{-3}$, which may reach $10–12 \, \text{g cm}^{-3}$ at the center. The silicate shell with $3 \, \text{g cm}^{-3}$ was less than $1/6^{\text{th}}$ of the entire radius. Roche's Earth model was influenced by meteorites, since he noted:

> "the central terrestrial mass is therefore in specific weight analogous to meteoritic iron, while the stratum that envelops it is comparable to aerolites of a stony nature, where iron enters only in a small proportion."

Roche's model satisfied the flattening and precession values available at that time, provided that "the nucleus of the globe solidified and has taken its definitive form under the influence of a rotation less rapid than that with which the Earth is now animated." Progressive cooling could explain the precession and that the observed flattening was higher than expected for a completely fluid interior. Roche noted that an increase in angular velocity must also follow from the contraction upon cooling.

Towards the beginning of the twentieth century, seismologists set out to prove the existence of the Earth's core. The interior structure of the Earth is traceable through the velocities of reflected and refracted seismic waves resulting from earthquakes, strong detonations and larger meteorite impacts. Emil Wiechert (1861–1928) devised valuable quantitative analytical tools to decipher the information about the Earth's interior in seismic waves. He called for the existence of an iron core because a density increase of silicates between 3.2 and $3.4 \, \text{g cm}^{-3}$ alone could not account for the density of $5.53 \, \text{g cm}^{-3}$ (which was one of the Earth's density values adopted at the time). To match the density and the Earth's moment of inertia, Wiechert (in 1897) found that a core of about 5000 km (0.78 that of the Earth's radius) with a density of ~ 8 g cm^{-3} was needed below a rocky mantle of ~ 1400 km thickness.

Oldham (1858–1936) thought along the same lines as Wiechert, and in 1899 derived a core radius of 0.55 that of the Earth's total radius, taking density and moment of inertia as constraints. By 1906, Oldham made an initial model for the seismic data and found a core of ~2500 km in radius, 0.4 times that of the entire Earth. By 1914, Beno Gutenberg (1889–1960) suggested that the discontinuity discovered by Oldham at 2900 km depth corresponds to a transition to a solid core of 0.54 times (~3500 km) the radius of the Earth, which is close to the currently established value.

Up to 1914, most researchers had modeled the Earth as entirely solid; the other extreme to an essentially liquid Earth favored in the older days. However, the truth is in-between. In 1926, Harold Jeffreys (1891–1989) demonstrated that the core must consist of a fluid metallic medium because seismic shear waves, in the absence of rigidity, are not transmitted through liquids and because the overall rigidity of the Earth is less than that of the silicate mantle. This could only be found after the rigidity of the Earth's mantle was determined from seismic data.

The solid inner core of 1220–1230 km in radius (~19% of the Earth's radius) was discovered by Inge Lehmann (1888–1993) in 1936. A solid inner core is required to account for the occurrence of certain transmitted shear waves, which was determined by Gutenberg and Richter in 1938. By 1946, Francis Birch and K. E. Bullen had clearly demonstrated that the inner core is indeed solid. For more on this subject, see for example, Brush,[7] Bullen[8] and Helffrich and Wood.[9]

4.2.2 Chemical Properties

Chemical information on planetary materials is plentiful, in particular for the Earth, where we have the best access to samples. Analysis of crustal and mantle derived rocks yield comparatively consistent estimates of the composition of the Earth's crust (and hydrosphere) and its mantle. From this, the representative element inventory of the "fertile," "primitive" or unfractionated silicate mantle – before it differentiated into the present mantle and the crust – can be derived. The elemental abundances of the siderophile and chalcophile elements in the primitive mantle provide indirect insights into core composition and core formation processes in the Earth.

The abundances of the elements in other planetary objects can be estimated when rocks from them are available for study. Table 4.2 gives an overview of the materials and the means by which they have been investigated.

4.2.2.1 Silicate Portion of the Earth. The composition for the entire silicate portion of the Earth, also called the "bulk silicate Earth (BSE)," is derived from the compositions of the crust, present mantle and

Table 4.2 Exploration of planetary materials.

	Surface/crustal rocks	*Mantle derived rocks*
Mercury	Spectroscopy	None
Venus	Space probe & landers	None
Earth	Field samples	Field samples
Moon	Spectroscopy, space probe & landers, field work, samples, lunar meteorites	Space probe, field samples, lunar meteorites
Mars	Spectroscopy, space probes & landers	SNC meteorites
Vesta	Spectroscopy	EHD meteorites

hydrosphere by computing a weighted average of the total of their compositions. For this purpose, the composition and the total mass fractions of the crust and mantle must be known. The largest lithophile element reservoir by mass is the silicate mantle. The BSE abundances have been evaluated by several authors and the data are within reasonable agreement among the different groups.

Igneous rocks found in the crust and mantle formed from magmas in two different settings, which lead to characteristic differences in their texture (grain size, shape and mineral arrangements). The first setting is volcanism, where completely molten or partially molten magma extruded onto the surface. This gives rocks with relatively small mineral sizes because of relatively rapid cooling. The rocks formed from intrusion of such melts into near surface rock may count as igneous volcanic (hypabyssal) rocks if crystallization is fast. The second setting is subsurface consolidation of melts and relatively slow crystallization. Rocks from melt intrusion possess larger mineral crystals and are called "intrusive" or "plutonic" rocks. Much of the ground-breaking work on determining rock compositions was done by Clarke and Washington at the US Geological Survey in the early 1900s. Their *Data on Geochemistry* published from 1924 onward remain a valuable reference of rock analyses.[10]

A first-order distinction of the different magmatic rocks is their content of lighter and darker minerals. The dark minerals, collectively called "mafic" minerals, include pyroxenes, olivine, magnetite, titanite, ilmenite, amphibole, biotite, garnet and apatite. The light-colored, "felsic" minerals are mainly feldspars and quartz. To sort magmatic rocks, the volume proportions of the mafic and felsic minerals (modal mineral content) are determined first. Rocks with more than ~85–90 vol.% mafic minerals are called "mafic" rocks, those with more than 90 vol.% are called "ultramafic" rocks.

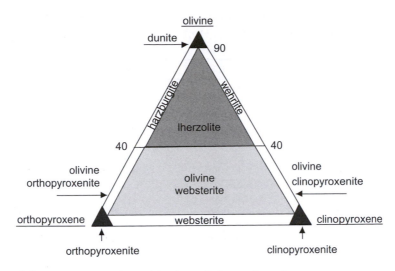

Figure 4.4 Compositional classification of ultramafic rocks.

The ultramafic rocks are the major constituents of planetary silicate mantles. These rocks are distinguished by their contents of olivine, orthopyroxene and clinopyroxene as shown in Figure 4.4. Ultramafic rocks are known from the Earth, the Moon and among the SNC meteorites, which likely represent samples from Mars.

Composition of the Earth's Crust. The crust is divided into the oceanic ($\sim 35.7\%$ by mass) and continental crust ($\sim 64.3\%$ by mass). The average composition of the crust is derived through geologic mapping of the volume of rock types present in different areas of the world and taking accordingly weighted averages of the rock compositions. Table 4.3 lists compositions for the crust and its components.

Only about a third of the continental volume – the upper ~ 12 km of the continental crust – is accessible, and the lower continental crust and the oceanic crust must be explored otherwise. The average thickness of the continental crust is ~ 35 km, that of the oceanic crust is 5–10 km. Ongoing crustal recycling by plate tectonics, re-surfacing by volcanism and erosion erased much of the earliest terrestrial rock record that had survived the late planetary bombardment period. In 1832, H. T. de la Beche (1796–1855), the first director of the British Geological Survey, noted to his reader:[15] " . . . indeed he may consider that no considerable portion of the Earth's surface has ever remained long, geologically speaking, in a state of rest; . . . " Recycling makes old crust material quite rare, but it is of interest to know how early the crust separated from the silicate mantle. For example, granitic rocks (pegmatites) in up to 200 km deep remnant crust fragments ("cratons") found on the

Table 4.3 Composition of the Earth's crust and its present silicate mantle (ppm by mass). Sources: Hofmann,[11] Wänke *et al.*,[12] Wedepohl,[13] and Taylor and McLennan.[14]

Element	Crust + hydrosphere	Continental crust	Oceanic crust	Present mantle
H	9 140	2 200		
Li	17	18	10	2.07
Be	2.2	2.4	0.5	0.1
B	10.6	11	4	0.8
C	1 870	2 000		24
N	68	60		
O	497 600	472 000	444 400	449 000
F	490	525		16.3
Na	22 760	23 500	20 335	2 745
Mg	20 640	22 000	46 045	222 200
Al	74 350	79 600	80 800	21 700
Si	269 000	28 8000	233 400	213 100
P	709	757		60
S	710	697		122[a]
Cl	1 720	472	174	0.5
K	20 000	21 400	1 067	127
Ca	36 000	38 400	80 800	25 000
Sc	15	16	39.685	16.9
Ti	3 750	4 010	9 340	1 320
V	92	98	250	81.3
Cr	120	126	270	3 010
Mn	670	716	1 000	1 016
Fe	40 350	43 200	81 335	58 600
Co	22	24	47.035	105
Ni	52	56	142.25	2 108
Cu	23	25	80.2	28.2
Zn	61	65	85	48
Ga	14	15	17	3.7
Ge	1.3	1.4	1.5	1.31
As	1.6	1.7	1	0.14
Se	0.11	0.12	0.16	0.012 6
Br	5.3	1	0.4	0.004 6
Rb	73	78	1.731	0.276
Sr	312	333	121.6	26
Y	22	24	33.91	2.9
Zr	190	203	92.12	7.8
Nb	18	19	2.8535	0.6
Mo	1	1.1	1	
Ru	0.000 1	0.000 1	0.001	
Rh	0.000 06	0.000 06	0.000 2	
Pd	0.000 4	0.000 4		
Ag	0.065	0.07	0.026	0.002 51
Cd	0.093	0.1	0.13	0.025 5
In	0.047	0.05	0.072	0.018 1
Sn	2.2	2.3	1.391	
Sb	0.28	0.3	0.017	0.004 5
Te	0.004 7	0.005	0.003	0.019 9
I	1.5	0.8	0.008	0.004 2

Table 4.3   (*Continued*).

Cs	3.2	3.4	0.022 05	0.001 44
Ba	545	584	19.435	2.4
La	28	30	3.7975	0.35
Ce	56	60	11.750 5	1.41
Pr	6.3	6.7	1.937	
Nd	25	27	10.589 5	1.28
Sm	5	5.3	3.526	0.49
Eu	1.2	1.3	1.3175	0.18
Gd	3.7	4	4.838 5	0.69
Tb	0.61	0.65	0.877 5	0.12
Dy	3.6	3.8	6.002	0.73
Ho	0.75	0.8	1.321	0.17
Er	2	2.1	3.921 5	0.44
Tm	0.28	0.3	0.580 5	0.047
Yb	1.9	2	4.5	
Lu	0.33	0.35	0.574 5	0.071
Hf	4.6	4.9	2.737	0.26
Ta	1	1.1	0.246	0.012 6
W	0.93	1	0.5	0.016 4
Re	0.000 4	0.000 4	0.000 9	0.000 23
Os	0.000 05	0.000 05	0.000 004	0.003 1
Ir	0.000 05	0.000 05	0.000 02	0.002 8
Pt	0.000 4	0.000 4	0.002 3	
Au	0.002 3	0.002 5	0.000 23	0.000 5
Hg	0.037	0.04	0.02	
Tl	0.49	0.52	0.012	
Pb	13.8	14.8	0.644 5	
Bi	0.079	0.085	0.007	
Th	7.9	8.5	0.203 5	0.056 7[b]
U	1.6	1.7	0.085 5	0.021

[a]See table 4.4 for S.
[b]A Th/U of 2.7 is adopted here.

Canadian shield, Australia, the south African continent or in India range from 2.7 to 3.4 Ga in age.

On average, the continental crust is 960 Ma old and the oceanic crust $\sim$60 Ma, with the oldest oceanic crust extending to $\sim$200 Ma. The oldest samples of the Earth's continental crust are detrital zircons (ZrSiO$_4$). These are about 100–200-micron-sized grains found in metamorphic gneiss rocks from Jack Hills in Western Australia. The zircons are 3.8 to 4.4 Ga old and are older than the gneiss, which formed not before 3.8 Ga ago.[16] Analyses of zircons from the Acasta Gneiss in the Canadian Northwest Territories gave ages up to 4.03 Ga.[17] Zircons are more resistant to weathering than other minerals in the ancestral rock and were left behind to be later incorporated into the gneiss. The formation of zircons required water. Therefore, liquid water and a continental crust must have been present within 150–200 Ma of the Earth's formation. This rapid development places constraints on the thermal history of the Earth.

Currently the oldest entire rocks (not only residual minerals) known are from the Nuvvuagittuq greenstone belt in northern Quebec. Their age is 4.28 Ga, from ^{146}Sm-^{142}Nd dating.[18] This age is under debate and may not date the rocks themselves but the magma from which the rocks formed. Conventional dating with long-lived radionuclide chronometers gives ages between 3.6 and 3.8 Ga, which places the formation of these Greenstone belt rocks at a time shortly after the peak of the large bombardment period.

The upper continental crust is mainly made of granitic rocks, and the lower continental crust and oceanic crust contain gabbroic and basaltic rock types. Characteristic types of intrusive rocks in the Earth's upper continental crust (in order of decreasing abundance) are granites (most common), quartz monzonites, granodiorites, gabbros, quartz diorites, diorites, anorthosites, peridotites, syenites and alkalic rocks (least common). Characteristic types of volcanic rocks in the upper crust are tholeiitic and alkali-olivine basalts, andesites, dactites, rhyolites and trachytes. Chemically, the basaltic oceanic crust contains more Mg, Ca, Fe and Ti when compared to the continental crust, which contains more Al, Si, alkali elements and Th and U (Table 4.3).

Composition of the Earth's Silicate Mantle. The chemical composition of the Earth's silicate mantle is derived from analyses of relatively unaltered rocks thought to come from the mantle, petrological models of peridotite-basalt melting and element ratios (*e.g.*, refs 19–22). Table 4.4 gives the composition of the silicate mantle obtained from the different geochemical modeling approaches. All accessible mantle samples come from the upper mantle and one still unresolved question is whether the upper mantle's chemical composition is representative of the entire mantle or if the lower mantle (which is about 68% of the entire mantle) is compositionally different. In the latter case, using upper mantle compositions to derive the composition of the silicate portion must be incorrect. Although the mineral phases present are different in the lower and upper mantle because of total pressure effects, any differences in chemical composition may become minimized over time through convection. Most studies adopt the upper mantle elemental composition as representative for the entire silicate mantle.

For the Earth, we have to distinguish the "present mantle" and the "primitive silicate mantle." The primitive mantle is the entire silicate portion (BSE) after core formation but before differentiation into

Table 4.4 Composition of the primitive and present-day silicate mantle of the Earth (ppm by mass).

Element	Primitive mantle (=bulk silicate Earth, BSE)				Present mantle
	PO03	*MS95*	*KL93*	*WDJ84*	*WDJ84*
H	120		54.7		
Li	1.6	1.6	1.71	2.15	2.07
Be	0.07	0.068	0.077 5		0.1
B	0.26	0.3	0.433		0.8
C	100	120	65	46.2	24
N	2	2	0.88		
O	443 300		444 200		
F	25	25	20.7	19.4	16.3
Na	2 590	2 670	2 932	2 890	2 745
Mg	221 700	228 000	220 100	222 300	222 200
Al	23 800	23 500	21 230	22 200	21 700
Si	212 200	210 000	216 100	214 800	213 100
P	86	90	79.4	64.5	60
S	124[a], 200	250	274	13.2	8,122
Cl	30	17	36.4	11.8	0.5
K	260	240	232.4	231	127
Ca	26 100	25 300	25 300	25 300	25 000
Sc	16.5	16.2	16.4	17	16.9
Ti	1 280	1 205	1 180	1 350	1 320
V	86	82	90	82.1	81.3
Cr	2 520	2 625	2 905	3 010	3 010
Mn	1 050	1 045	1 057	1 020	1 016
Fe	63 000	62 600	62 690	58 900	58 600
Co	102	105	104.6	105	105
Ni	1 860	1 960	1 948	2 110	2 108
Cu	20	30	31.2	28.5	28.2
Zn	53.5	55	53.9	48.5	48
Ga	4.4	4	4.29	3.8	3.7
Ge	1.2	1.1	1.15	1.32	1.31
As	0.066	0.05	0.17	0.152	0.14
Se	0.079	0.075	0.044 3	0.013 5	0.012 6
Br	0.075	0.05	0.035 2	0.045 6	0.004 6
Rb	0.605	0.6	0.598	0.742	0.276
Sr	20.3	19.9	20.7	27.7	26
Y	4.37	4.3	3.91		2.9
Zr	10.81	10.5	11.47		7.8
Nb	0.588	0.658	0.765		0.6
Mo	0.039	0.05	0.059 3		
Ru	0.004 55	0.005	0.004 23		
Rh	0.000 93	0.000 9	0.001 31	0.001 18	
Pd	0.003 27	0.003 9	0.004 63		
Ag	0.004	8×10^{-8}	0.008 45	0.002 92	0.002 51
Cd	0.064	0.04	0.031 7	0.026 1	0.025 5
In	0.013	0.011	0.006 9	0.018 5	0.018 1
Sn	0.138	0.13	0.28		
Sb	0.012	0.005 5	0.008	0.005 7	0.004 5
Te	0.008	0.012	0.016 1	0.019 9	0.019 9
I	0.007	0.01	0.010 7	0.013 3	0.004 2

Table 4.4 (*Continued*).

| Element | Primitive mantle (= bulk silicate Earth, BSE) | | | | Present mantle |
	PO03	MS95	KL93	WDJ84	WDJ84
Cs	0.018	0.021	0.013 1	0.009 14	0.001 44
Ba	6.75	6.6	6.41	5.6	2.4
La	0.686	0.648	0.622	0.52	0.35
Ce	1.786	1.675	1.592	1.73	1.41
Pr	0.27	0.254	0.242		
Nd	1.327	1.25	1.175	1.43	1.28
Sm	0.431	0.406	0.36	0.52	0.49
Eu	0.162	0.154	0.145	0.188	0.18
Gd	0.571	0.544	0.529	0.74	0.69
Tb	0.105	0.099	0.095 5	0.126	0.12
Dy	0.711	0.674	0.656	0.766	0.73
Ho	0.159	0.149	0.146	0.181	0.17
Er	0.465	0.438	0.43	0.46	0.44
Tm	0.071 7	0.068	0.062 7		0.047
Yb	0.462	0.441	0.437	0.49	
Lu	0.071 1	0.067 5	0.064 8	0.074	0.071
Hf	0.3	0.283	0.3	0.28	0.26
Ta	0.04	0.037	0.041 2	0.025 6	0.012 6
W	0.016	0.029	0.012	0.024 1	0.016 4
Re	0.000 32	0.000 28	0.000 293	0.000 236	0.000 23
Os	0.003 4	0.003 4	0.004 05	0.003 106	0.003 1
Ir	0.003 2	0.003 2	0.003 25	0.002 81	0.002 8
Pt	0.006 6	0.007 1	0.011 2		
Au	0.000 88	0.001	0.001 03	0.000 524	0.000 5
Hg	0.006	0.01	0.001 8		
Tl	0.003	0.003 5	0.010 4		
Pb	0.185	0.15	0.149		
Bi	0.005	0.002 5	0.004 2		
Th	0.083 4	0.079 5	0.078 2		0.056 7[b]
U	0.021 8	0.020 3	0.022	0.029 3	0.021

[a]Recommended here. PO03: Palme and O'Neill,[23] MS95: McDonough and Sun,[61] Kargel and Lewis,[21] WDJ84: Wänke *et al.*[12]
[b]A Th/U of 2.7 is adopted here.

the present mantle and crust. The primitive mantle composition may still be represented by samples that have not experienced loss and fractionations of the elements by larger amounts of melt extraction. The criterion for such "unaltered" primitive mantle samples is based on the idea that the Earth (and other planets) contains refractory lithophile elements in chondritic proportions. Searches were carried out to find fertile mantle rocks (peridotites, spinel-lherzolites and garnet lherzolites) with approximate chondritic abundance ratios for refractory elements. Not all refractory elements have the same compatibility in the major mantle minerals olivine and pyroxene, and partial melting or preferential re-crystallization would have fractionated these elements from their originally chondritic ratio. The Ca/Al, Mg/Si and Al/Si abundance ratios and the rare earth element (REE) abundances are particularly diagnostic for such fractionations. Only a few such mantle nodules with chondritic Ca/Al ratios and weakly fractionated heavy REE are known. These have been studied in detail for their major and trace element contents to "directly" derive the primitive mantle concentrations for a larger suite of compatible elements. A detailed description has been given by Palme and O'Neill.[23]

To derive the mantle and planetary bulk silicate compositions from mantle derived rocks with fractionated abundances, the method of "element correlations" or "element ratios" has been applied to terrestrial, lunar and martian rocks, and achondrites. The element ratio method is based upon Goldschmidt's observations that elements with similar ionic radii and similar silicate/melt partition coefficients (*i.e.*, similar compatibility) fractionate to about the same extent during igneous processing. The absolute concentrations of a pair of elements may be different in a mantle source and in the igneous rock formed from it, but the relative concentrations of the two elements in the rock and mantle source remain approximately constant. If the absolute mantle concentration of one element in the pair can be determined independently, the mantle abundance of the second element can easily be calculated from the element ratio using the abundance of the first element of the pair.

The element ratio concept (*e.g.*, Hofmann[62]) utilizes two elements (1 and 2), with initial (c°) and final (c) concentrations, in a mantle source and melt. For example, we assume that an igneous rock formed by crystallization of the melt. If F is the fraction of melting (ranging from 0 for no melting to 1.0 for complete melting) and D_i is the silicate/melt partition coefficient by mass (see below) for either element in each phase, we can write:

$$\frac{c_1}{c_2} = \frac{c_1^0 F + D_2(1 - F)}{c_2^0 F + D_1(1 - F)} \qquad (4.3)$$

for the elemental concentrations in mantle source (c°) and melt (c). As long as $D_1 \approx D_2 < F$, then:

$$\frac{c_1}{c_2} = \frac{c_1^0}{c_2^0} \qquad (4.4)$$

and the concentration ratio in the melt and mantle source remain approximately constant. For example, highly incompatible elements that reside preferentially in the melt, have $D_i \approx 0.001$ and the approximation holds for all F values greater than about 0.01 (that is 1% melting or more). Larger degrees of melting are required for this approximation to be valid if concentration ratios of more compatible elements with larger silicate/melt partition coefficients are considered. For elements with $D_i \approx 0.01$, only 10% melting is needed for element ratios to remain constant in the source and melt.

Geochemical analyses and modeling of terrestrial rocks, lunar samples and basaltic achondrites (eucrites) have shown that several element pairs such as K/La, Ba/La and W/U display good positive correlations that imply relatively constant concentration ratios during melting and crystallization. This method allows us to derive the concentrations for volatile elements if these pairs are of elements with significantly different condensation temperatures but with similar geochemical properties. The K/La and K/U ratios can be used as a diagnostic of planetary volatile element depletion. The CI-chondritic (solar) K/U mass ratio is $\sim 67\,000$; in other carbonaceous chondrites and ordinary chondrite groups the K/U ratio ranges from $20\,000$ to $60\,000$. In contrast, terrestrial samples have K/U $\sim 10\,000$, lunar samples about ~ 2000. These differences point to incomplete accretion or loss of volatile K during the formation of these objects. Other element pairs with one volatile and one refractory element are Rb/La, Cs/La, Na/Ti, Zn/Fe, or Cl, Br and I relative to La.

Another diagnostic ratio is the Fe/Mn ratio, or better still the FeO/MnO ratio, which is useful to determine the amount of Fe as oxide from the total planetary Fe, which can be oxide, sulfide and metal. Both elements readily substitute for each other in olivine and pyroxene because they have the same valence and similar ionic radii. One problem with this ratio is that Fe and Mn have different condensation temperatures and the Fe/Mn ratio can be affected by volatility.

The element ratio method is ideal for deriving the abundances of the siderophile elements in the silicate portion of the Earth and of other objects. Elements like Ni, Co, Mo, W, Cr, Cu, Ag, Au, Ga, In, Tl, P, As, Sb and the platinum group elements can alloy with the metallic iron that is removed into the core. If present, iron sulfide can remove the chalcophile elements S, Se and Te.

The extraction of siderophile and chalcophile elements into the core is not complete for those elements that are more readily oxidized so that some fraction of the total element inventory remains in the silicate portion. Elements behaving this way are the moderately siderophile elements Ni, Co, Cu, Ga, Ge, P, Mo and W. Their concentrations can be derived from element ratios such as Ni/Mg, Co/Mg, Cu/Ti, Ga/Al, Ge/Zr, W/La, Mo/Nd and P/Tb. Several of these element ratios are often employed; however, some element ratios show more scatter than others, and an element ratio that is well defined in one object may show more scatter in samples of another.

The composition for the bulk silicate Earth evaluated by different authors and methods are in reasonable agreement (Table 4.4). In addition to the element concentrations derived from fertile mantle nodules, the composition of the entire silicate portion of the Earth, or primitive silicate mantle, is obtained from the weighted sum of the concentrations in the silicate mantle and crust.

The present silicate mantle represents 99.37% by mass of the entire silicate portion and is the overwhelming reservoir for most elements of the bulk silicate Earth. For the incompatible elements, the crust (including oceans) is the most important reservoir. Lithophile elements with high valences and/or large atomic radii such as K, Zr, Nb, Th and U preferentially enter the Si and Al-rich minerals in the continental crust. The heavy alkalis and the halogens are depleted in the present mantle, and these elements are readily dissolved in the oceans (one may call these elements aquophile). Figure 4.5 shows the distribution of some incompatible elements. The crust contains more than 50% of the U, K, Ba, Rb and Cs of the silicate portion of the Earth, and oceans

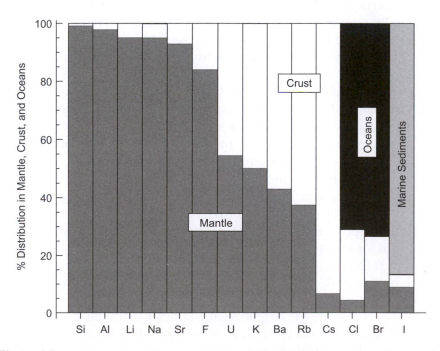

Figure 4.5 Distribution of alkalis, halogen and other lithophile elements between the oceans, crust and silicate mantle of the Earth.

are the major reservoir of Cl and Br. Most iodine is sequestered in marine sediments, which one may count as part of the oceanic reservoir. If we neglected the oceans and only use mantle and crust compositions to calculate the overall elemental abundances for the Earth, these abundances would come out too low. The hydrosphere may be unique to Earth now but oceans (of undefined size) on Mars and/or Venus may have existed in the past. Any elements that had been dissolved in such Martian or Venusian oceans still should reside on the surface. Interestingly, larger concentrations of halogens (Cl, Br) were detected in the Martian soil by the 1975 Viking missions, and about 0.5% Cl was confirmed by surface soil analysis during the 1997 Pathfinder mission. The high halogen content could be interpreted as salts left behind from an evaporated ocean. On the other hand, one would obtain salt deposits from volcanic emanations reacting with surface rocks, which is a likely case for Venus since HCl and HF are well-known constituents of terrestrial volcanic gases and the Venusian atmosphere (see below).

The concentrations of volatile and siderophile elements in the silicate portion of the Earth are diagnostic for volatility fractionations and siderophile element removal into the Earth's core. To evaluate bulk silicate Earth abundances for these fractionations, the data are conventionally normalized to their respective abundances in CI-chondrites. Occasionally, the silicate/CI-chondritic concentration ratio of an element is further normalized to the respective ratio of another element such as Mg or Si.

The reason for normalizing to CI chondrites is that the chemical composition of CI-chondrites is representative for the abundances of the rock-forming elements in the solar system (Chapter 1). All solar system objects somehow formed from the solar mixture and incorporated varying amounts of the elements from it. The normalized abundances then can be interpreted for the element fractionations that may have taken place in the solar nebula, during accretion and during planetary differentiation. The normalization to CI-chondrites, however, should by no means imply that the Earth has an overall composition like these meteorites, or that the Earth formed from exactly those phases that are found in these meteorites.

The bulk silicate Earth abundances normalized to CI-chondrites are shown in Figure 4.6 as a function of 50% condensation temperatures. The symbols indicate the geochemical character – lithophile, siderophile and chalcophile – of the individual elements. If all elements were present in relative CI-chondritic abundances, they would plot on a horizontal line. These fractionations describe the extent of volatility fractionations

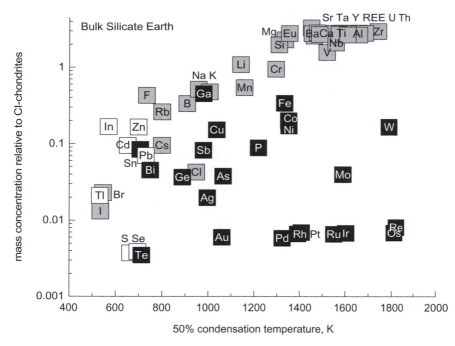

Figure 4.6 Elemental abundances in the bulk silicate Earth normalized to CI-chondrites are plotted as a function of the 50% condensation temperatures at 10^{-4} bar. Symbol shades indicate the geochemical affinities of the elements: grey, lithophile; black, siderophile; and white, chalcophile.

of the elements, removal of siderophile elements from the silicates during core formation, or both.

Many lithophile elements with condensation temperatures above ~1400 K are present in ~2.8 times CI-chondritic abundances, indicating that they are not fractionated through either incomplete condensation or evaporation processes. The abundances of the lithophile elements with lower condensation temperatures such as Cr, Mn, Cl, Na, K and Zn are lower. At first glance, the abundances decline relatively smoothly with decreasing condensation temperatures, suggesting incomplete condensation or retention of these elements. The abundances of F, Cl and Cs show more scatter from the volatility trend, but the condensation temperatures of the halogens and heavy alkalis are not well known, and their abundances have larger uncertainties. The abundances of lithophile Si, V, Cr and Mn are lower than expected from the abundance trend defined by the other lithophile elements, which has drawn particular interest from geo- and cosmochemists. Silicon and Mg have very similar condensation temperatures so one would expect a similar Mg/Si ratio for the Earth's silicates as in CI chondrites (Mg/

Si = 0.9) instead of the observed Mg/Si = 1.0, if there was no Mg/Si fractionation in the materials accreting to Earth, and if Si was not lost from the terrestrial silicates later. The Earth's silicates can be depleted in Si by either loss through volatility, for example, during the giant impact that led to the formation of the Moon, or through extraction into the core (see below).

The low abundances of V, Cr and Mn cannot be ascribed to volatility. Vanadium, among the refractory elements, is clearly less abundant than Ca or Sr with similar volatility. Manganese is more volatile than Cr, and the relative abundances are consistent with a volatility fractionation. However, both Cr and Mn plot below the volatility trend established by other lithophile elements, suggesting either different conditions for the volatility fractionations or yet another depletion mechanism. The condensation temperatures are for a gas of solar composition, but can change under different oxidation states, which changes the potential loss (or retention) of these elements. However, the volatilities of other elements would change, and the overall abundance trend with the condensation temperatures for a solar gas may argue against this possibility. Another interpretation is that V, Cr and Mn changed their geochemical affinity because the very early Earth differentiated under more reducing conditions than observed now as, for example, postulated in the two-component accretion models. If so, Cr and Mn behave as siderophile and/or chalcophile elements and were removed from the silicates into the metallic core. The occurrence of Cr and Mn as sulfides is known from the highly reduced enstatite meteorites, which contain Si alloyed in their FeNi metal. Models considering this possibility suggest 5820–7000 ppm Mn and 6800–7790 ppm Cr in the Earth's core (Wänke and Dreibus[35]; Allegre et al.[22]).

In addition to volatility, the abundances in the silicate portion of the Earth, crust plus mantle, were influenced by core formation. During core formation, denser metal and possibly sulfide segregated from the silicates and depleted the silicates in siderophile and chalcophile elements. Gallium, often classified as a siderophile element, fits right in with the lithophile elements establishing the volatility trend for the Earth (Figure 4.6). However, the classification of "siderophile" depends on the oxidation state. Gallium is more lithophile when oxygen fugacities are high enough to favor "Ga_2O_3" incorporation into silicates instead of "Ga-metal" incorporation into FeNi-alloys.

The partition coefficient or distribution coefficient is a quantitative measure of how elements are distributed among different mineral phases and/or melt according to their compatibility during igneous differentiation processes.

The Nernstian equilibrium partition coefficient, or bulk distribution coefficient, D_i is given by the mass concentration ratio of i between a solid mineral phase and coexisting liquid:

$$D_i = C_{i,s}/C_{i,\text{liq}} \tag{4.5}$$

In a binary system the equilibrium concentrations of an element in the solid and liquid are related to the total concentration by:

$$C_{i,\text{tot}} = C_{i,s}X_s + C_{i,\text{liq}}X_{\text{liq}} = C_{i,s}\left[1 - X_{\text{liq}}(1 - 1/D_i)\right] \tag{4.6}$$

where the last part of the equation makes use of the condition $X_s + X_{\text{liq}} = 1$ and the definition of the distribution coefficient.

The partition coefficient between metal and silicate is a measure of the extent of the lithophile and siderophile nature of an element. These partition coefficients are a function of temperature, total pressure and oxygen and/or sulfur fugacity (fO_2, fS_2). Partition coefficients are related to thermodynamic properties of the exchange reaction of an element between two phases. For example, the exchange reaction for partitioning of an element between metal and silicate is:

$$\text{M (in metal)} + (y/2)O_2 = MO_y \text{ (in silicate)} \tag{4.7}$$

for which the equilibrium constant K_{eq} is defined as:

$$\log K_{\text{eq}} = \log\left(\frac{X_{\text{MO}y}}{X_M}\right) + \log\left(\frac{\gamma_{\text{MO}y}}{\gamma_M}\right) - \frac{y}{2}\log fO_2 \tag{4.8}$$

where the activity a_i is replaced by the product of mole fraction and activity coefficient ($X_i \cdot \gamma_i$) of species i in the respective phases. Molar metal/silicate partition coefficients can be parameterized by:

$$\log K = a + b \log fO_2 + c/T \tag{4.9}$$

which in isothermal cases is:

$$\log K = a + b \log fO_2 \tag{4.10}$$

where the coefficient "a" contains the constant temperature term.

The parameter b is related to the valence state ($2y$) of the oxide in the silicate as:

$$2y = -4b \tag{4.11}$$

and the ratios of the activity coefficients are related to parameter "*a*:"

$$\log\left(\frac{\gamma_{MOy}}{\gamma_M}\right) = \log K_{eq} - a \tag{4.12}$$

Similar equations can be obtained to describe the Nernstian partition coefficients (mole fractions must be converted into mass concentrations, and log K is replaced by log D). Often, experimental partition coefficient data are measured as a function of oxygen fugacity and temperature and are fitted to:

$$\log D = a + b \log f O_2 + c/T \tag{4.13}$$

These fits provide information about the valence state (coefficient b for the fO_2 term) of an element in the silicate phase, which is of interest for transition metals with different possible oxidation states. Under oxygen fugacities usually assumed for planetary differentiation processes, moderately siderophile elements such as Fe, Ni, Ga, Ge, Mo and W have metal/silicate partition coefficients $D \ll 10\,000$, while highly siderophile elements, such as the noble metals, have $D > 10\,000$ (see Lodders and Fegley[24] and Walter *et al.*[25] for data summaries).

Figure 4.6 shows that the moderately siderophile elements are less abundant than the lithophile elements of similar volatility. For example, Mo and W are refractory elements but they are ~ 17 times less (W) and ~ 70 times less (Mo) abundant than the lithophile refractory elements. Assuming that the entire Earth has the same complement of refractory lithophile and siderophile elements as found in chondrites, the only explanation for the low amounts of Mo and W in the silicate portion is that they were removed into the metal phase that segregated to form the core. Other siderophile elements with similar condensation temperatures as their lithophile counterparts are also less abundant in the silicate portion of the Earth (*e.g.*, moderately siderophile Fe, Co, Ni and lithophile Mg and Si). The abundances of these elements in the silicate portion were likely established by volatility and partitioning into the core.

There are several problems with this simple scenario if one tries to perform quantitative modeling. The abundances of the highly siderophile elements are much higher in the silicate portion of the Earth than expected from simple core–mantle equilibration. Noticeably, the relative abundances are chondritic, *i.e.*, they plot at a constant concentration ratio that is independent of condensation temperature. If there was core–mantle equilibration, the silicate portion of the Earth should be devoid of all these noble metals because of their high affinity for metal (high metal–silicate partition coefficients). This leads to the idea that the highly siderophile element abundances in the Earth's silicates came from a late accreting "veneer" after core-formation was essentially complete. This veneer was broadly chondritic in composition, but it cannot have been CI-chondritic because the observed abundances of the chalcophile elements S, Se and Te are even lower than that of the noble metals. This points to a chondritic source depleted in the more volatile S, Se and Te.

Like the highly siderophile elements, the abundances of the moderately siderophile elements (Figure 4.7) are not consistent with simple core–silicate mantle equilibration. The easiest example to illustrate this is a comparison of the Ni and Co abundances. Their abundances are essentially at the same chondritic level in the BSE. Both elements have the same condensation temperature, so volatility fractionations between these two elements can be ruled out. However, the metal/silicate partition coefficient of Ni is about a factor of ten larger than that of Co under all plausible oxygen fugacities that prevailed during metal–silicate equilibration in the early Earth. Therefore, partitioning of Ni and Co into a metallic phase leads to a larger depletion in Ni than Co, which is not observed.

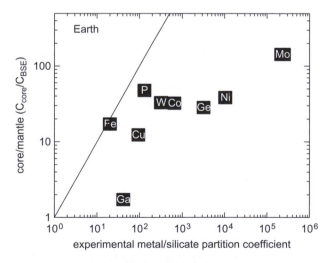

Figure 4.7 Terrestrial core/silicate concentration ratio of moderately siderophile elements compared to the experimental metal/silicate partition coefficients at an oxygen fugacity (fO_2) 2.3 log-units below the Fe–Fe$_{0.947}$O buffer ($\log fO_2 = $ IW-2.3). At this fO_2, the corresponding ratios match for Fe. If there was equilibrium between metal and silicates, all other elements should plot along the 1 : 1 line for the corresponding experimental and observed ratios. Calculation of the core concentrations takes volatility depletions of siderophile elements into account.

This mismatch extends to other siderophile elements. Figure 4.7 shows the core/silicate concentration ratios of the moderately siderophile elements as a function of their metal/silicate partition coefficient at oxygen fugacities 2.3 log-units below the iron-wüstite, IW, buffer. This buffer fixes the oxygen fugacity as long as pure iron coexists with wüstite, $Fe_{0.947}O$. Using the $Fe-Fe_{0.947}O$ equilibrium as a reference takes the temperature dependence of the oxygen fugacity into account. The oxygen fugacity of IW-2.3 for the Earth is that required to obtain the current "FeO" content of the mantle silicates and the Fe-metal composition of the core. The line in Figure 4.7 indicates where experimental and observed data would be equal. Except for Fe, which falls onto the line by definition, none of the ratios for the other moderately siderophile element ratios match the line. A comparison with the partition coefficients measured under different oxygen fugacities also cannot explain the observed elemental abundances of all moderately siderophile elements. There is no unique oxygen fugacity for which experimental partition coefficients explain these abundances by metal–silicate equilibration. These conclusions were based on experimental partition coefficients determined at 1 bar total pressure but these pressures are not relevant for metal–silicate partitioning in the deeper Earth. Attempts to explain the abundances by pressure-dependent element partitioning into the core were inconclusive. Another possibility is that the prevailing oxygen fugacities changed from very reducing to oxidizing conditions during accretion of the Earth. In that case, siderophile element abundances are dependent on the rates at which differently oxidized material accreted to the Earth. More about different accretion models postulated to explain the observed abundances are given below.

4.2.2.2 Composition of the Earth's Core.
The presence of metallic Fe as a constituent of the Earth's core is no longer under debate. Nickel and Co are other very likely components and well-known constituents of iron meteorites. The Ni/Fe of ~ 0.06 of the core is taken to be similar to that of chondrites adjusted for Fe and Ni that is retained in the silicate mantle.

Most researchers agree that the solid inner core is mainly made of Fe and Ni. The temperature of the inner core is above the Curie temperature and cannot hold a permanent magnetic field. The magnetic dipole field of the Earth is created in the fluid convective outer core through dynamo action of the rotating Earth. Seismic velocity measurements show that the outer core must contain 5–15% (by mass) of a light element (or elements) because the outer core is less dense than an FeNi alloy at the same temperatures and pressure.

Considering the abundances of the elements and the amount of the light element required, the most popular choices include C, O, S, P, Si and H. The presence of these elements alters the physical properties such as thermal and electrical conductivities, and lowers the melting point of the FeNi alloy. Knowing which lighter element is present and its phase relation with FeNi metal at the total pressures of the boundary of the inner and outer core may provide a temperature estimate.

Choices for the light element included the more abundant elements S, Si and O. Sulfur was suggested as the light element in the Earth's core by Goldschmidt as early as 1922.[26] The amount of S in the entire core has been estimated as high as 9–12%, but from cosmochemical constraints discussed below the amount of S in the core is less than 2%. Iron sulfide was long suspected as a core component because it is ubiquitous in iron meteorites, and has a larger density than silicate melts. Sulfide is expected to settle with metallic iron to the center of a planet. If metal and sulfide are present, metallic melt formation begins at the eutectic of 1260 K (at 1 bar total pressure; 1423 K at 100 kbar), much lower than the melting point of pure Fe (1809 K). Melting at lower temperatures fosters density segregation of metallic melts during accretion. Melting relations of Fe–FeS at higher total pressures have been investigated in detail to gain insights into the possible nature of the Earth's core and that of other planetary objects.

Sulfur, like siderophile Ni and Co, is depleted in the silicate mantle relative to CI-chondrites. This would be consistent with a removal of S into the core if the Earth ever had an overall CI-chondritic composition for refractory, moderately volatile and volatile elements. Sulfur is a volatile element like the lithophile volatile elements, and the relative depletion of S in the silicate mantle may have been caused by volatility, and not solely by loss of sulfides into the core.

An estimate of how much S was lost by volatility and how much is in the core is carried out by comparing the relative abundances of sulfur with lithophile elements of similar volatility, *e.g.*, Zn. Dreibus and Palme[27] suggested that as a whole the Earth has the same abundance ratios for elements of similar volatility as chondrites, such as, for example, the S/Zn ratio. The assumption is that the initially accreted amounts of the elements only depended on their relative volatility, for which condensation temperatures are used as a measure (Chapter 3). The absolute amounts of Zn and S in different chondrites and entire planets (silicates plus core) may vary, but the abundance *ratios* of elements with similar volatility will remain the same if these were only controlled by volatility.

This estimate for the S content of the core relies on the S/Zn from CI-chondrites, which represents solar composition for rock-forming

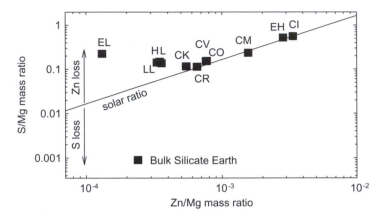

Figure 4.8 Abundance ratios of S/Mg and Zn/Mg in chondrites and in the bulk silicate Earth. The line corresponds to the CI-chondritic S/Zn ratio. Arrows indicate how changes from the solar S/Zn ratio can be achieved. The ratios above the CI chondrite value in ordinary and EL chondrites require lower Zn abundances (*e.g.*, from incomplete Zn condensation or selective Zn loss). Lower ratios such as observed in the silicate portion of the Earth can indicate either S depletion through volatility fractionations or S removal into the Earth's core, or both.

elements. If fractionations are solely controlled by volatility in a solar composition (H_2, He-rich) gas, the S/Zn ratio should be the same in all chondrite groups and, by analogy, the same in other objects that accreted in the inner solar system. The carbonaceous chondrites and the EH chondrites have S/Zn ratios close to the CI-chondrite ratio (indicated by the line in Figure 4.8). However, the ordinary chondrites (H, L and LL) and the EL chondrites have higher S/Zn ratios and plot above the CI-chondrite ratio line. These higher ratios must be ascribed to lower Zn concentrations in these objects and may be explained by the total pressure dependence of the Zn and S volatilities. The sulfur condensation temperature is pressure-independent, but that of Zn decreases with decreasing total pressure, which would increase the S/Zn ratio.

In contrast, the bulk silicate Earth has a smaller S/Zn ratio than most chondrites and plots far below the CI chondrite ratio line. This lower ratio must be ascribed to loss of sulfur. Since we are only considering the S/Zn ratio of the silicate portion of the Earth, it is plausible to assume that the "missing" sulfur is in the Earth's core.

A second assumption for estimating the S content of the core is that, during differentiation, Zn quantitatively remains in the silicates whereas S partitions between the metallic core and the silicate mantle. The Zn concentration of the Earth's mantle silicates is 50 ppm by mass, and the entire Earth contains $50\,ppm \times 0.675 = 33.8\,ppm$ Zn (the 0.675 is the mass fraction of total silicates in the Earth). From this and a CI-chondritic S/Zn mass ratio of 166, the amount of S in the entire Earth is $33.8\,ppm\times166 = 5600\,ppm$ S. Sulfur is not very abundant in the present Earth's mantle (122 ppm, Table 4.4) and in the crust (697 ppm, Table 4.3), which leads to a S concentration of 124 ppm in the entire silicate fraction of the Earth (using mass fractions of 0.9962 for the mantle and 0.0038 for the crust). This is not a considerable reservoir for S. Taking the estimate for the entire Earth from the chondritic S/Zn ratio, only 1.5% ($124\,ppm\times0.675/5600\,ppm$) of the entire Earth's S is in the silicate portion and the remaining S must be in the core. This gives a concentration of $(5600 - 124 \times 0.675)/0.325 = 17\,000\,ppm$, or 1.7 mass% S for the Earth's core (here 0.325 is the mass fraction of the core). This percentage of S is much lower than required by seismic density constraints for the core composition, and the presence of another light element is needed.

Silicon is another favorite choice for the light element in the core. First, it is another abundant light element, and it can alloy with FeNi under reducing conditions. The relatively low Si/Mg ratio in the Earth's primitive mantle supports the idea that some Si is in the Earth's core. Like sulfur, silicon in the outer core depresses the melting point of FeNi and enables formation of convective flows in the outer core to drive the dynamo. Freezing of Fe out of liquid Si-bearing melt leaves excess Si at the inner to outer core boundary. This creates a concentration gradient in the outer core and introduces flows that may be faster than the less efficient thermally driven convective flows. Estimates for the Si content in the core are up to 6% by mass. Together the amounts of S and Si make $\sim 8\%$ of the light element contribution in the core. Adding the amounts of Cr and Mn mentioned above gives about 9%.

The possibility of oxygen in the core was considered by Goldschmidt in 1922, who thought of an Fe-Ni-O-S core. Oxygen becomes more soluble in NiFe alloys with increasing temperatures, but this increase in solubility is counteracted by high total pressures that are found at the core–mantle boundary (CMB, $\sim 1.4\,Mbar$). The limit for the O content of the core is taken as about 2% by mass.[28] However, the presence of O in the core is inconsistent with the presence of metallic Si in the core. Oxidation leads to SiO_2, which would move from the outer core into the silicate mantle. The presence of O in the core is not discussed further here.

The presence of carbon in the core, either alloyed or in the form of carbides, is plausible if one makes the argument that core formation occurred under reducing conditions and that larger amounts of highly volatile C were accreted to the early Earth. Accretion of carbon would naturally lead to more reducing conditions that may allow removal of Si, but also Cr and Mn into the core. There are no good cosmochemical constraints for the amount of carbon that may have been accreted. Carbon is a highly volatile element and there is no suitable lithophile element with similar volatility that can be used to estimate the C concentration from an abundance ratio in a similar manner as done above for S and Zn. An upper limit estimate for the C content of the core is about 0.2% by mass.

Some of the less abundant elements such as K or U have been suggested as radioactive heat sources in the core to drive the geomagnetic dynamo. Estimates for the energy requirements to drive the dynamo through electric current flow in the core are $\sim 10^8$ W. Under very reducing conditions, K can enter sulfide minerals, which is observed in the highly reduced enstatite meteorites. Thus it is possible to incorporate K together with Cr and Mn if the Earth started to accrete from reduced material as postulated in the two-component models.

4.2.2.3   Composition of the Silicate Portion of Mars.   Direct chemical information about Mars comes from the numerous space missions, landers and rovers that explored the surface mineralogy and chemistry, which is discussed later. The surface rocks and soil on the up to 50 km thick Martian crust are the products of more extensive fractionation processes and their measured compositions show the effects of aqueous and aeolian weathering. There is only limited information that can be derived from these measurements for the composition of the entire silicate portion and, by extension, all of Mars. Another problem is that measurements for many diagnostic but minor elements – volatile and siderophile – are not accessible so that similar modeling as done for the composition of the Earth's mantle is difficult to do.

The sample record for Mars is extended by the SNC meteorites – the shergottites, nakhlites and chassignites (*e.g.,* McSween[29]). These achondrites are linked together through their O-isotope compositions, which form a fractionation curve parallel to the terrestrial fractionation curve with an offset of $\Delta^{17}O = +0.30$ per-mil (see Figure 2.2). The SNC meteorites are igneous rocks that underwent varying degrees of differentiation and elemental fractionation (*e.g.,* McSween[29]). The shergottites (named after the Shergotty meteorite fall in 1865) are basaltic achondrites and until the late 1960s were not distinguished from eucrites (see below) with which they share several similarities in mineralogy. However, the occurrence of maskelynite, a glassy form of the feldspar labradorite (Table 2.4), and the absence of brecciation sets shergottites apart from eucrites. Only two shergottites were known (Shergotty and

Figure 4.9   Pieces of the Zagami shergottite, a basaltic rock believed to have come from Mars. The largest piece is about 2 cm long. (Photograph by the authors.)

Zagami, fallen in 1962) until the Antarctic and desert finds made shergottites a larger group of achondrites. Figure 4.9 shows a few pieces of the Zagami shergottite.

The lherzolitic shergottites contain relatively Fe-rich olivine (Fa$\approx$27%), orthopyroxene and smaller amounts of pigeonite, augite, maskelynite, chromite and the phosphate whitlockite. The olivine grains show a preferential crystallographic orientation that suggests that these are cumulative rocks.

Nakhlites are clinopyroxenite or wehrlite rocks (Figure 4.4). They have $\sim$75% diopside, $\sim$15% relatively Fe-rich olivine (Fa$\approx$65%) and $\sim$10% plagioclase (An$\approx$35%). Magnetite, pyrite, pyrrhotite and Cl-apatite are present in minor amounts. Veins of the Nakhla meteorite contain iddingsite deposits, which are pre-terrestrial in origin. These and other mineral deposits required water to form as alteration products from olivine, which requires that these meteorites originated on an object where water circulation and brine mobilization was possible. Chassignites, named after the meteorite from Chassigny, France, consist almost entirely of cumulative olivine rich in Fe ($\sim$33% fayalite) and minor amounts of pyroxene, maskelynite, chromite and sulfides such as troilite, pentlandite and marcasite. One member of the SNC meteorite makes a category of its own. The Antarctic meteorite ALH84001, found in the Allan Hills, contains mainly orthopyroxene, and minor minerals, including magnetite, phosphates, pyrite, Fe-sulfides, sulfates and carbonates. McKay *et al.* have suggested that some small sub-micron size magnetite and sulfide formations associated with carbonates in this meteorite are remnants of Martian microfossils.[30] This claim is debated because the formation of such mineral associations also can be achieved through inorganic processes. The ALH 84001 meteorite has a significantly older crystallization age (closer to $\sim 4\,Ga$) than the shergottites with ages of $\sim 180\,Ma$ and the Nakhlites and chassignites with $\sim 1300\,Ma$.

Figure 4.10 shows the various ages of the SNC meteorites. The crystallization ages refer to the ages when the rocks formed by crystallization from a magma. The ejection age is the time when the rock left the planet and was in space and on Earth. The time spent in space is measured from the amounts of cosmogenic nuclides that were produced while the meteorites were exposed to cosmic rays. The terrestrial age refers to the time when the meteorites landed on Earth. This age is measured from the amounts of short-lived radioactive nuclides that were produced when the meteorites traveled in space.

The young crystallization ages require a larger object that could still provide heat for magmatic processing about 180 Ma ago. This points to a planetary rather than asteroidal origin. A close association to Mars comes from measurements of gas inclusions found in the shergottite EET 79001. The relative abundances of the inert gases are remarkably similar to those measured in the near-surface Martian atmosphere

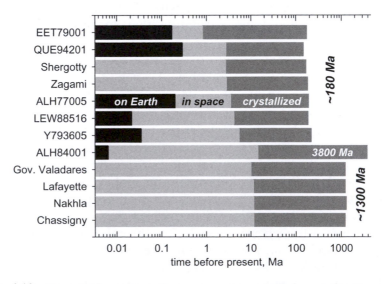

Figure 4.10 Terrestrial ages, cosmic ray exposure ages and crystallization ages of some SNC meteorites.

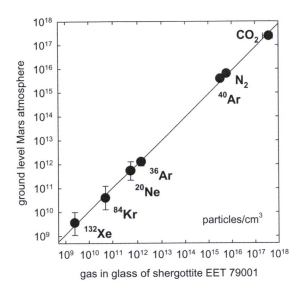

Figure 4.11 Relative concentrations of inert gases in the ground-level Martian atmosphere and released from the SNC meteorite EET 79001.

(Figure 4.11). The gas inclusions in the meteorite are thought to be trapped Martian atmosphere that got into the rock when it was ejected.

The elemental compositions of the SNC meteorites are related to, but are not identical to, the elemental composition of the Martian mantle. In a similar manner as done for the silicate portion of the Earth, the element ratio method is used to derive the elemental composition of the Martian mantle from chemical analyses of the SNC meteorites and the assumption that refractory elements are present in relative chondritic proportions. Table 4.5 summarizes the composition of the Martian silicate portion obtained by this approach.[31,32] Further modeling of the siderophile element abundances in the silicate portion provides estimates for the core composition and the overall elemental abundances of Mars.

Figure 4.12 shows the data for the Martian silicate portion normalized to CI chondrites, which can be compared to the composition of the terrestrial silicate portion in Figure 4.6.

The lithophile element abundances are volatility controlled because they decrease relatively smoothly with decreasing condensation temperatures.

In comparison to the primitive silicate mantle of Earth, the primitive Martian mantle received higher abundances of volatile elements from the planetesimals that accreted to Mars. The likely cause is more efficient condensation of volatile elements further from the Sun. On the other hand, the Earth may have lost volatiles during the giant impact that led to the formation of the Moon (see below).

Table 4.5 Elemental composition of the silicate portion of Mars and composition as major oxides.[a]

Element	Composition (ppm)	Element	Composition (ppm)	Element	Composition (ppm)	Oxide	Composition (mass%)
Li	0.9	Cr	5 720	In	0.016	Na$_2$O	0.349
F	28	Mn	3 520	Sb	0.005	MgO	29.89
Na	2 590	Fe	136 400	I	0.016	Al$_2$O$_3$	3.02
Mg	180 250	Co	62.5	Cs	0.073	SiO$_2$	44.40
Al	16 000	Ni	345	La	0.44	P$_2$O$_5$	0.148
Si	207 540	Cu	4.2	Nd	0.85	K$_2$O	0.036
P	645	Zn	61	Eu	0.082	CaO	2.45
S	250	Ga	6.45	Hf	0.224	TiO$_2$	0.138
Cl	29	As	0.04	Ta	0.035	Cr$_2$O$_3$	0.836
K	300	Se	0.06	W	0.0835	MnO	0.454
Ca	17 500	Br	0.13	Au	0.0015	FeO	17.55
Sc	8.2	Rb	1	Tl	0.0037	Sum	99.27
Ti	828	Mo	0.086	Th	0.068		
V	52	Ag	0.007	U	0.0165		

[a]Derived from element correlations and assuming that SNC meteorites are from Mars (see Lodders and Fegley[32] and references therein).

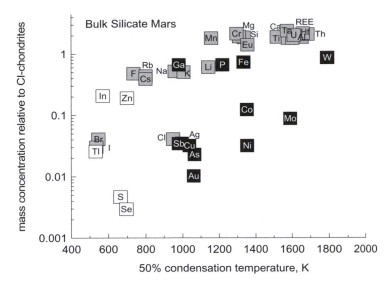

Figure 4.12 Composition of the silicate portion of Mars relative to CI-chondrites as a function of the 50% condensation temperatures at 10^{-4} bar. Symbol shades indicate the geochemical affinities of the elements: grey, lithophile; black, siderophile; and white, chalcophile.

The abundances of siderophile elements in the silicate portion of Mars depend on the relative volatility and the efficiency of their removal into the metallic core. The extent of the volatility fractionation is easily estimated from the abundance trend of lithophile elements with similar condensation temperatures. These are the expected abundances ($C_{expected}$). Any further decrease of the concentration of a siderophile element in the silicates ($C_{silicate}$) below the volatility trend (Figure 4.12) is then due to core–mantle partitioning, so $C_{missing} = C_{expected} - C_{silicate}$. The "missing" amount of a siderophile element in the silicate mantle relates to its concentrations in the core (C_{core}) by $C_{core} = (C_{expected} - C_{silicate})X_S/X_C$.

To calculate the elemental abundances in the core, we need the relative mass fraction of the silicate portion (X_S) and the core (X_C) of Mars. Considering the relative solar Fe abundance and that in the silicate mantle, the core mass comes as ~20% of total Mars. Further consideration of the volatile element inventory and the moment of inertia suggest that the Martian core is mainly Fe, Ni, smaller amounts of Co, P and other siderophile elements plus about 10–14 mass% S.

With an estimated core size, the siderophile element concentrations in the core are easily obtained, and, with those, we have the core/mantle ratios. These are compared to experimental metal/silicate partition

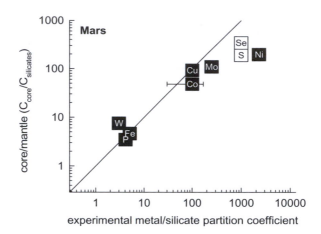

Figure 4.13 The Martian core/silicate concentration ratios of moderately siderophile elements show a relatively smooth correlation with experimental metal/silicate partition coefficients at oxygen fugacities about 1 log unit below the Fe-wüstite equilibrium buffer (log $fO_2 = IW-1$). Calculation of the core concentrations takes volatility depletions of siderophile elements into account. The core contains about 29 mass% sulfide, and the partition coefficients are for the corresponding metal-sulfide composition. This figure is similar to Figure 4.7 for the Earth.

coefficients in Figure 4.13. A relatively good correlation is found for (metal + sulfide)/silicate partition coefficients at oxygen fugacities about 1 log-unit below the Fe-wüstite equilibrium buffer. The existence of such a correlation suggests that the distribution of siderophile elements between mantle silicates and the metal-(sulfide) core approached an equilibrium state. This is quite different from what is observed for the Earth.

This may simply suggest that metal–silicate equilibrium is more likely to be attained in a smaller object, which is apparently corroborated by the metal/silicate equilibrium for the small asteroid Vesta (see below). In objects that are only a tenth of the Earth's mass or less, a change in the redox state during accretion may not leave a pronounced change in siderophile abundances as is observed for the Earth's silicate mantle. The timescales of planet accretion and core formation are also different for Earth and Mars.

The timing of core formation can be determined from the abundance of ^{182}W relative to other W isotopes in rocks of planets and asteroids. The short-lived nuclide ^{182}Hf (half-life of 9 Ma) was present in the early solar system and decayed into ^{182}W. Both Hf and W are refractory elements but they behave differently during planetary differentiation. Hafnium is a lithophile element and is retained in the silicates but W is moderately siderophile and is readily removed into the metallic core. The terrestrial and Martian silicate portions are depleted in W when compared to the refractory lithophile elements. If W is (partially) removed from the silicates into the core and core formation is completed before all ^{182}Hf has decayed, there will be a build-up of ^{182}W in the silicate portion later on. This leads to an increase in the ratios of ^{182}W to the other stable W isotopes in the silicate portion. For example, the ^{182}W/^{183}W ratio in a planetary silicate will be larger than the ratio in chondritic meteorites or their oldest components, if core formation predated complete decay of ^{182}Hf to ^{182}W. If the initial solar system ^{182}W/^{183}W ratio is known, the time difference between solar system formation and the time when the ^{182}W/^{183}W ratio was established can be derived.

W-Hf dating[53-55] shows that the formation of the Earth's core was completed within 30 Ma after the oldest known solids in the solar system had formed (the refractory inclusions, or CAIs, see Chapter 3). Core formation on Mars and Vesta was faster and was completed after 13 Ma (Mars) and within 3–5 Ma (Vesta).

If metal–silicate fractionation was rapid in small planetesimals, the larger planets most likely accreted from a mixture of differentiated (like Vesta) and undifferentiated (like chondrite parent bodies) planetesimals. Thus there was already some metal–silicate redistribution of siderophile elements before the planetesimal merged with the major planet mass, where the final distribution of siderophile elements was determined.

4.2.2.4 Differentiated Asteroid Vesta. The eucrites and the related diogenites and howardites (EHD meteorites) are thought to come from the asteroid 4 Vesta and its fragments, the Vestoids, which share the orbit and spectral characteristics of Vesta. The relation to the EHD meteorites was established by the good match of reflectance spectra from Vesta with those taken of the basaltic eucrites.[33] Vesta has a basaltic surface and that shows it is differentiated. It has a relatively high density (3.5–4.0 g cm^{-3}) compared to other asteroids, which suggest it is a compact object with a small metallic core.

Eucrites are basaltic rocks that contain mainly Ca-rich pyroxene (pigeonite) and plagioclase feldspar (An $\sim$80–95). The minor minerals are magnetite, chromite, ilmenite, whitlockite, troilite and kamacite, and the three SiO_2 modifications quartz, tridymite and cristobalite.

Eucrites show brecciated structures, indicating that their parent body experienced extensive bombardment over time. Several eucrites contain rock fragments from other meteorite types, such as CM chondrites. The mineralogy of the diogenites is that of orthopyroxenites. Diogenites are

Ca-poor and contain mainly bronzite-hypersthene pyroxene (Fs 23–27). Other components are olivine (Fa $\sim$28), plagioclase feldspar (An 85–90), and minor chromite, troilite and kamacite ($\sim$3% Ni), and glass. Most diogenites are monomict breccias that contain orthopyroxene rock clasts in fine to coarse grained matrices. In some polymict brecciated diogenites, material from eucrites is mixed in. The howardites were named by Gustav Rose in honor of the British chemist Edward Howard (1774–1818), who performed the first detailed meteorite analyses. Howardites are mainly orthopyroxene (Fs<40%), and pigeonite (Fs >45%). Less abundant are plagioclase feldspar (An >75%), olivine of variable composition (Fa$\approx$8–40%), kamacite ($\sim$4% Ni), taenite ($\sim$40% Ni) and troilite. The howardites can be understood as products from mixing eucrite and diogenite fragments, which makes howardites polymict breccias.

The mineral fabric of eucrites is two-fold: non-cumulate eucrites formed from magmas extruding on the surface of their parent body, and cumulate eucrites as intrusive rocks near the surface. Cumulative eucrites are reminiscent of terrestrial gabbros, which formed out of minerals that accumulated by slow growth in a magma chamber. The cumulate eucrites have large, well developed and oriented crystals whereas non-cumulate eucrites have fine grained textures. The non-cumulate eucrites resemble rocks formed near the surfaces of a lava flow, which sometimes show vesicles from escaping gas. Figure 4.14 shows

Figure 4.14   Ibitira, a basaltic meteorite with vesicles from the asteroid Vesta. The width of the piece on the right is about 1.5 cm on its widest side. Photograph by the authors.

pieces of the non-cumulate eucrite Ibitira, which is a rare example of a vesicular structure in achondrites.

Table 4.6 and Figure 4.15 show the estimated chemical composition of the silicate portion of Vesta derived from the element ratio method that was used to estimate the composition of the silicate portions of Earth (Figure 4.6) and Mars (Figure 4.12). As in previous Figures, the abundances are normalized to CI-chondrites. The silicate portion of Vesta is highly depleted in volatile elements, and their abundances show a correlation with condensation temperatures. The alkali abundances are orders of magnitude lower than in Mars, which is unexpected. If Mars accreted larger amounts of volatile elements from volatile-rich planetesimals that formed in the orbital region of Mars (1.52 AU), Vesta located at 2.36 AU should have higher volatile abundances, or at least similar ones as observed from Mars, but not significantly less. Therefore, the low abundances of volatiles are not only controlled by the amounts present in the accreting planetesimals, but volatile loss during accretion must be considered. As noted above, loss of volatiles from a low-mass object is more likely when heated sufficiently to allow melting and metal–silicate separation.

The distribution of moderately siderophile elements between core and silicate mantle on Vesta is derived in the same manner as before for Earth and Mars. These ratios are compared to the experimental metal/silicate partition coefficients in Figure 4.16. Considering the Fe and lithophile element concentrations, Vesta has a metal core of about 20 mass%. This is about the same percentage as for the Martian core. However, there is an important difference between the cores of Mars and Vesta. Mars is a volatile element-rich planet and therefore some fraction of its core consists of sulfide (about 6–9 mass%, depending on model details), but Vesta is volatile element poor, so its core should be very sulfide poor. This is consistent with the lower oxygen fugacity for Vesta that is required to match the observed core/silicate distributions to experimental values than that for Mars. A larger fraction of total Fe has to remain in the metallic state on Vesta, therefore less iron is in the form of "FeO" (the Fe content of the silicate portion is 11.3% for Vesta, and 13.8% for Mars).

At an fO_2 about 1.5 log-units below the iron-wüstite buffer the observed Fe distribution matches reasonably well, and more detailed modeling suggests an $fO_2 = IW-1.4$. This is about 2.5-times more reducing than the fO_2 estimated for the Martin core–mantle equilibrium.

4.2.3 The Earth's Moon

Of the large objects in the inner solar system, the Moon is the only one orbiting a planet. If it were not for its orbital properties, the Moon's

Table 4.6 Elemental composition of the silicate portion of Vesta and composition as major oxides. [a]

Element	Composition (ppm)	Element	Composition (ppm)	Element	Composition (ppm)	Oxide	Composition (mass%)
Na	757	Mn	3 520	Cs	0.002	Na$_2$O	0.10
Mg	190 950	Fe	115 100	Ba	4.8	MgO	31.67
Al	17 150	Co	19	La	0.484	Al$_2$O$_3$	3.24
Si	215 100	Ni	42	Nd	0.942	SiO$_2$	46.02
P	79	Cu	0.6	Sm	0.304	P$_2$O$_5$	0.018
Cl	4.6	Zn	0.36	Eu	0.116	K$_2$O	0.008
K	66	Ga	0.54	Yb	0.338	CaO	2.58
Ca	18 420	Rb	0.06	Lu	0.051	TiO$_2$	0.155
Sc	12	Sr	15.6	Hf	0.212	Cr$_2$O$_3$	0.778
Ti	930	Y	3.1	W	0.013	MnO	0.455
V	97	Zr	7.2	Th	0.066	FeO	14.81
Cr	5 320	Mo	0.005	U	0.017	Sum	99.82

[a]Assuming EHD meteorites are from Vesta. Data mainly derived from element correlations (see also ref. 34)

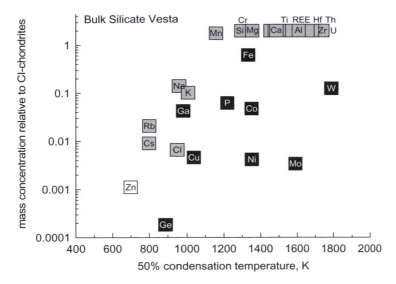

Figure 4.15 Composition of the silicate portion of Vesta as derived from eucrites relative to CI-chondrites as a function of the 50% condensation temperatures at 10^{-4} bar. Symbol shades indicate the geochemical affinities of the elements: grey, lithophile; black, siderophile; and white, chalcophile.

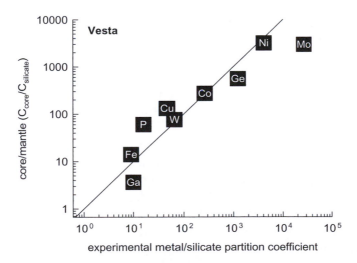

Figure 4.16 The core/silicate concentration ratios of moderately siderophile elements in Vesta show a relatively smooth correlation with experimental metal/silicate partition coefficients at oxygen fugacities about 1.5 log-units below the Fe-wüstite equilibrium buffer ($\log fO_2 = \text{IW-}1.5$). Calculation of the core concentrations takes volatility depletions of siderophile elements into account. The core is assumed to be mainly FeNi without sulfides. Note the larger core/silicate ratios for Vesta than for Mars (Figure 4.13) despite the fact that they have similar relative core sizes (about 20% of the objects mass).

physical properties would qualify it as a planet. The Moon is ~ 1800 km in diameter, which is about 1/3 of the Earth's, and its mass is 80 times less than the Earth's. It has enough self-gravity to force its shape into a near-spherical ellipsoid, which is one criterion for the IAU definition of a planet or dwarf planet.

For comparison, the two small Moons Phobos and Deimos orbiting Mars are much smaller and irregular in size ($13.5 \times 10.8 \times 9.4$ km and $7.5 \times 6.1 \times 5.5$ km, respectively). They are most likely asteroids that were gravitationally captured from the near-by asteroid belt. However, the origin of the Earth's Moon cannot be separated from the history of Earth's origin.

We limit our discussion of lunar chemistry to a few geochemical aspects and some historical accounts on the origin of the lunar craters and the Moon itself. Much chemistry that happened during the formation of the Moon remains to be explored. Lunar exploration in the future will bring much more data that will allow us decipher more about the Moon's origin and the current processes that are operating on the lunar surface that is exposed to the perils of space weather.

4.2.3.1 Chemical Signatures of the Moon. The composition of the lunar silicate portion has been assessed by several authors using similar methods as discussed previously for the other differentiated objects. Table 4.7 lists data that are mainly derived from element correlations of lunar rocks.[35–38]

It is useful to compare the composition of the silicate portions of the Earth and the Moon. Overall, the dry lunar silicates are very poor in volatile lithophile elements such as the alkalis. The lunar silicates have about 12–13 mass% FeO (although estimates range from 6–18%) whereas the terrestrial inventory of FeO is about 8 mass%. This is the first hint that the Moon is not just made of the primitive Earth's mantle after the terrestrial core had formed. A favorite theory for the origin of the Moon is that it formed from debris in orbit after the early Earth encountered a giant impact after the Earth's core had formed (see below). Because the lunar silicates have a higher FeO content, one could argue that the Earth's core was not yet completely done and some metallic Fe was still mingled in with the silicates that were ejected and formed the Moon. However, this required an oxidizing agent to introduce "FeO" into the lunar silicates, whose origin is not clear. Overall, the reasons for the different FeO contents of the silicate portions of the Earth and Moon are not well defined yet.

Table 4.7 Elemental composition (in ppm by mass) of the silicate portion of the Moon and composition as major oxides (mass%).[a]

Element	Composition (ppm)	Element	Composition (ppm)	Element	Composition (ppm)	Oxide	Composition (mass%)
F	1.4	Mn	1 245	Cs	~0.000 17	Na_2O	0.057
Na	420	Fe	97 300	Ba	5.1	MgO	35.49
Mg	214 000	Co	85	La	0.59	Al_2O_3	3.76
Al	19 900	Ni	470	Eu	0.14	SiO_2	44.28
Si	207 000	Cu	2.1	Yb	0.4	P_2O_5	0.0046
P	20	Ga	0.66	W	0.018	K_2O	0.0049
K	41	Ge	~0.006	Re	0.000 006	CaO	3.15
Ca	22 500	Rb	0.11	Os	0.0004	TiO_2	0.18
Sc	14.3	Sr	21	Ir	0.000 06	Cr_2O_3	0.356
Ti	1 100	Mo	0.0022	Pt	0.0004	MnO	0.161
V	84	Ru	0.0001	Au	0.000 06	FeO	12.52
Cr	2430	Pd	0.0001	U	0.02	Sum	99.96

[a]Data mainly derived from element correlations of lunar rocks.[35-38]

Siderophile and Chalcophile Elements. Moderately siderophile and/
or chalcophile elements such as Ni, Cu, Ga, Ge, As, Pb and Bi are less
abundant in lunar than in terrestrial silicates (Figures 4.6 and 4.17). The
highly siderophile elements Ru, Pd, Os, Ir, Pt, Au and Re are present at
about 0.0002 times the CI-chondrite abundances, which is significantly
less than the ~ 0.008 times chondritic abundance ratio in the bulk silicate Earth. Interestingly, both the Moon and the Earth have flat
chondrite-normalized abundances of the highly siderophile elements,
which suggests contamination by accretion of a late veneer of chondrite-
like composition (Figures 4.6 and 4.17).

Figure 4.18 shows the concentrations of Ni and Au plotted against
the Ir concentration in lunar rocks collected during the Apollo missions,
in meteorites originating from the Moon, and average ranges of values
(grey boxes) for the Earth's crust, Earth's mantle, and chondritic
meteorites. The Apollo and Luna samples plus the lunar meteorites
represent samples from different locations with the lunar meteorites
presenting a much broader range of geologic settings, for which,

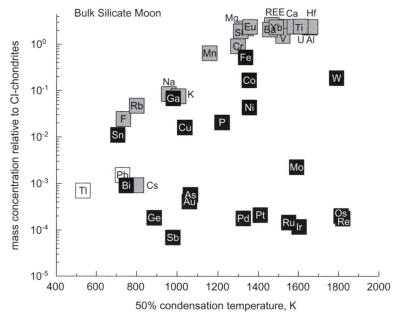

Figure 4.17   Composition of the silicate portion of the Moon as derived from samples
collected during the Apollo and Luna missions and from Lunar me-
teorites. Data are relative to CI-chondrites and are plotted as a function
of the 50% condensation temperatures at 10^{-4} bar. Symbol shades indicate the geochemical affinities of the elements: grey, lithophile; black,
siderophile; and white, chalcophile.

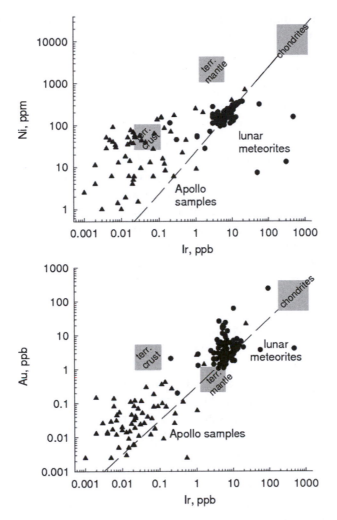

Figure 4.18   Concentrations of Ni and Au plotted against the Ir concentration in lunar rocks collected during the Apollo missions (filled triangles), in meteorites originating from the Moon (filled circles), and average ranges of values (grey boxes) for the Earth's crust, Earth's mantle, and chondritic meteorites. The Apollo samples generally have lower Ir, Au and Ni contents than the lunar meteorites. Also note the clustering of the data for the lunar meteorites along the line for the chondritic ratio.

however, the exact locations on the Moon are unknown (see Korotev for a comprehensive review[39]). The Apollo samples generally have lower Ir, Au and Ni contents than the lunar meteorites. Also note the clustering of the data for the lunar meteorites along the line for the chondritic ratio, which is indicative for meteoritic contamination. This strongly suggests that lunar meteorites incorporate significant amounts of the material from the impacting objects that are responsible for ejecting these rocks from the Moon, which can complicate the use of lunar meteorites for deriving the overall lunar siderophile element abundances and their interpretation.

Many of the moderately siderophile elements are also volatile and smaller relative abundances of these elements on the Moon than in the terrestrial mantle could come from volatility loss in the aftermath of the giant impact, in addition to removal into the small lunar core. The relative abundances of the refractory moderately siderophiles Co and W are similar in the silicates of the Earth and Moon. This suggests that late formation of a lunar core cannot have influenced moderately siderophile element abundances dramatically if the Moon had only inherited the siderophile elements that were left after core formation in the ejected terrestrial silicates. However, considering that the projectile was about Mars-size (about 11% of Earth's current mass) and that core formation can be expected to be complete when the Earth's core had reached maturity, one has to ask what was the siderophile element inventory brought to the Moon by the silicates from the impacting object? It is safe to assume that the Mars-size object was differentiated into a silicate portion and a metallic core by the time Earth was largely differentiated. The dynamical impact simulations suggest that the impactor's core "shot" through the Earth's mantle and combined with the terrestrial core. Thus, no significant contribution of siderophile elements, if any, would come from the impactor's core to the silicate portion of either Earth or Moon.

There is some uncertainty about how much of the Moon's silicates were derived from evaporated terrestrial silicates, and how much was contributed by the projectile. Several chemical models of the Moon suggest that most of the Moon stems from ejected and vaporized terrestrial silicates. In contrast, dynamical impact models lead to the conclusion that most of the Moon is made of the silicate portion of the projectile and that only a smaller portion originated from evaporated terrestrial silicates.

A differentiated object impacting the Earth and combination of its core with the terrestrial core can explain the absence of a significant core as indicated by the low density of the Moon. The lunar core cannot be much more than 2% of the total lunar mass; estimates by different authors run from 2–5 mass%, and the presence of the lunar core is also necessary to explain the more recent Lunar Prospector magnetometer measurements. The abundances of the siderophile elements in the lunar silicates are also consistent with the presence of a small lunar core.

Compared to the bulk silicate Earth, the bulk silicate Moon has lower siderophile element abundances. Low abundances of volatile lithophile elements such as the alkalis and halogens are also a typical lunar characteristic. These elements are not likely to be moved into the lunar core if metal or sulfide separation from lunar silicates ever took place. Therefore, these can be used to estimate how much depletion of the volatile, siderophile elements is volatility related. The loss of volatile elements must have happened quite early because the lunar silicates have a very low $^{87}Sr/^{86}Sr$ ratio, often called LUNI (lunar initial ratio). Over time, the ratio is affected by decay of ^{87}Rb to ^{87}Sr. However, if only little of the volatile Rb ever accreted or was lost early, the amount of ^{87}Sr remains low, and thus the $^{87}Sr/^{86}Sr$ ratio remains low.

Defining a trend of volatility as a function of condensation temperatures as performed for Earth, Mars and Vesta can be done. This would assume, though, that the Moon formed from the same building blocks formed within the solar nebula so that the condensation temperatures computed for a H_2 and He rich gas of solar composition are applicable. However, if there was volatile element loss from the ejected terrestrial and the impactor silicates, the condensation temperatures do not really apply because this volatilization happened in an essentially H_2 and He-gas free environment. Overall, the evaporation of the terrestrial mantle produces much more oxidizing conditions than those present in the solar nebula. Considering this, oxidation of smaller amounts of metallic Fe that had escaped core formation appears as a possible explanation of the FeO content of the lunar silicates. A change to more oxidizing conditions also could explain the differences in moderately siderophile element abundances in the terrestrial and lunar silicates: Under higher oxygen fugacities, gallium becomes more refractory as incorporation of Ga_2O_3 is favored into silicates. The moderately siderophile Mo and W and possibly Ge become more volatile because their gaseous di- and trioxides become more stable. (However, for easy comparison to other objects shown in Figures 4.6, 4.12 and 4.15, we kept Figure 4.17 in a similar format and plotted the normalized concentrations of the lunar silicate *versus* the 50% condensation temperatures.)

The terrestrial and lunar silicates share several abundance peculiarities that are not observed in silicates of Mars or Vesta. Most notable are the increasing depletions of V, Cr and Mn in the lunar and terrestrial silicates when compared to the chondritic abundance ratios (Figure 4.6 and 4.17). Similarly, Si appears to be depleted when compared to Mg, Ca, Ti or other lithophile and non-volatile elements.

Assuming that the Earth started accreting and differentiated under very reducing conditions, the depletion of V, Cr, Mn and, possibly, Si could have been caused by removal of these elements into the Earth's core. This would leave silicates with relative depletions in V, Cr, Mn and Si, which would remain if a larger portion of the Moon was derived from Earth's silicates. This is one of the major arguments made from chemistry that a larger portion of the Moon must come from the silicates of the early Earth. There is no doubt that there has to be a contribution from the impacting object as well because the lunar silicate portion is less depleted in Mn than the bulk silicate Earth (the lunar to terrestrial Mn ratio is 1.2). Another argument for a larger portion of the Moon stemming from the Earth is the similar oxygen isotopic composition, which is distinctly different than that of Mars and Vesta (see Figure 2.2). Overall, the chemical signatures of the lunar silicates are not yet fully understood. In part, modeling is difficult because well-defined bulk abundances of the Moon require more measurements of lunar samples with more sensitive methods so that the bulk composition derived from such data become more robust. The other difficulty in deriving the formation history of the Moon from the chemical imprints in its silicates is that the formation history of the Moon is quite complex – as is the history of the proposed formation models itself.

4.2.3.2   Origin of the Moon.   There are at least five proposed theories for the origin of the Moon:

1. Fission from the Earth, George H. Darwin, 1879,
2. gravitational capture, H. Gerstenkorn, 1955,
3. co-accretion with the Earth, O. Schmidt, 1959,
4. disruptive capture, E. Opik, 1972,
5. formation of the Moon after a giant impact on the Earth, Daly 1946.[40]

Currently many researchers favor the impact origin of the Moon as the most plausible hypothesis. It postulates that the Moon formed in the aftermath of a single, glancing collision of a large, Mars-size body with the early Earth. Such a blow can transfer enough angular momentum to satisfactorily explain the observed angular momentum density of the Earth–Moon system when compared to that of the other planets. Other models have serious problems to explain the angular momentum density. The mass of the Earth lost to space during the impact can remove much of the original angular momentum.

George Darwin had derived the distribution of the angular momentum of the Earth–Moon system and discovered that the Moon must have been in much closer proximity to the Earth (as close as ∼3 Earth radii). Tidal friction increases separation of the Moon from Earth by about 4–5 cm per century at decreasing speed. The angular momentum of the Moon orbital motion now is about 4.82 times that of the angular momentum of the rotating Earth. Since angular momentum has to be preserved, the rotation of the Earth has slowed down correspondingly over time, which led to the "lengthening of day." Extrapolating back in time then would lead to an arrangement of the Moon being close to the Earth. This is where Darwin's idea about the formation of the Moon – tidal fission – came from: forming the Moon from the silicate portion of the Earth. This naturally explains the absence of a lunar metallic core, as implied by the Moon's lower density when compared to the higher densities of the Earth, Venus, Mars or Mercury.

However, the problem with direct separation of the Moon out of the Earth is that the Moon would be within the Roche limit of the Earth. Darwin himself pointed out that

"if the Moon were to revolve at a distance of less than 2.86 [Earth] radii, or 11 000 miles [the Roche limit], she would be torn to pieces by the Earth's tidal force."

Being of about Martian mass, the impacting body was very likely differentiated into a silicate mantle and a metallic core. The Moon formed from the silicates of the impacting body and ejected and evaporated terrestrial silicates. The core of the impactor would have penetrated though the Earth's silicate layer and combined with the Earth's metallic core. This scenario explains some similarities of several trace element abundances in the lunar and terrestrial silicates. The absence of a large metallic core on the Moon is also plausible if the Moon-forming impact happened after core formation in the Earth was essentially complete. However, any model that derives the Moon out of material from the Earth (such as fission theories) could lead to the same conclusions; such theories fail, though, the angular momentum test.

Recent reviews and research papers often refer to Hartmann and Davis[41] and Cameron and Ward[42] for proposals on the giant impact hypothesis for the origin of the Moon. There is no doubt that these works were the incentive for research on this topic carried out in the past three decades. However, the idea of an impact origin of the Moon was already proposed around 1946. Perhaps too buried in the literature, the paper describing this idea seems to be unknown to most researchers and historians on this topic since it is rarely, if at all, quoted in more recent reviews or books discussing the Moon's origins.

It may even be more of a surprise to researchers in the field that the idea of an impact origin of the Moon goes back to Henry Norris Russell, a pioneer in several astronomical studies (*e.g.*, see solar composition discussion in Chapter 1) and the American–Canadian geologist Reginald A. Daly (1871–1957).

In 1946, 30 years before the idea was re-proposed, Daly followed up on the impact origin of the Moon that Russell had suggested to him in a personal letter:

"He [Russell] suggested that it might be worth while to study the question whether the main part of the Moon's substance represents a planetoid, which, after striking the Earth with a glancing, damaging blow, was captured."

Daly[40] included this possibility as a "catastrophe hypothesis" in his paper on the lunar origin, and further postulated

"... that at the epoch concerned the Earth had practically its present mass and was liquid at the surface; that the planetoid had direct motion and struck the Earth at or near the equator; and that fragments were torn off directly by the visitor, along with others ejected because of an explosion resulting from the collision..."

In his summary, Daly already thought about several aspects that remain of interest today:

"...a 'planetoid', captured because of tangential, slicing, collision with the liquid Earth, brought with it so much angular momentum as to ensure its perpetuation as a separate, revolving body – the Moon we know." "...initially liquid fragments were exploded out of the planet [Earth] well beyond Roche's limit. Many of these were gravitatively [*sic*] aggregated by the pull of master fragment or captured 'planetoid' to make the substance of our Moon, and the somewhat diminished Earth felt a prolonged rain of other Earth-fragments, large and small."

Hartmann and Davis[41] proposed "Collision of a large body with Earth could eject iron-deficient crust and upper mantle material, forming a cloud of refractory, volatile-poor dust that could form the Moon."

Cameron and Ward[42] suggested:

"... a collision with a major secondary body in the late stages of accumulation of the Earth, with the secondary body adding its mass to the remainder of the protoEarth [*sic*]. ... Consider the consequence of this tangential impact [of a Mars-size object]. ... most silicates vaporize upon shock-unloading but the metallic core of the smaller body probably would not. ..."

Condensed dust grains "form a thin disk of the refractory particles. ...Beyond the Roche limit near three Earth radii, a collective gravitational instability can [...] produce gravitational clumping of particles. ... We estimate that the Moon can form very quickly from this disk."

Daly's catastrophe hypothesis "implies the absence of a volumetrically important iron core in the Moon – a fact demonstrated by the mean density of the body." Daly also notes "surface material of the Moon, exposed to the low gravitational pressure in our satellite and made rapidly viscid through free radiation of heat, to be vesicular to this day." Daly speculated that the silicate material of the Moon was initially hot and liquid at the surface, which is an early version of the magma ocean model that saw a revival in recent years. Gas held in solution could be freed during pressure crystallization inside the lunar globe. Thus, the lunar surface should show signs of moderate volcanism, which is seen in the dark Mare basins.

However, most of the lunar craters are products of impacts and not all are of volcanic origin. The impact origin for the craters was already entertained by Gilbert 1895, and an early review on the impact origin of craters is by Putnam Beard in 1917.[48] Some of clear impact signatures are the large diameters of the mainly circular craters, the light-colored striations of debris emanating from the large craters, the frequently occurring central peaks, and the relatively smooth crater floors. An impact origin of the craters requires impacting objects. Baldwin,[43,44] a dedicated proponent of the impact origin of the craters, assumed that the impacting objects are of terrestrial origin or smaller planetoids. Similarly, Daly[40] thought that these craters were made from remnant fragments of terrestrial material from the giant impact and/or from impacting asteroids.

If the craters were mainly due to impacts of other asteroidal or meteoritic material, the composition of the lunar soil should have signs of meteoritic contamination such as high Ni and Ir contents, which are the well-known signatures used to identify terrestrial impact structures. Figure 4.18 shows that these signatures – relatively high abundances of Ir and correlating abundances along the chondritic ratio) are present in the Apollo samples and the lunar meteorites.

We close this section with a little bit about the debate about the origin of the craters on the lunar surface. The term "crater," derived from the Greek bowl or cup, was originally associated with volcanoes, familiar features on the Earth, leading to the notion that craters on the Moon also have volcanic origins. As early as 1665, Robert Hooke (1635–1703) proposed the two competing theories for the origin of the lunar craters in his *Micrographia*. One was the volcanic origin driven by "Earthquakes and a hot interior" and the other the impact origin. Hooke tested his theories experimentally. He assumed that craters could form by steam explosions and he checked this Cyclopean gas-bubble theory by observing the behavior of boiled alabaster as it cooled down. Gas bubbles bursting through the solidifying surface left features that produced a small brim around the center of explosion looking not unlike lunar craters. Hooke preferred this model for their origin.

Hooke tested the impact theory by shooting bullets into watery tobacco pipe clay and noticed the similarity of the resulting structures to the lunar craters. His problem was to identify the projectiles hitting the Moon – meteorites were recognized only about 150 years later. Therefore, Hooke did not favor this theory, although the resemblance of impact craters and lunar craters was much closer.

The fight between scientists about which mechanism is responsible for the lunar craters continued into the middle of the twentieth century. Theories around the 1880s and 1890s proposed craters originated by volcanism driven by tidal stresses; a mechanism that is responsible for driving volcanism on the Galilean Moon Io. Franz von Gruithuisen (1774–1852) seems to be the first to revive the impact origin of the lunar craters, but since his ideas also included more exotic ones, such as life on the Moon or rapidly growing jungles on Venus, he was not taken very seriously.

Grove K. Gilbert brought credible considerations to the impact origin theory because experiments such as bombarding smooth clay surfaces with projectiles made of various materials reproduced the structural details of the lunar craters.[45,46] One needs to account for crater locations in the low laying surface plains, their large sizes, the presence of central peak that never exceed the height of the crater rims, and the dark inner crater plains (called maria). The impact theory also explains rugged and precipitous craters, and streaks radiating from craters such as Tycho and Copernicus. These structures are not characteristic of volcanic craters. Overall, Gilbert put forward detailed, logical foundations for the impact origin of lunar craters in his work *The Moon's Face; A Study of the Origin of its Features* in 1893:

"The impact theory applies a single process to the entire series. . ., correlating size variation with form variation in a rational way. . . . It brings to light the history of a great cataclysm, whose results include the remodeling of vast areas, the flooding of crater cups, the formation of irregular maria and the conversion of mere cracks to rills with flat bottoms. It explains the straight valleys and the white streaks. In fine, it unites and organizes as a rational and coherent whole the varied strange appearances whose assemblage on our neighbor's face cannot have been fortuitous."

More about the contributions by Hooke, Gilbert, as well as Alfred Wegener, better known for his contribution on continental drift, to this subject are given by E. T. Drake and P. D. Komar.[47] An early review by Putnam Beard (1917)[48] summarized the arguments in favor of an impact origin of lunar craters, concluding:

"In that remote aeon [*sic*] when the Earth was racked by the volcanic tumult of her early geologic birth, the Moon was bombarded and gashed by colossal meteoritic missiles from the sky; everlastingly sealing her doom as the abode of intelligent life."

4.2.4 Comparison of Planetary Compositions

The silicate portions of Earth, Moon, Mars and Vesta contain the refractory, lithophile elements in approximately chondritic abundances, but are all notably deficient in volatile lithophile elements when compared to chondritic meteorites. Here "chondritic" means not strictly CI-chondritic composition but encompasses the range of refractory element ratios found in different chondritic meteorite groups (Chapter 2). Figure 4.19 gives a comparison of some characteristic element ratios of the planetary objects.

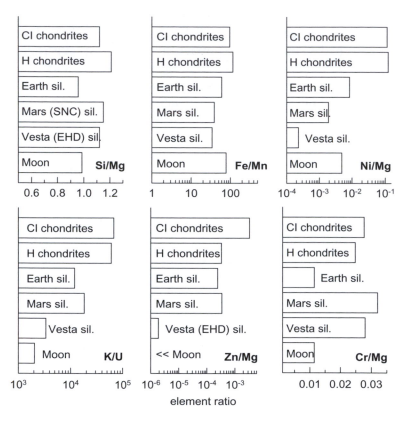

Figure 4.19   Characteristic element abundance ratios in the silicate portion of planetary objects and chondrites.

A low Si/Mg ratio is characteristic for the Earth and Moon whereas Mars and Vesta have chondritic ratios. The Cr/Mg ratios of the Earth and Moon are about the same and unique, and a similar conclusion can be drawn for V/Mg and Mn/Mg ratios (not shown). The Fe/Mn ratios of the Earth and Moon are between those of CI (or H) chondrites and Mars and Vesta. This ratio reflects the extent of Fe removal into the core (or the oxidation state of the silicates), but only if it can be shown that Mn is not depleted from chondritic values through volatilization or extraction into the core under very reducing conditions. This is important when the Earth and Moon are considered because their Mn/Mg, as well as Cr/Mg, ratios (Figure 4.19) are lower than in other objects.

All these previous ratios involve more or less refractory elements. However, there are large differences between chondritic and planetary abundances for volatile elements such as the alkalis, halogens or Zn. For example, they span about two orders of magnitudes for K/U, and more than three orders for Zn/Mg (Figure 4.19). These depletions do not correlate with an object's distance from the Sun, because Vesta – being more distant – should have more or the same relative amounts of volatiles as Mars. The volatile abundances may depend on object mass because there are notably lower volatile element inventories in the smaller objects (moon and Vesta). Volatile elements may be lost during accretion and differentiation through hydrodynamic escape, which is easier from objects of low-mass. The mechanism of the strong volatile element depletions in differentiated objects is not yet well understood.

A planet with about chondritic refractory lithophile element abundances is likely to have chondritic abundances of refractory siderophile elements. This applies to the high-density objects Earth, Mars and Vesta but not to the Moon, which lacks a large metallic core that would host most of the refractory siderophile elements. Figure 4.19 shows that the Ni/Mg (and respective ratios for other moderately siderophile elements) are all lower in the differentiated objects than in chondrites due to extraction of Ni into the core. The extent of siderophile element removal depends on the oxygen fugacity during planetary differentiation and most likely on the size of the object as metal–silicate equilibration may not be achieved.

The abundances of siderophile elements in the planetary silicates and experimental metal/silicate partition coefficients are roughly consistent with silicate mantle and core equilibration in the smaller objects Mars and Vesta. This suggests that these objects accreted homogenously (see below). Judging from the available low- and high-pressure partition coefficients and the impossibility of matching all siderophile element abundances in the Earth by element partitioning we have to conclude that the Earth's core is not anywhere near equilibrium with its silicate mantle. This suggests either incomplete silicate–metal equilibration or that the Earth accreted heterogeneously.

Mercury and Venus are lacking in our discussion of the terrestrial planets' composition because we simply do not have sufficient chemical information on their rock chemistry. The results from *in situ* measurements on space craft landers are not yet sufficient for reliable models of Venus' "geo" chemical evolution. There are only very limited measurements of the Venusian crust from Soviet Venera missions, which are not very informative of the silicate mantle chemistry. The composition of the dense Venusian atmosphere is better known and understood, and surface interactions with the hot atmosphere are described later.

The situation for Mercury is not much better but will be improved by the ongoing Messenger mission. Ideally, better analytical *in situ* tools such as employed during the various Mars missions may yield data to better understand Mercury, Venus and the asteroids. The best way to analyze samples for their major- and trace element chemistry and their isotopic compositions is in the laboratory, which requires sample return. The lunar samples are the only extraterrestrial samples where we know the parent object for sure. Although the inference is relatively strong that the SNC meteorites are from Mars and the EHD meteorites from Vesta, we cannot state this as a fact with 100% certainty until samples have been returned from these objects.

4.3   ACCRETION SCENARIOS

Possible planet compositions depend on the available amounts of the elements, the thermal stability of the phases that they form and the planets' accretion locations within the solar nebula. The amount of gas, rock and ice in the accretion disk and the accretion efficiencies determine a planet's possible mass. Once accreted, differentiation processes in the planetary environment determine the phases that are present within a planet and which then determine the planet's structure and size.

4.3.1   Rock, Ice and Gas

A simple approach to making terrestrial planets is to assume that there are no elemental fractionations other than volatility based fractionations. The condensation temperatures (Chapter 3) provide useful guidelines for this. This leads to three principal types of compounds, loosely referred to as "rock," "ice" and "gas," that can go into making planets. The distinction of "rock" from "ice" is based on the thermal stability of the different phases included in these components. Here "rock" is the assortment of high temperature phases made from the major elements like Si, Mg, Ca, Fe, Ni, Al, Ti, *etc.* that form oxides and silicate minerals, FeNi metal alloys and iron sulfide. For example, usually the most thermally stable solid containing most of the Fe is an FeNi metal alloy; for Mg, it is the Mg-silicate forsterite, (Mg_2SiO_4), and for sulfur, it is troilite, FeS. A good working definition of "rock" is a component containing the approximate elemental abundances and consisting of the mineral phases that are observed in undifferentiated, chondritic meteorites. All the rocky compounds are relatively refractory and require high temperatures for evaporation or condensation in a gas of solar system composition (Chapter 3). Rocky material is the last material to evaporate or the first to condense, so it is always present among solids in an accretion disk. The solar system abundance of the elements limits the amount of rocky material to $\sim 0.5\%$ of the total mass in the solar nebula.

Low temperature phases that sequester C, N and O more or less quantitatively are collectively called "ice." Icy materials only become stable below ~ 200 K in the low pressure solar nebula. Water ice is the most important ice because O is the third most abundant element in the solar system, and water ice is the most refractory ice among these ices. (The maximum oxygen in rock bound to abundant cations is $\sim 20\%$ of all available solar system oxygen.) If completely condensed, the ices make about $\sim 1\%$ of the total mass (water is $\sim 0.6\%$) and, together, the rocky and icy components (noble gases inclusive) make up $\sim 1.5\%$. The "gas" component is mainly H_2 and He and includes all the remaining mass that is left after rocky and icy condensates are removed.

The terminology rock, ice and gas is often used to characterize the element content of phases that accreted to a planet, and not necessarily the nature of phases that occur in a planet today. For example, Jupiter's putative core is often described as a "rocky and icy core," but this is only descriptive for the elemental make-up, and does not mean that there are actually rocky and icy phases present as we normally known them.

To first order, the radial distribution of the rocky and icy components depends on the temperature and total pressure gradients in the solar nebula. In absence of turbulent redistribution of these components, the relative amounts of the solid planetary building blocks vary with radial distance. This should have affected the masses, types and number of resulting planets.

In the terrestrial planets and asteroids, the "rocky" component is dominant and contributions from the icy and gas components are rather small. During accretion, heating may break down the individual phases in the rocky planetesimals, and the elements react and redistribute into phases that are stable under conditions of the planetary environment. The separation of the "rocky" elements into the silicate and metallic portions in the Earth and other terrestrial planets is a well-known example. Another result of planetary differentiation is the redistribution of rocky elements within the Earth's silicate portion into a crust, the upper mantle of mainly olivine, and a lower mantle containing larger amounts of a high-pressure modification of $(Mg,Fe)SiO_3$ called silicate perovskite (not to be confused with the Ca-Ti mineral perovskite, which has a similar structure at normal pressure). By analogy to meteorites, the elements in the metallic cores in terrestrial planets are the "rocky" elements Fe, Ni and S, which transform into high pressure metal and sulfide phases in planetary interiors; phases that are in quite different states than those encountered in the low-pressure mineralogy of chondritic meteorites.

The solar abundances of the rocky elements permit a range of metal and sulfide to be present depending on the oxidation state. Assuming that the Fe/Mg ratio in the planet is the same as in solar composition but allowing for S abundances from solar S/Fe to zero S, it is relatively simple to calculate the range in "cosmochemically constrained" core sizes that can result for terrestrial planets. This is shown in Figure 4.20.

The proportions of silicates and metallic phases are not the same in each of the terrestrial planets, which is easily seen from a comparison of the planet densities and estimated core sizes. In Figure 4.20, two end members are shown. One assumes that sulfur is present in relative solar abundance. If all Fe is in metallic form and the silicates are FeO-free, the sum of Fe + S represents the maximum metallic portion (38%) that can be obtained for a planet that contains the rock-forming elements in solar proportions. On the other hand, if a core were solely made of FeS and all the other Fe is in silicates, one obtains a pure sulfide core of 21% of the total planetary mass. The second case is that the planet has no volatile elements such as sulfur. Then the core size is only determined by the distribution of Fe between the metal core and FeO and NiO in

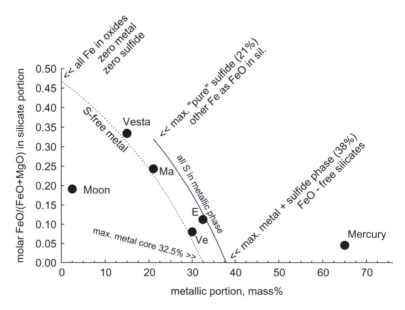

Figure 4.20 Inferred core sizes for the terrestrial planets, the Moon and Vesta are compared to the possible compositions derived from the solar abundances of elements.

silicates. If all Fe is oxidized, there is no core and the silicates have a molar $FeO/(FeO + MgO)$ of $\sim 47\%$. If the silicates were FeO-free, the maximum core size is 32.5% of the planet's mass. This is a little smaller than in the previous case because there is no sulfur contributing to the core. The core sizes with varying FeO contents of the silicates fall along the lines drawn for S-free planets and planets that have a solar complement of S in their cores. Planets with varying S contents from zero to solar can have core sizes that fall between the two lines given by the end-member compositions.

The estimated core sizes of the Earth, Mars, Venus and Vesta are consistent with what one can expect for core sizes from solar compositions and varying S contents. We have seen above that the Earth's core is not very likely to contain as much S as would be implied by Figure 4.20 (the Earth plots close to the S-rich core line). It more likely contains larger amounts of Si instead. An additional element as a core constituent is not considered in the modeling of Figure 4.20 and would add another dimension to the models.

Figure 4.20 illustrates that the estimated core sizes of the Moon and Mercury are inconsistent with the "cosmochemical" core sizes from solar composition. The Moon has only a small core, and we have seen that the formation of the Moon is unlike that of the larger planets Earth, Venus and Mars. Mercury, on the other hand, has a rather larger metallic core.

As mentioned earlier, Mercury's small size and high bulk density of $5.44 \, \text{g cm}^{-3}$ show that it has an iron to silicate mass ratio of about 70% to 30%, which is about twice as high as any other terrestrial planet, the Moon or the eucrite parent body. So Mercury either accreted from iron-rich, silicate-poor material or accreted from a mixture of iron and silicates, from which silicates were lost later. Our discussion of nebular chemistry in Chapter 3 showed that Fe metal condenses prior to ferromagnesian silicates at pressures above 10^{-4} bar in solar composition material. The temperature difference between the Fe and olivine condensation curves increases with increasing total pressure, but only amounts to a small radial distance separation between metal and silicate (for thermal profiles with T inversely proportional to radial distance R in the nebula). A very narrow planetary accretion zone is required for Mercury to accrete Fe-rich and silicate-poor material. The accretion zone is widened if condensation took place at higher C/O ratios, like those required for formation of the enstatite chondrites. In this case the separation between the metal and olivine condensation curves is larger. However, if Mercury accreted from material formed at high C/O ratios it would contain large amounts of carbides, nitrides and sulfides that condense at high temperatures at high C/O ratios.

Alternatively, Mercury lost most of its silicates after its accretion. This could have occurred due to extensive vaporization of its silicate mantle or to a large impact that shattered the silicate mantle. The two scenarios lead to compositional differences such as depletions of U, Ce and other easily oxidized rare earth elements during vaporization but not during impacts. Whether or not this is the case remains unknown, but may be testable from the elemental analysis measurements being made by the MESSENGER spacecraft.

4.3.2 Accretion Models

The general idea is that the larger planetary objects grew from a hierarchy of small, km-size planetesimals in the solar nebula disk. The overall elemental composition of the planet-building planetesimals was controlled by the abundances of the elements and their cosmochemical behavior under the conditions at their formation locations in the inner solar nebula. Taking such a simple approach, there could be trends in physical properties and chemical compositions of the planets as a function of their orbital distance from the Sun. If so, this may relate planet composition back to the thermal and density structure of the solar nebula. On the other hand, the final composition and make-up of a planet as seen today was influenced by its mass, its distance from the Sun and its accretion and post-accretion history.

Dynamical models of planetesimal accretion suggest that the terrestrial planets mainly formed from rocky planetesimals in the inner solar system. These planetesimals formed within similar orbital regions that the planets and asteroids – the remnant planetesimals in the asteroid belt – occupy today. A widely used concept is that the planets accreted material from certain "feeding zones" extending from inside to the outside of their orbits. If so, the overall composition of the terrestrial planets should have retained the memory of the local composition of the solar nebula from which the materials for the planetesimals and planets were derived. To a first approximation, such expected compositions can be obtained from equilibrium and non-equilibrium condensation calculations considering the radial density variation of the solar nebula and its thermal structure.[56–58]

If the inner solar nebula was too hot to allow full condensation of moderately volatile and volatile elements, one should find volatility-controlled abundance gradients in planetary compositions. Planets closer to the Sun should be rich in refractory elements, and planets closer to Jupiter should be rich in volatile elements. An individual planet would accrete relatively more reduced and refractory element-rich materials from the inner edge of its feeding zone and volatile rich and oxidized materials from the outer edge of the feeding zone. A complication of this simple scenario is that cooling of the nebula increased the thermal stability of volatile-bearing compounds. We can expect that the planetesimal composition changed over the lifetime of the solar nebula. However, of all planets, Mercury is the most likely planet that accreted the least volatiles, such as alkali elements or water. The amount of water, for example, influences the oxidation state and therefore the FeO content of planetary silicates and the size of the metallic core.

There is some evidence that compositional classes of planetesimals dominated as a function of radial distance within the solar system. The asteroids, the principal meteorite parent bodies, are remaining planetesimals believed to have formed between about 2 and 5 AU (Figure 4.21). Reflectance spectroscopy reveals distinct populations categorized by similar spectral features of asteroid surfaces at different locations in the asteroid belt. In the inner belt, asteroid types with "dry" surface mineralogies are common (*e.g.*, the S type asteroids) whereas the outer asteroid belt populations have evidence for varying amounts of hydrated silicates as well as organic materials and ices (*e.g.*, the C type asteroids). The different meteorite classes represent at least one type of asteroid and, considering the diversity of meteorites, the solar nebula had a pretty inhomogeneous composition over a relatively short heliocentric distance ($\sim$2–4.8 AU); from extremely reducing materials with Si in metal alloys (*e.g.*, enstatite meteorites) to extremely oxidized meteorites with large amounts of Fe in the magnetite (*e.g.*, CI-chondrites). One is tempted to conclude that the rest of the solar nebula was very inhomogeneous in composition as well.

It is possible that planetesimals accreting to a terrestrial planet originated from solar nebula materials outside a planet's feeding zone. A planetesimal formed outside a planet's feeding zone may be scavenged by it after being dynamically scattered into the planet's orbit. This introduces changes to the planetary composition because material with different amounts of volatile elements and oxidation state can be added to the solids at the planet's orbital region.

For example, water ice accretion by planetesimals can only occur in cool regions of the nebula beyond $\sim$2.5 AU. Consequently, planetesimals formed in the inner solar nebula are expected to be bone-dry. This would have led to equally dry terrestrial planets and one has to ask where the Earth, Venus and Mars got their water from. On the other

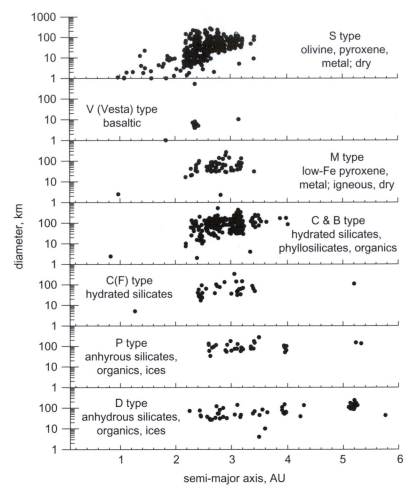

Figure 4.21 Some of the different asteroid classes as a function of radial distance from the sun.

hand, planetesimals from the outer asteroid belt contain hydrated silicates, and the small objects in the outer solar system usually contain water ice and other volatile compounds (such as icy satellites and comets). The accretion of water-bearing planetesimals that originally formed beyond the orbits of the terrestrial planets has long been taken as a conceivable explanation for the presence of water on Earth and Mars and for the very likely presence of water on early Venus. These planetesimals are also plausible contributors of other volatile substances such as organics.

Even today, large-scale movements with changes in orbital parameters occur for the asteroids. Collisions break asteroids apart and smaller asteroids are dynamically scattered into new orbits. Such events are needed to replenish the near-Earth asteroid populations from which we derive our meteorites. Similar events happen in the Kuiper belt and the Oort cloud, where the remnants of icy planetesimals are still scattered into new orbits to produce comets. The cratering record on the Moon, Mercury, Venus, Mars, the asteroids and cometary nuclei shows that large impacts were common in the early solar system about 4500–3900 Ma ago. Towards the end of this "heavy bombardment period" the planets accreted to their final sizes; for example, the Earth accreted up to 5% of its mass, corresponding to a 30–120 km thick layer covering the Earth. It is believed that most of the Earth's water and other volatile compounds were delivered during this period, by comets or "icy" planetesimals. Thus, the gravitational forces between planetesimals removed objects with planet-crossing orbits by impacts onto planets, or moved them into more dynamically stable orbits in the asteroid belt.

The Earth was bombarded in the same manner as the other terrestrial planets and the Moon, although plate tectonics, volcanic resurfacing and erosion obliterated the cratering record on Earth over time. Larger impact structures from the not-so distant past remain. The 170 km diameter Chicxulub impact crater on the Yucatan peninsula stems from a projectile that caused an environmental disaster that wiped out the dinosaurs at the end of the Cretaceous about 66 million years ago. The 1.2 km diameter Barringer meteorite crater in Arizona is only about 50 000 years old. The occurrence of meteorites with masses up to several tens of tons on the Earth's surface attests that smaller objects continue to accrete to the Earth, and we are still seeing the final trickles of accretion.

4.3.2.1 Component Models. In absence of better evidence, many accretion models for the terrestrial planets postulate compositionally different components to reproduce characteristic planetary properties (silicate oxidation state, core composition, volatile content, density, core size). The relative abundances of refractory lithophile elements (but not of volatile elements) in planetary silicates are comparable to those in chondritic meteorites, suggesting a kinship between the relatively unaltered, primitive solar nebula material sampled by chondrites and the planets.

The principle of using meteorites as analogs for the make-up and building blocks of the terrestrial planets goes at least back to Chladni in 1794. Some accretion models for the Earth, Mars and Vesta invoke chondritic matter in more or less modified form as building blocks.

Meteorites are the only available records of what the rocky planet-esimals accreting to the terrestrial planets may have looked like.

The diversity of meteorite compositions requires different formation conditions in the solar nebula. The observations of variable volatile and refractory element contents, different oxidation states and different oxygen isotopic compositions that are seen in meteorites extend to planetary compositions. However, each planetary object has a charac-teristic oxidation state, metal and sulfide content, volatile element con-tent and oxygen isotopic composition, and no planet as a whole matches a particular chondrite group. No single group of chondrites is repre-sentative for the matter accreted by a planet, or as noted by Hart and Zindler (1986):[20]

> "In other words, the Earth is not like any chondrite (in major elements) but is composed of its own blend of accretion products, with a bias toward a higher proportion of the high temperature refractory components (or, equivalently, a lower proportion of the volatiles and partially refractory components)."

Successive planet build-up occurs from the assembly of smaller km-sized planetesimals that accrete into larger and larger objects. Objects of lunar and Martian size are occasionally called "planetary embryos." The basic schemes of the homogeneous and heterogeneous planet accretion models have very different assumptions but may lead to the same final planet. The "truth" of planet formation is probably in-between, and, as always, there are more considerations necessary to get the planet formation story right.

Homogeneous Accretion Model. The homogenous accretion model postulates that planetesimals contain silicates, sulfides and metal similar to (but not necessarily identical to) chondrites. Relatively slow accretion of these planetesimals does not produce much heating to melt silicates. Slow accretion without heating to silicate melting temperatures is necessary to explain larger chondrite parent planetesimals, since there are still chondritic meteorites around.

The type of asteroids providing the chondrites it is not known, let alone how large chondrite parent bodies were initially before they were shattered in the asteroid belt. It is plausible that these objects must have reached several 100-km in diameter. With ongoing asteroid collisional breakdown, only material from initially larger objects can survive in smaller objects and finally be delivered as meteorite fragments to present-day times.

Chondrites tell us that larger, undifferentiated planetesimals formed. Furthermore, observations of the largest asteroid Ceres suggest that this asteroid did not undergo metal–silicate differentiation, and not all "large" objects are differentiated into a silicate mantle and metal core. However, Ceres is "differentiated" in the sense that aqueous alteration led to the formation of hydrated minerals such as phyllosilicates. However, the term "differentiated" is usually not applied to such alterations.

To form a planet that differentiated into a metallic core and silicate mantle, an efficient heat source is required. The available heat sources are gravitational energy released during accretion and radioactive heating from short-lived radionuclides (Chapter 3).

Heterogeneous Accretion Models. There are two variations of the heterogeneous accretion model in the literature, which is unfortunate as it only serves to confuse new students in the field. The original version of the heterogeneous accretion model, going back in its theme at least to William Thomson (Lord Kelvin), postulates that the first accreting planetesimals are metal-rich and lay down a "core" onto which later on silicate-rich planetesimals will accrete. This would naturally explain the layer structure of the terrestrial planets and some differentiated meteorite parent bodies. However, it leaves the question where such metal-rich or only silicate-rich planetesimals were produced in the first place. One possibility is condensation of metal before silicates at relatively large total pressures in the inner solar nebula (see Figure 3.8). Other mechanisms of metal–silicate fractionation within the solar nebula before larger planetesimal accretion are aerodynamic density sorting or magnetic separations. While attractive, these may not have worked in the inner solar nebula that is occupied by the terrestrial planets and asteroids today. For example, the existence of chondrites with their intimate mixtures of fine silicates, metal and sulfides requires some flexibility in the model to explain why metal–silicate fractionation did not occur in certain nebular regions.

A consequence for a heterogeneously accreted planet is some pre-differentiation into a core and a silicate shell. In other words, one would not expect a well-developed equilibration of silicate and metal, which would leave particular signatures in trace element distributions between the core and mantle silicates. Equilibration of metal and silicate would be restricted to any residual metal that came with the silicate-rich planetesimals, if there is a sufficient heat source to allow melting and density separation of such metal from the silicates.

Modified heterogeneous accretion models were proposed.[49] In these models, silicate-metal(-sulfide) bearing planetesimals of different oxidation states and with different amounts of low-temperature condensates accreted in succession. This variation of heterogeneous accretion model is also referred to as a "two-component model." This terminology is adopted here to distinguish this from the original heterogeneous "layer" accretion model for the terrestrial planets.

In addition to these major accretion themes, there are several other variants of accretion models. All these models have their merits, but also their failures in explaining the composition of the terrestrial planets and differentiated asteroids.

4.3.2.2   Heat Sources for Differentiation.   The Earth as an initially hot and molten mass was already considered by Descartes and Leibnitz. If several rocks on the Earth's surface were still forming by igneous processes, Leibnitz argued, the Earth interior must have been and still be hot enough to melt them. He thought that the liquid Earth first cooled at the crust, and that the interior still is liquid but, as described above, this view had to be revised. The energy necessary for planetary differentiation comes from three principle sources: release of heat generated during formation of the Earth; release of latent heat from crystallization of the inner core; and heating from decay of radioactive nuclides.

Early and fast accretion of planetesimals and planets provide the best conditions to heat and melt planetary materials. Heating by gravitational energy release is effective when planetesimal and planet mass-build-up is faster than radiative surface cooling, which is the only heat loss mechanism for any planet. Latent heat release from crystallization of the inner core may support the fluidity of the outer core driving the magnetic field. The other widely discussed heat source is the energy released during radioactive decay of short-lived radioactivities (Chapter 3).

Gravitational Heating.   The Earth and other terrestrial planets (Mercury, Venus and Mars) formed by accretion of smaller rocky planetesimals ranging in size from boulders to Mars-size objects. The accretion of the terrestrial planets released large amounts of gravitational potential energy as thermal energy. The accretion heat caused geochemical reactions between molten metal and molten silicate, vaporized part of the accreting planets to form atmospheres, and was radiated away to space. We know the most about accretion of the Earth and focus on this in the discussion below.

The thermal energy (E) released by the accretion of the Earth or another planet is:

$$E = \frac{GM_P^2}{R_P} \tag{4.14}$$

where G is the universal constant of gravitation (6.6726×10^{-11} m^3 kg^{-1} s^{-2}), M_P is the planetary mass in kilograms, R_P is the planetary radius in kilometers, and the thermal energy (E) is in joules. For example, accretion of the Earth ($M_P \sim 6 \times 10^{24}$ kg, radius $R_P \sim 6370$ km) releases 4×10^{32} J of heat. Thermodynamic calculations show that this is more than enough energy to heat up and vaporize Earth's silicate lithosphere.

As discussed above and in Chapter 6, the lithosphere ($M_L \approx 4.01 \times 10^{24}$ kg) is dominated by the mantle (99.4% by mass). The mantle is mainly composed of forsterite Mg_2SiO_4 ($\mu \approx 140$ g mol^{-1}) with smaller amounts of other minerals. To a good first approximation the Earth's lithosphere is forsterite. The energy required to heat up, melt and vaporize one mole of forsterite from room temperature (298 K) to its 1-bar boiling point (3540 K) is about 1180 kJ. This is the enthalpy of vaporization ($\Delta_{vap}H$) of forsterite, and most of this energy goes into melting solid forsterite and vaporization of molten forsterite. The energy (E_{vap}) to vaporize the Earth's lithosphere is:

$$E_{vap} = \frac{M_L}{\mu}\Delta_{vap}H = \frac{4.01 \times 10^{27}\,\text{g}}{140\,\text{g mol}^{-1}} \times 1.18 \times 10^6\,\text{J mol}^{-1} \sim 3 \times 10^{31}\,\text{J} \quad (4.15)$$

This is about ten-times less than the thermal energy released by accretion of the Earth, so vaporization of the entire lithosphere could occur if all energy were released at one time. How likely is this?

Computer models of terrestrial planet accretion show that impacts between large objects were frequent in the early solar system. For example, the impact of a Mars size object with the Earth during its accretion is thought to have formed the Moon. Objects accreted by the Earth had impact velocities of about 10 km s^{-1}. The kinetic energy (E_K in Joules) from each impact is:

$$E_K = \tfrac{1}{2}Mv^2 = 5 \times 10^7 M \quad (4.16)$$

The impactor mass (M) is in kilograms in this equation. The kinetic energy from each impact is used to heat up solids to their melting points, melt the solids, heat the resulting melts to their boiling points, vaporize the melts and then heat the gases produced. Table 4.8 lists the kinetic energy and thermal effects from impacts of increasingly larger objects on the Earth during its accretion. As mentioned earlier, part of the thermal energy produced is radiated away to space and cools the Earth. Radiative cooling times vary from about 10 years for a Pallas size

Table 4.8   Impact energy.

ΔE (J)	Size of impactor	Thermal effects for Earth
7×10^{27}	1.4×10^{20} kg ($\sim$ mass of asteroid 2 Pallas)	Boil oceans and heat to 2000 K
5×10^{28}	1×10^{21} kg ($\sim$ mass of asteroid 1 Ceres)	Melt crust and heat to 2000 K
2×10^{29}	4×10^{21} kg ($\sim$ 5% mass of Earth's moon)	Vaporize crust and heat to 3200 K
3×10^{31}	6.8×10^{23} kg ($\sim$ mass of Mars)	Vaporize silicate Earth and heat to 3540 K

impact to about 1000–10 000 years for the Moon forming impact of a Mars-size object.

Radioactive Heat Sources. The significance of short-lived radio-nuclides as heat sources for planetoid and planetary differentiation was realized by Urey (1955),[59] who suggested ^{26}Al, and Kohman and Robison (1980),[60] who proposed ^{60}Fe. These are isotopes of more abundant elements in rocky planets and therefore are expected to be more efficient heat sources. The most productive heat sources are short-lived radio-nuclides that have higher abundances and higher decay energies.

In 1955, Urey considered eight known radioactive nuclides as heat sources for driving differentiation processes on planetesimals and planets. The reason for considering this was that other work had suggested the possibility of star and planetesimal formation within a few million years. Urey estimated the heat generated by ^{26}Al, ^{36}Cl, ^{93}Zr, ^{107}Pd, ^{129}I, ^{135}Cs, ^{236}U and ^{237}Np and concluded that of these known short-lived nuclides at the time only ^{26}Al and ^{36}Cl could have been abundant enough to be considerable heat producers. He estimated that the initial abundance of the radionuclides might have approached that of their neighboring isotopes. In that case, and considering the solar elemental abundance distribution (Chapter 1), nuclides from elements beyond Fe can be expected to be much lower in abundances, which only left ^{26}Al and ^{36}Cl in his list. Because the half-life of ^{36}Cl is only half that of ^{26}Al, Urey concluded that ^{26}Al, which had just been discovered to have a half-life of the order of around a million years, is the best chance for a short-lived radioactive heat source in the early solar system.

The potential of ^{60}Fe as a heat source in the early solar system and use as a chronometer was investigated by Kohman and Robison (1980). At the time, one problem in the modeling was the uncertain half-life of ^{60}Fe, which is currently taken as twice that of ^{26}Al. The ^{60}Fe half-life of 1.49 Ma still has an uncertainty of $\sim$ 18%, and new evaluations suggest the half-life is longer. The decay of ^{60}Fe over ^{60}Co to stable ^{60}Ni causes relative ^{60}Ni excesses, and meteoritic metal and especially Fe-rich but relatively Ni-poor sulfides are good targets for searches of such excesses, which have been found a few years ago.

Heating by short-lived nuclides was more important for the young Earth. The major contribution of long-term radiogenic heat comes from ^{40}K, ^{232}Th, ^{235}U and ^{238}U. The radiogenic heat production rate per unit volume is calculated from:

$$dE/dt(\mathrm{J\,cm^{-3}\,yr^{-1}}) = \rho_{\mathrm{sample}}\Sigma(\varepsilon_i \cdot f_{ij} \cdot C_i) \times 10^{-6} \qquad (4.17)$$

where ρ_{sample} is the sample density in $\mathrm{g\,cm^{-3}}$, C_i is the concentration of element i in the sample in $\mathrm{\mu g\,g^{-1}}$ (ppm), f_{ij} is the present-day mass fraction of radioisotope j for element i and ε_i is the decay energy per unit mass and time in $\mathrm{J\,g^{-1}\,yr^{-1}}$. Table 4.9 gives the decay energies of the four long-lived isotopes and their respective mass fractions of K, Th and U.

Inserting the values in Equation (4.17) gives the heat production rate for present-day radionuclide concentrations:

$$dE/dt\,(\mathrm{J\,cm^{-3}\,yr^{-1}}) = \rho_{\mathrm{sample}}(1.05 \times 10^{-4}C_{\mathrm{K}} + 0.8368C_{\mathrm{Th}}$$
$$+ (0.1295 + 2.9489)C_{\mathrm{U}} \times 10^{-6} \qquad (4.18)$$

Heat production rates from these radionuclides were larger in the past since their initial concentrations have been reduced through decay. The concentration and the heat production rate of a radionuclide with radioactive decay constant λ_i is reduced by a factor of $\exp(-\lambda_i t)$ over time t. The mean heat production rate over 4.5 Ga is described by:

$$dE/dt\,(\mathrm{J\,cm^{-3}\,yr^{-1}}) = \rho_{\mathrm{sample}}\Sigma\varepsilon_i \cdot f_{ij} \cdot g_{ij} \cdot C_i/\exp(-\lambda_i t) \qquad (4.19)$$

Here g_{ij} is a weighing factor for each radionuclide; values are listed in Table 4.9. Inserting values gives:

$$dE/dt\,(\mathrm{J\,cm^{-3}\,yr^{-1}})$$
$$= \rho_{\mathrm{sample}}\left[3.05 \times 10^{-4}C_{\mathrm{K}} + 0.903\,C_{\mathrm{Th}} + (1.460 + 3.849)C_{\mathrm{U}}\right] \qquad (4.20)$$

Table 4.9 Mass contributions of radionuclides to given elements and their decay energies, and weighing factors for mean heat production.

Radionuclide	f_{ij} [g (g-element)$^{-1}$]	ε_i $(\mathrm{J\,g^{-1}\,yr^{-1}})$	g_{ij}
^{40}K	0.000119	0.879	0.272
^{232}Th	1	0.8368	0.860
^{235}U	0.0072	17.99	0.135
^{238}U	0.9927	2.971	0.645

Using the mantle and crust concentrations in $10^{-6}\,\mathrm{g\,g^{-1}}$ (ppm) in Tables 4.3 and 4.4, the characteristic present-day heat production rate for mantle rocks is $0.67\,\mathrm{J\,cm^{-3}\,yr^{-1}}$ and 39.33 and $1.46\,\mathrm{J\,cm^{-3}\,yr^{-1}}$ in the continental and oceanic crust, respectively. Corresponding values for the mean generated heat over 4.5 Ga are 1.17, 62.8 and $2.59\,\mathrm{J\,cm^{-3}\,yr^{-1}}$. The significantly larger heat production rate in the continental crust stems from the large incompatibility of K, Th and U in mantle rocks and their accumulation in the continental crust. The oceanic crust and (upper) mantle have lower heat production rates.

4.4 ATMOSPHERES AND SURFACES OF VENUS AND MARS

4.4.1 Overview

Venus and Mars are the two planets nearest to Earth. Mars is about 52% farther away from the Sun than the Earth (1.52 *versus* 1 AU) and receives about 43% less solar energy. The amount of solar radiation received by Mars is 45% greater at perihelion (where it is closest to the Sun) than at aphelion (where Mars is farthest from the Sun) because of its large orbital eccentricity (0.093 *versus* 0.017 for Earth and 0.007 for Venus). Venus is about 28% closer to the Sun than Earth (0.723 *versus* 1 AU) and receives about 1.9 times as much solar insolation as Earth.

However, Venus absorbs only about 62% as much solar energy as the Earth because of its high albedo (reflectivity) of 0.76. This is higher than the albedo of either Earth (0.31) or Mars (0.16) and is due to the reflectivity of the global aqueous sulfuric acid droplet ($H_2SO_4 \cdot 2H_2O$) clouds on Venus. Most of the solar radiation absorbed by Venus is deposited in the upper atmosphere and clouds, in contrast to the Earth where 66% of solar energy is deposited at the surface. On average, clouds cover only 52% of Earth's surface and have albedos ranging from 0.15 to 0.80.

On Mars, atmospheric silicate dust, when present in even small amounts, is the major absorber of solar radiation, which is otherwise absorbed by CO_2. Mars is noted for its periodic regional global dust storms, which occur annually, and for planet-wide dust storms that occur about once every three Martian years. In general atmospheric dust plays a central role for absorption of solar radiation in the Martian atmosphere and causes changes in its thermal structure and dynamics over diurnal, seasonal and inter-annual time periods. Condensate clouds formed from water ice (*e.g.*, over the northern polar regions in wintertime) and CO_2 ice (*e.g.*, during nighttime in the polar winter or at 60–100 km high in the atmosphere) also occur on Mars.

The average surface temperatures are 740 K on Venus and 220 K on Mars (*versus* 288 K on Earth). The greenhouse effect (discussed in Chapter 6) raises the surface temperature of Venus far above the temperature it would otherwise have, which is 229 K (*versus* 254 K for the Earth). The CO_2, SO_2 and H_2O in Venus' atmosphere produce about 511 K of greenhouse warming *versus* 34 K on Earth from the CO_2 and H_2O in the terrestrial atmosphere. Greenhouse warming on Mars is negligible in comparison (about 4 K) because of its thin CO_2 atmosphere.

The average surface pressures are 95.6 bar on Venus and $\sim 6 \times 10^{-3}$ bar on Mars *versus* about 1.013 bar on Earth. The high surface pressure on Venus arises from its massive CO_2-rich atmosphere, which has a molecular column density of 1.40×10^{27} CO_2 molecules per cm^{-2}.[i] The amount of CO_2 in Venus' atmosphere is about twice the amount of carbon in Earth's crust, and is 256 000 times larger than the amount of CO_2 in Earth's atmosphere (5.46×10^{21} molecules cm^{-2}). Even though the Martian atmosphere is thin, it is mainly CO_2 and it contains 2.26×10^{23} CO_2 molecules cm^{-2}, about 40 times larger than that in the terrestrial atmosphere.

Venus' lower atmosphere (0–60 km altitude, z) is convective with a temperature gradient (dT/dz) of 7.7 K km^{-1}, which is close to the dry (condensation cloud free) adiabatic gradient for a binary 96.5% CO_2–3.5% N_2 mixture, *i.e.*, the same proportions as in Venus' atmosphere. The temperature and pressure are 660 K and 48.0 bar at the top of Maxwell Montes, which is the highest point on Venus (10.4 km above the modal radius where $T = 740$ K).

The lowermost part (0–10 km) of the Martian atmosphere is nearly isothermal and the rest of the troposphere has a temperature gradient of 2.5 K km^{-1}, which is lower than the dry adiabatic gradient of 4.5 K km^{-1}. Temperatures at the surface of Mars vary from ~ 147 K over the polar caps to ~ 300 K at the subsolar point (where the Sun is directly overhead) at perihelion in Mars' orbit.

Tables 4.10 and 4.11 give the chemical compositions of the atmospheres of Venus and Mars, respectively. These data come from Earth-based and spacecraft spectroscopic observations and from *in situ* measurements by analytical chemistry experiments, gas chromatographs, mass spectrometers and optical spectrometers on atmospheric

---

[i] The column density of a gas is the number of gas molecules or atoms throughout an atmospheric column from the surface to space. It has units of atoms (or molecules) per unit area and can be calculated from $P_i N_A / g\mu$ where P_i is the partial pressure of the gas, N_A is the Avogadro constant, g is the altitude-dependent gravitational acceleration on the planet and μ is the formula weight of the gas.

Table 4.10 Chemical composition of the atmosphere of Venus.

Gas	Abundance	Source(s)	Sink(s)
CO_2	$96.5 \pm 0.8\%$	Outgassing	UV photolysis, carbonate formation
N_2	$3.5 \pm 0.8\%$	Outgassing	NO_x formation by lightning
$SO_2{}^a$	150 ± 30 ppm (22–42 km) 25–150 ppm (12–22 km)	Outgassing & reduction of OCS, H_2S	H_2SO_4 formation, $CaSO_4$ formation
H_2O^a	30 ± 15 ppm (0–45 km)	Outgassing and cometary impacts	H escape, Fe^{2+} oxidation
^{40}Ar	$31 {}^{+20}_{-10}$ ppm	Outgassing (^{40}K)	—
^{36}Ar	$30 {}^{+20}_{-10}$ ppm	Primordial	—
CO^a	28 ± 7 ppm (36–42 km)	CO_2 photolysis	Photooxidation to CO_2
$^4He^b$	0.6–12 ppm	Outgassing (U, Th)	Escape
Ne	7 ± 3 ppm	Outgassing, primordial	—
^{38}Ar	5.5 ppm	Outgassing, primordial	—
OCS^a	4.4 ± 1 ppm (33 km)	Outgassing, sulfide weathering	Conversion into SO_2
H_2S^a	3 ± 2 ppm (<20 km)	Outgassing, sulfide weathering	Conversion into SO_2
HDO^a	1.3 ± 0.2 ppm (sub-cloud)	Outgassing	H escape
HCl	0.5 ppm (35–45 km)	Outgassing	Mineral formation
H_2SO_4	0.1–10 ppm (35–50 km)	SO_2 photolysis	Cloud formation
^{84}Kr	$25 {}^{+13}_{-18}$ ppb	Outgassing, primordial	—
SO^a	20 ± 10 ppb (cloud top)	Photochemistry	Photochemistry
$S_{1-8}{}^a$	20 ppb (<50 km)	Sulfide weathering	Conversion into SO_2
HF	4.5 ppb (35–45 km)	Outgassing	Mineral formation
NO	5.5 ± 1.5 ppb (sub-cloud)	Lightning	Conversion into N_2
^{132}Xe	<10 ppb	Outgassing, primordial	—
^{129}Xe	<9.5 ppb	Outgassing (^{129}I)	—

^{a}Abundances are altitude dependent.
^{b}He abundance in Venus' upper atmosphere is $12 {}^{+24}_{-6}$ ppmv.[50] The value listed above is a model-dependent extrapolation to lower altitudes.

entry probes and landers. However, the atmospheric composition below 22 km altitude on Venus, which is about 80% of the atmospheric mass, is poorly constrained. Several experiments on US and Russian spacecraft failed below this altitude and Earth-based spectroscopic observations of this region are extremely difficult.

Both atmospheres are dominantly CO_2 (96.5% on Venus, 95.32% on Mars) and N_2 (3.5% on Venus, 2.7% on Mars), both atmospheres contain variable amounts of water vapor ($\sim$30 parts per million by

Table 4.11 Chemical composition of the atmosphere of Mars.

Gas	Abundance[a]	Source(s)	Sink(s)
CO_2	95.32%	Volcanic outgassing, sublimation of CO_2 ice	Condensation to CO_2 ice, and photolysis to $CO + O_2$
N_2	2.7%	Outgassing	Escape as N
Ar	1.6%	Outgassing (^{40}K), primordial	—
O_2	0.13%	CO_2 photolysis	Photoreduction
CO	0.08%	CO_2 photolysis	Photooxidation
H_2O^b	~140 ppm	Evaporation, desorption	Condensation & adsorption
NO	~100 ppm (at 120 km)	Photochemistry (N_2, CO_2)	Photochemistry
H_2	15 ± 5 ppm	H_2O photolysis	Escape
He	10 ± 6 ppm	Solar wind, outgassing,	Escape
Ne	2.5 ppm	Outgassing, primordial	—
HDO^b	0.85 ± 0.02 ppm	Evaporation, desorption	Condensation & adsorption
Kr	0.3 ppm	Outgassing, primordial	—
Xe	0.08 ppm	Outgassing, primordial	—
O_3^b	~0.04–0.2 ppm	Photochemistry (CO_2)	Photochemistry
$H_2O_2^b$	10–20 ppb	H_2O photolysis	Photochemistry
HD	11 ± 4 ppb	H_2O photolysis	Escape
CH_4	10 ± 3 ppb	Outgassing	Photochemistry

[a]The mixing ratios, but not the column densities, of non-condensable gases are seasonally variable as a result of the annual condensation and sublimation of CO_2.
[b]Spatially and temporally variable.

volume (ppmv) on Venus, ~140 ppmv on Mars), and Ar is the most abundant noble gas in each atmosphere (0.007% on Venus, 1.6% on Mars). But there the similarity ends. Venus' atmosphere contains significant amounts of SO_2, OCS, HCl and HF that result from the high temperatures at Venus' surface. None of these gases are present in the Martian atmosphere and as discussed below; their abundances on Venus are orders of magnitude greater than those in Earth's atmosphere. The Martian atmosphere contains small amounts of O_2 (0.13% = 1300 ppmv), while the spectroscopic upper limit for O_2 is <0.3 ppmv in the well-mixed region (0–100 km) of Venus' atmosphere. Some O_2 is present in Venus' upper atmosphere (100–130 km) and produces visible and IR emissions (Herzberg II nightglow at 400–800 nm and infrared nightglow at 1.27 µm), but its abundance is not known. The Martian atmosphere also contains traces of O_3, H_2O_2 and CH_4 – none of which are present on Venus. Dioxygen, O_3, CO and H_2O_2 are present in the Martian atmosphere because of solar UV photochemistry. The CH_4

source for the Martian atmosphere is unknown, and various atmospheric, biological and geological mechanisms have been proposed.

The atmospheres of Venus and Mars are dramatically different than that of Earth (see Chapters 5 and 6), which is dominantly N_2 ($\sim 78\%$), with a large amount of O_2 ($\sim 20.9\%$). The Earth's atmosphere contains over 10 000 times more O_2 than Venus' atmosphere and over 15 000 times more O_2 than in the Martian atmosphere. The large amount of O_2 in Earth's atmosphere and of reduced carbon (*e.g.*, coal, gas and oil) in Earth's crust is due to photosynthetic conversion of CO_2 and H_2O into O_2 and organic matter. Biogenic emissions are the source of other important minor and trace gases such as CH_4, N_2O, OCS and CH_3Cl in the terrestrial atmosphere (see Tables 5.1 and 5.2). The atmospheres of Venus, Earth and Mars are three natural laboratories for studying the effects of thermochemistry, biochemistry and photochemistry on atmospheric chemistry, composition and structure.

4.4.1.1 Major Gases. The CO_2 abundance ($96.5 \pm 0.8\%$) is constant throughout Venus's atmosphere (0–100 km region). The $\pm 0.8\%$ uncertainty is due to the $\pm 0.8\%$ uncertainty in the N_2 abundance. The mass spectrometers on the US *Pioneer Venus* and Russian *Venera 11-12* spacecraft gave different N_2 abundances than the gas chromatographs on the same spacecraft. The apparent disagreements remain unresolved and the large uncertainty in the N_2 and CO_2 abundances persists. Volcanic outgassing is probably the ultimate source of CO_2 in the atmospheres of Venus and Mars by analogy with Earth where CO_2 is generally the second most abundant species (after steam) in volcanic gases. Sublimation of CO_2 ice (dry ice) is also a seasonal source of CO_2 on Mars. The annual sublimation of CO_2 from (and condensation of CO_2 back into) the polar caps causes an annual pressure change of 37% (2.4 millibar, mbar) relative to the global mean pressure of 6.36 mbar. This is much larger than pressure changes on Earth, which are typically 1% of the mean. Mars is volcanically inactive and volcanic outgassing of CO_2 is probably not taking place at present. In contrast, on Venus maintenance of the global sulfuric acid clouds requires ongoing volcanism at a rate that is about the same as the average rate of sub-aerial volcanism on Earth (1 km^3 magma per year). If the amount of CO_2 dissolved in the magma is similar to that on Earth, Venusian volcanism may provide $\sim 6.6\,Tg\,C\,yr^{-1}$ to the Venusian atmosphere (1 Tg $= 10^{12}$ g $= 10^9$ kg $= 10^6$ metric tons).

Volcanic outgassing is probably the major source of N_2 on Venus and Mars, by analogy with the Earth where atmospheric N_2 originated from volcanism and volcanic outgassing of N_2 continues today at a rate of < 1

Tg-N_2 per year (Chapter 6). Small amounts of nitric oxide NO are observed in the atmospheres of Venus and Mars, and NO formation by lightning (Venus) or upper atmospheric photochemistry (Mars) is a sink for N_2. Nitrate (NO_3^-) formation is another possible sink for nitrogen. The Wet Chemistry Laboratory and the Thermal and Evolved Gas Analyzer on the *Mars Phoenix Lander* provided evidence for perchlorate (ClO_4^-) salts on Mars. Photochemistry at the Martian surface provides an oxidizing environment that forms perchlorates and possibly nitrates. Nitrogen atom escape may also be a small sink for N_2 on Venus and Mars.

Photolysis of CO_2 is an important sink on Venus and Mars. Ultraviolet sunlight continually converts CO_2 into O_2 and CO on both planets. The central problem is why both atmospheres are still CO_2-rich and O_2-, CO-poor. On Venus, photolysis of CO_2 occurs throughout the upper atmosphere down to the cloud tops at ~ 60 km altitude. Ultraviolet sunlight would destroy all CO_2 above the clouds in $\sim 14\,000$ years and all CO_2 in Venus' atmosphere in ~ 5 million years unless CO_2 reforms in some manner. Furthermore, observable amounts of O_2 (above the spectroscopic upper limit of 0.3 ppmv) would be produced in about 5 years. On Mars, photolysis of CO_2 occurs throughout the *entire* atmosphere down to the surface. This is not too surprising when one considers that conditions at the Martian surface and in Earth's stratosphere are qualitatively similar [220 K, 6.36 millibar (mbar) at the Martian surface versus 235 K, 6.35 mbar at 35 km in Earth's stratosphere]. The lifetime of CO_2 on Mars is even less than that on Venus – about 7740 years to destroy all CO_2 in the Martian atmosphere.

Ultraviolet sunlight ($\lambda < 227$ nm) photolyzes CO_2 into O atoms and CO molecules:

$$CO_2 + h\nu \rightarrow O(^3P) + CO \quad (\lambda = 167 - 227.5\,\text{nm}) \tag{4.21}$$

$$CO_2 + h\nu \rightarrow O(^1D) + CO \quad (\lambda = 129 - 167\,\text{nm}) \tag{4.22}$$

Ultraviolet light from 167 to 227 nm produces oxygen atoms in their electronic ground state (^3P) while shorter wavelength light produces electronically excited O atoms in the electronically excited ^1D (129–167 nm) or ^1S (129–167 nm) states. Collisions of the electronically excited oxygen atoms with unreactive atoms or molecules (denoted as M in the reaction below) removes the excess energy forming oxygen atoms in their ground electronic states, for example:

$$O(^1D) + M \rightarrow O(^3P) + M \tag{4.23}$$

At first glance, it seems that CO_2 could be reformed *via* the direct recombination of its photolysis products:

$$O(^3P) + CO + M \rightarrow CO_2 + M \qquad (4.24)$$

However, this is not possible. The rate constant for this reaction is:

$$k = 1.6 \times 10^{-32} \exp\left(\frac{-2184}{T}\right) cm^6 s^{-1} \qquad (4.25)$$

Reaction (4.24) is spin forbidden by quantum mechanics and is significantly slower than the reaction of O atoms with one another to form O_2:

$$O(^3P) + O(^3P) + M \rightarrow O_2 + M \qquad (4.26)$$

$$k = 1.1 \times 10^{-27} T^{-2.0} cm^6 s^{-1} \qquad (4.27)$$

For example, at 220 K, oxygen atom recombination to O_2 is 29 500 times faster than CO_2 formation from O atoms plus CO. This temperature is the average surface temperature on Mars and at 73 km in Venus' atmosphere. Carbon dioxide photolysis occurs at the 220 K level on both planets. In the absence of any other compensating reactions, the net result of CO_2 photolysis on Venus and Mars would be production of O_2 and CO on a fairly short timescale.

Compensating reactions occur and restore CO_2 lost by photolysis. In the Martian atmosphere hydroxyl radicals are produced by UV sunlight photolysis of atmospheric water vapor:

$$H_2O + h\nu \rightarrow OH + H \quad (\lambda < 212 \, nm) \qquad (4.28)$$

and by the reaction of electronically excited O atoms with water vapor:

$$O(^1D) + H_2O \rightarrow OH + OH \qquad (4.29)$$

The OH radicals then enter into catalytic cycles that recombine O atoms and CO to CO_2:

$$OH + CO \rightarrow CO_2 + H \qquad (4.30)$$

$$H + O_2 + M \rightarrow HO_2 + M \qquad (4.31)$$

$$HO_2 + O \rightarrow OH + O_2 \qquad (4.32)$$

$$Net\,reaction : CO + O \rightarrow CO_2 \qquad (4.33)$$

The net effect of Reactions (4.30)–(4.32) is Reaction (4.33). Reactions (4.30)–(4.32) are a catalytic cycle because OH radicals, H atoms and hydroperoxyl (HO_2) radicals are conserved; that is, Reaction (4.30) destroys OH while (4.32) reforms OH. Reaction (4.30) produces H atoms and reaction (4.31) destroys H atoms. Reaction (4.31) produces HO_2 radicals and (4.32) destroys HO_2 radicals. We will encounter other catalytic cycles below, and in Chapter 5 when we discuss the destruction of Earth's ozone layer by chlorofluorocarbon (CFC) gases and other species.

Hydrogen peroxide vapor H_2O_2 is involved in another catalytic cycle that restores CO_2 and is more important at higher H_2O mixing ratios:

$$HO_2 + HO_2 \rightarrow H_2O_2 + O_2 \qquad (4.34)$$

$$H_2O_2 + h\nu \rightarrow OH + OH \qquad (4.35)$$

$$OH + CO \rightarrow CO_2 + H \qquad (4.30)$$

$$H + O_2 + M \rightarrow HO_2 + M \qquad (4.31)$$

$$H + O_2 + M \rightarrow HO_2 + M \qquad (4.31)$$

$$Net\,reaction : 2CO + O_2 \rightarrow 2CO_2 \qquad (4.36)$$

On average, the Martian atmosphere contains 140 ppmv water vapor, but about 4.5 times less H_2O is present in the lower Venusian atmosphere, and only a few parts per million by volume at the cloud tops where CO_2 photolysis occurs. The solar flux on Venus is about 4.4 times higher than on Mars, which is 2.1 times further from the Sun. Thus, other catalytic cycles are required to restore CO_2 on Venus because not enough OH radicals may be present in its upper atmosphere.

One set of catalytic cycles thought to be important on Venus involves chlorine. Venus' atmosphere contains about 0.5 ppmv HCl. This is photolyzed by UV light with $\lambda < 237.5$ nm and produces Cl atoms that enter into catalytic cycles exemplified by:

$$HCl + h\nu \rightarrow H + Cl \quad (\lambda < 237.5\,nm) \qquad (4.37)$$

$$Cl + CO + M \rightarrow COCl + M \qquad (4.38)$$

$$COCl + O_2 + M \rightarrow ClCO_3 + M \qquad (4.39)$$

$$ClCO_3 + O \rightarrow CO_2 + O_2 + Cl \qquad (4.40)$$

$$Net\ reaction : CO + O \rightarrow CO_2 \qquad (4.33)$$

$ClCO_3$ is the peroxychloroformyl radical and is sometimes written as $ClC(O)OO$ to denote that it has a C=O bond and a $C-O_2$ bond. Another catalytic cycle proposed to reform CO_2 on Venus also produces the sulfuric acid clouds and is discussed below.

4.4.1.2 Minor and Trace Gases. Tables 4.10 and 4.11 list several minor and trace gases that occur in the atmospheres of Venus and Mars. The abundances of many of these gases vary with altitude, time and/or latitude because of their sources and sinks. For example, SO_2, OCS, SO and CO vary with altitude on Venus because photochemistry destroys SO_2 and OCS and produces SO and CO. Water vapor varies with altitude on Venus because it is consumed by formation of aqueous sulfuric acid cloud droplets. Carbonyl sulfide and sulfur vapor vary with altitude below Venus' clouds because they are produced by thermochemical reactions that become more important at the higher temperatures and pressures in Venus' lower atmosphere and surface. Likewise, O_2, O_3, H_2O_2 and NO vary with altitude on Mars because they are produced photochemically. Water vapor varies seasonally and with latitude because it is transported from pole to pole over the course of the Martian year. We discuss some of these minor and trace gases in more detail below.

Water vapor is an important minor gas in the atmospheres of Mars and Venus. As discussed earlier, photolysis of H_2O is a source of OH radicals that play a key role in recycling CO_2 destroyed by photochemistry. The OH and HO_2 radicals from photolysis of water vapor produce H_2O_2 and destroy O_3, whose abundance is anti-correlated with H_2O. The amount of H_2O in the Martian atmosphere is often expressed in terms of precipitable micrometers of water (pr µm). One precipitable micrometer is equivalent to a mass density of $10^{-4}\,g\,cm^{-2}$ or a column density of 3.35×10^{18} H_2O molecules cm^{-2}, which is equal to 14 ppmv (relative to a total atmospheric column density of $2.4 \times 10^{23}\,cm^{-2}$). The global annual average H_2O content is 10 ± 3 pr µm (140 ± 47 ppmv). The water vapor content is largest at high latitudes during summer in either hemisphere, but the maximum values differ (100 pr µm, northern hemisphere; 50 pr µm, summer hemisphere). The water vapor content is smallest (< 5 pr µm) at mid and high latitudes of either hemisphere during fall and winter. Water ice clouds form in the cold Martian atmosphere when the H_2O partial pressure is equal to the vapor pressure of H_2O over water ice. Cloud condensation occurs at different altitudes (10–15 km altitude during aphelion and > 30 km altitude during perihelion) due to seasonal changes in the atmospheric thermal structure. The seasonal behavior and latitudinal profile of water vapor differs between the northern and southern hemispheres. For example, water vapor is transported from the northern to southern hemisphere as the summertime maximum in the north decays away, but the converse behavior does not occur.

On average, the total amount of atmospheric water vapor on Mars is equivalent to 1–2 km^3 liquid water. In contrast, the volumes of the northern and southern polar layered deposits are 1.14×10^6 and 1.6×10^6 km^3, respectively, which are comparable to the ice sheet covering Greenland (2.6×10^6 km^3). The north and south polar layered deposits are composed of water ice and silicate dust, *e.g.*, the south layered polar deposits are 85% ice and 15% dust by volume and have an average density of about 1220 kg m^{-3}. The total amount of water in the polar caps and deposits and buried elsewhere on Mars are important questions for modeling the Martian hydrologic cycle.

Earth-based IR observations show that HDO is present in the Martian atmosphere. The abundance varies seasonally and with latitude, like that of H$_2$O. Observations near perihelion gave an abundance of $6.06 \pm 0.5 \times 10^{16}$ HDO molecules cm^{-2}, which corresponds to a D/H ratio of $\sim 9 \times 10^{-4}$ taking the global average value given above for water vapor. The Martian D/H ratio is 5.8 ± 2 times higher than the terrestrial value of 1.56×10^{-4} for standard mean ocean water (SMOW). The high D/H ratio could be primordial, *i.e.*, Mars formed from material with a higher D/H ratio than Earth, or it could point to the escape of water over time from an atmosphere originally having a lower D/H ratio, which is assumed to be the terrestrial D/H ratio in models.

As mentioned earlier, H$_2$O$_2$ takes part in one of the catalytic cycles to restore CO$_2$ destroyed by UV sunlight. Hydrogen peroxide was finally observed in the Martian atmosphere in 2003 when it was detected by microwave spectroscopy. Since then it has also been observed by IR spectroscopy. The H$_2$O$_2$ mixing ratio is about 10–20 ppbv (parts per billion by volume) (in agreement with photochemical model predictions) and varies spatially and temporally with the amount of water vapor in the Martian atmosphere. Hydrogen peroxide is formed from hydroperoxyl radicals *via*:

$$HO_2 + HO_2 \rightarrow H_2O_2 + O_2 \qquad (4.34)$$

The HO$_2$ radicals are formed by Reaction (4.31). In turn, the H atoms in Reaction (4.31) are formed from water vapor by photolysis (4.28) or reaction with electronically excited oxygen atoms (4.29). Thus, the abundance of H_2O_2 correlates with that of water vapor. Hydrogen peroxide is destroyed by photolysis:

$$H_2O_2 + h\nu \rightarrow OH + OH \tag{4.35}$$

Although the reaction pathways and rates are unquantified, H_2O_2 could be destroyed by reactions with the material on the surface of Mars. For example, H_2O_2 may have given the positive response in the Labeled Release Life Science Experiment on the 1976 US Viking Lander.

Molecular hydrogen is also present in the Martian atmosphere. It is produced by reaction of H atoms with hydroperoxyl radicals:

$$H + HO_2 \rightarrow H_2 + O_2 \tag{4.41}$$

The major sinks for H_2 are its reactions with OH radicals and O (^{1}D) atoms:

$$OH + H_2 \rightarrow H + H_2O \tag{4.42}$$

$$O(^1D) + H_2 \rightarrow H + OH \tag{4.43}$$

The observed H_2 abundance is 15 ± 5 ppmv, which is several times larger than the H_2 mixing ratio of 0.55 ppmv in Earth's atmosphere. The HD abundance on Mars is 11 ± 4 ppbv, which corresponds to a D/H ratio of $\sim 4 \times 10^{-4}$, about 40% of that in water on Mars. The difference is due to photochemical fractionation of D between HDO and HD.

Ozone in the Martian atmosphere has spatially and temporally variable abundances of 6–170 ppbv and is produced by solar UV photolysis of O_2 (see also Chapter 5):

$$O_2 + h\nu \rightarrow O + O \quad (\lambda = 180 - 240 \, \text{nm}) \tag{5.59}$$

$$O + O_2 + M \rightarrow O_3 + M \tag{5.37}$$

The M in Reaction (5.37) is any third body, and is most often CO_2, N_2 or Ar, the three most abundant gases in the Martian atmosphere. Ozone is destroyed by solar UV photolysis and by reaction with monatomic oxygen atoms:

$$O_3 + h\nu \rightarrow O(^1D) + O_2 \quad (\lambda = 200 - 300 \, \text{nm}) \tag{5.35}$$

$$O(^1D) + M \rightarrow O + M \tag{4.23}$$

$$O + O_3 \rightarrow O_2 + O_2 \tag{5.61}$$

The reactions responsible for ozone production and loss are discussed in detail in Chapter 5 because they were first proposed for terrestrial atmospheric chemistry. As on Earth, O_3 is efficiently destroyed by H atoms, OH and HO_2 radicals that are interconverted but not destroyed during the catalytic cycle and function as gas phase catalysts, for example:

$$HO_x \text{ Cycle I} \quad H + O_3 \rightarrow HO + O_2 \tag{5.64}$$

$$HO + O \rightarrow H + O_2 \tag{5.65}$$

$$\textit{Net reaction}: \quad O + O_3 \rightarrow O_2 + O_2 \tag{5.61}$$

As discussed above, the H, OH and HO_2 are formed by photolysis of water vapor. The O_3 abundance is high when that of H_2O is low and *vice versa*. For example, during northern winter O_3 is high while H_2O is low. Typical O_3 abundances are 0.5–$15\,\mu\text{m-amagats}$ ($1\,\mu\text{m-amagat} = 2.69\times10^{15}\,O_3\,\text{cm}^{-2} = 11.2\,\text{ppbv}$ relative to a total column density of $2.4\times10^{23}\,\text{cm}^{-2}$), or 6–$170\,\text{ppbv}$.

Carbon monoxide is the second most abundant C-bearing gas in the atmospheres of Venus (~ 17–$30\,\text{ppmv}$ CO below the clouds) and Mars ($\sim 0.08\%$ CO). The CO mixing ratio varies with altitude on both planets (see below), and in principle it should vary on a timescale of a few years in the Martian atmosphere. Photolysis of CO_2 produces CO in both atmospheres and it is reconverted back into CO_2 *via* the catalytic cycles described above. Thermochemical reactions in the lower atmosphere and at the surface of Venus are a sink for CO, and its mixing ratio decreases with decreasing altitude toward Venus' surface ($45 \pm 10\,\text{ppmv}$ at $64\,\text{km}$, $30 \pm 18\,\text{ppmv}$ at $42\,\text{km}$, $20 \pm 3\,\text{ppmv}$ at $22\,\text{km}$ and $17 \pm 1\,\text{ppmv}$ at $12\,\text{km}$). Carbon monoxide is also present in the terrestrial atmosphere, but with a smaller average mixing ratio of $0.12\,\text{ppmv}$. It is produced from anthropogenic and biogenic sources and is destroyed by reaction with OH radicals, as in the catalytic cycles described previously.

Earth-based IR spectroscopy shows that CH_4 occurs in the Martian atmosphere with spatially and temporally varying mixing ratios of a few to a few tens of ppbv. For example, during northern Martian summer 2003, CH_4 mixing ratios reached $33\,\text{ppbv}$ over the Nili Fossae region, where hydrated silicates are observed. The sources and sinks for Martian

CH_4 are of great interest, but are not known with certainty. Possible sources include (i) low temperature fumaroles, (ii) hydrothermal systems, (iii) formation of hydrous silicates (serpentinization), (iv) decomposition of CH_4–CO_2 clathrate hydrates $(CH_4,CO_2) \cdot 7H_2O$ (s), (v) meteoritic infall and (vi) biological sources such as methanogenic bacteria. Low temperature fumaroles favor production of CH_4 because gas phase equilibria such as:

$$CO_2 + 4H_2 = CH_4 + 2H_2O \qquad (4.44)$$

proceed toward CH_4 in lower temperature fumarolic gases. A similar reaction involving CO_2 dissolved in water (denoted by aq) and H_2 occurs in hydrothermal systems and during formation of hydrous minerals such as serpentine, $(Mg,Fe)_3Si_2O_5(OH)_4$:

$$CO_2 \text{ (aq)} + 4H_2 = CH_4 + 2H_2O \text{ (liq)} \qquad (4.45)$$

The H_2 gas in these reactions is produced by reduction–oxidation (redox) equilibria involving Fe oxides and minerals. For example, H_2 in magmatic systems (that vent fumarolic gases) is produced *via* redox reactions between water, Fe^{2+}, Fe^{3+} and hydrogen dissolved in the magma. Schematically this is represented by:

$$H_2O \text{ (magma)} + 2FeO \text{ (magma)} = Fe_2O_3(magma) + H_2(magma) \qquad (4.46)$$

In hydrothermal systems and during formation of hydrous minerals, H_2 is produced by the oxidation of Fe^{2+}-bearing anhydrous silicates such as olivine, for example:

$$(Mg, Fe)_2SiO_4 \text{ (olivine)} + H_2O \text{ (liq)}$$
$$= Mg_3Si_2O_5(OH)_4 \text{ (serpentine)} + Mg(OH)_2 \text{ (brucite)} \qquad (4.47)$$
$$+ Fe_3O_4 \text{ (magnetite)} + H_2 \text{ (g)}$$

Methane–CO_2 clathrate hydrates are cage compounds in which CH_4 and CO_2 are trapped in the water ice crystal lattice in a gas to water ratio of about 1 : 7. Models indicate that CO_2-rich clathrate hydrates exist at shallow depth on Mars (<100 m in dry unconsolidated dry soil and ~ 10 m in icy soils). When heated the clathrate hydrate decomposes to gas plus water, releasing the enclathrated gas:

$$(CH_4, CO_2) \cdot 7H_2O \text{ (clathrate hydrate)} = 7H_2O \text{ (liq)} + CO_2 \text{ (g)} + CH_4 \text{ (g)} \qquad (4.48)$$

The CH_4 to CO_2 ratio in the clathrate depends on its formation conditions, which can vary. One cubic meter of pure CH_4 clathrate hydrate contains 116 kg CH_4. Clathrate deposits >100 m thick extending over large areas occur on Earth.

Meteoritic or cometary infalls are potential sources of CH_4 due to CH_4 in cometary ices, along with pyrolysis of organic material in meteorites. Grain-catalyzed conversion of CO into CH_4 (*via* Fischer–Tropsch type reactions) may occur in the cooling fireballs produced by cometary and meteoritic impacts. However, models indicate that neither cometary nor meteoritic infall can supply the observed amounts of methane.

Biological sources produce CH_4 on Earth and this exciting, but speculative, possibility has been suggested for Mars. Methanogenic bacteria (*e.g.*, *Methanobacterium*, *Methanosarcina*, *Methanococcus* and *Methanospirillum*) occur in many anaerobic (*i.e.*, O_2-poor) environments, including swamps, sewage and rice paddies. Methane is produced *via* bacterially mediated reduction of CO or CO_2 and the microbe derives energy from the process. As discussed in Chapter 5, methanogenic bacteria in wetlands and landfills, rice paddies and in cattle and other animals are the major source of CH_4 in Earth's atmosphere. In contrast, fumarolic gases, hydrothermal systems, serpentinization, clathrate hydrate decomposition and meteoritic infall are negligible CH_4 sources. The situation on Mars is possibly very different than on Earth. If Martian CH_4 is produced biologically it may be enriched in ^{12}C because of enzymatic discrimination in favor of lighter carbon as occurs during photosynthesis on Earth (Chapter 6). Such isotopic measurements are planned for future Mars lander missions.

The major sinks for CH_4 in the Martian atmosphere are photochemical reactions. Direct photolysis of CH_4 by solar Lyman alpha photons (121.6 nm) is important in the upper atmosphere of Mars (~ 80 km altitude). At lower altitudes, CH_4 is primarily destroyed by reaction with OH radicals:

$$CH_4 + OH \rightarrow CH_3 + H_2O \tag{5.31}$$

Reactions with O atoms (ground state and electronically excited) are also important sinks:

$$CH_4 + O \rightarrow CH_3 + OH \tag{4.49}$$

These processes destroy CH_4 in about 340 years. It is also possible that H_2O_2 (or other peroxides) absorbed on airborne dust grains destroy CH_4, but this sink is hard to quantify.

Sulfur dioxide is the major S-bearing gas on Venus, the third most abundant gas (after CO_2 and N_2), and one of the three most important greenhouse gases (CO_2, SO_2, H_2O). The mole fraction (mixing ratio) of SO_2 below the clouds is 150 ± 30 ppmv in the 22–42 km altitude range. The SO_2 mixing ratio decreases to about 25 ppmv in the 22–12 km region, where gas-phase thermochemical reactions reduce some SO_2 to OCS, for example:

$$SO_2 + 3CO = OCS + 2CO_2 \tag{4.50}$$

Over much longer times (1.9 Ma) gas–solid thermochemical reactions with Ca-bearing minerals on Venus' surface convert SO_2 into $CaSO_4$ (anhydrite), for example:

$$SO_2 + CaCO_3 \text{ (calcite)} = CO + CaSO_4 \text{ (anhydrite)} \tag{4.51}$$

Anhydrite formation irreversibly removes SO_2 from Venus' atmosphere and it must be replenished by volcanism at the same average rate to maintain the global sulfuric acid cloud cover. The average S/Si ratio of Venus' surface at the Venera 13, Venera 14 and Vega 2 landing sites shows that the rate of volcanism is $1 \text{ km}^3 \text{ yr}^{-1}$, if the erupted material has the same average composition as Venus' surface. The estimated volcanism rate is the same as the average rate of sub-aerial volcanism on Earth.

In contrast to Venus no SO_2 is observed on Mars and the upper limit is <1 ppbv ($<1.8 \times 10^{14}$ SO_2-molecules cm^{-2}). On Earth SO_2 is a trace gas that is less abundant than OCS and has a column density of $\sim 4 \times 10^{15}$ SO_2-molecules cm^{-2}. As discussed in Chapter 6, anthropogenic sources supply most SO_2 in Earth's atmosphere and volcanic emissions are $\sim 10\%$ of anthropogenic emissions. The major sink for SO_2 on Earth is oxidation to sulfite (SO_3^{2-}) and sulfate (SO_4^{2-}) via photochemical and thermochemical reactions in the gas phase, in cloud droplets and on particulates.

Photochemical oxidation to aqueous sulfuric acid cloud droplets is the major sink for SO_2 on Venus. Ultraviolet sunlight photolysis efficiently converts SO_2 into H_2SO_4 in the cloud-forming region (~ 40–70 km). In situ and Earth-based observations show the decrease in SO_2 throughout and above the clouds on Venus. Eventually, the SO_2 mixing ratio drops to very low values, e.g., it is ~ 10 ppbv above the clouds, which is about 15 000 times less than below the clouds. One of the photochemical schemes involves chlorine and regenerates CO_2 lost by photolysis:

$$Cl + CO + M \rightarrow COCl + M \tag{4.38}$$

$$COCl + O_2 + M \rightarrow ClCO_3 + M \qquad (4.39)$$

$$ClCO_3 + Cl \rightarrow CO_2 + ClO + Cl \qquad (4.52)$$

$$SO_2 + h\nu \rightarrow SO + O \quad (\lambda < 210\,nm) \qquad (4.53)$$

$$SO + ClO \rightarrow SO_2 + Cl \qquad (4.54)$$

$$SO_2 + O + M \rightarrow SO_3 + M \qquad (4.55)$$

$$SO_3 + H_2O + M \rightarrow H_2SO_4 + M \qquad (4.56)$$

$$Net\,reaction : CO + O_2 + SO_2 + H_2O + h\nu \rightarrow CO_2 + H_2SO_4 \qquad (4.57)$$

The sulfuric acid droplets evaporate to H_2SO_4 vapor at the cloud base, for example:

$$H_2SO_4 \quad (liquid) = H_2SO_4 \quad (gas) \qquad (4.58)$$

Sulfuric acid vapor thermally dissociates to sulfur trioxide and water vapor:

$$H_2SO_4 = SO_3 + H_2O \qquad (4.59)$$

so a mixture of H_2SO_4, SO_3 and H_2O in equilibrium with one another is expected below the sulfuric acid clouds. Microwave observations from spacecraft (*Pioneer Venus, Magellan*) and from Earth show about 12 ppmv sulfuric acid vapor below the clouds, but SO_3 has not yet been observed on Venus. Although H_2SO_4, SO_3 and H_2O are in equilibrium with one another, they are less stable than SO_2 plus H_2O in Venus' lower atmosphere. As a result SO_3 is reduced to SO_2 *via* reaction with CO:

$$CO + SO_3 = CO_2 + SO_2 \qquad (4.60)$$

As SO_3 is destroyed H_2SO_4 dissociates to make more SO_3, and both gases are removed.

Carbonyl sulfide (OCS) is the second most abundant S-bearing gas in Venus' atmosphere with a mixing ratio of 4.4 ± 1.0 ppmv at 33 km altitude. Earth-based IR spectroscopic observations show that the OCS mixing ratio is inversely correlated with altitude (*e.g.*, 2 ppbv at 70 km, 14 ppbv at 64 km, 4400 ppbv at 33 km). Extrapolation of this gradient, the decreasing SO_2 mixing ratio with decreasing altitude and chemical models suggest that OCS has a mixing ratio of several tens of ppmv at Venus' surface.

Water vapor has an average mixing ratio of 30 ppmv below the clouds, which is 20% of the average water vapor mixing ratio in the Martian atmosphere. However, Venus' atmosphere above the clouds is much drier and contains only a few ppmv water vapor. Meteorology causes spatial and temporal variations in the water vapor abundance above and below the clouds. As mentioned above, the cloud droplets are composed of aqueous sulfuric acid. Water vapor reacts with the sulfuric acid cloud droplets to form hydronium (H_3O^+) and bisulfate ions (HSO_4^-):

$$H_2O + H_2SO_4 = H_3O^+ + HSO_4^- \qquad (4.61)$$

Consequently, the equilibrium concentration of H_2O in the concentrated sulfuric acid cloud droplets is very small and the equilibrium vapor pressure of water vapor over the sulfuric acid droplets is very small. The water vapor partial pressure at the cloud tops is less than that over pure water ice at the same temperature.

Venus' atmosphere is very dry, and contains less water vapor than on Mars or Earth. Nevertheless, H_2O is the major H-bearing gas on Venus and is important for the greenhouse effect. As discussed below, it is also involved in gas–solid reactions that control the partial pressures of HF and HCl in the Venusian atmosphere. Over geologic timescales of 10^8–10^9 years water loss *via* hydrogen escape to space and oxidation of Fe^{2+}-bearing igneous rocks to Fe^{3+}-bearing rocks, *e.g.*, *via* schematic reactions such as (4.62) and (4.63) regulated the oxidation state of Venus' atmosphere and surface:

$$H_2O + 2FeO \text{ (in rock)} = Fe_2O_3 \text{ (hematite)} + H_2 \qquad (4.62)$$

$$H_2O + 3FeO \text{ (in rock)} = Fe_3O_4 \text{ (magnetite)} + H_2 \qquad (4.63)$$

The FeO in the equations above represents Fe^{2+}-bearing minerals such as olivine $(Mg,Fe)_2SiO_4$ and pyroxene $(Mg,Fe)SiO_3$ in rocks.

The large amount of HDO in Venus' atmosphere supports this picture. The Venusian D/H ratio ($\sim 2.5\%$) is about 160 times higher than the average D/H ratio on Earth (1.56×10^{-4} in SMOW). If Venus initially had a lower D/H ratio, *e.g.*, like that on Earth, a large amount of water was lost to give its high D/H ratio. Models based on this assumption show that Venus initially had water equivalent to a global layer 4–530 m thick over the entire planet. For comparison, the total amount of water on the Earth is equivalent to a global layer 3300 m thick over the entire planet. On the other hand, Mars and Titan also have D/H ratios higher than Earth. Some meteorites and interplanetary dust particles have D/H ratios up to 50 times higher than that of the Earth. If Venus formed with a higher D/H ratio it could have lost less water (or none at all) to get to the present day value.

4.4.2 Surface Composition and Chemistry

To a first approximation, the surface of Venus is a hot, dry desert and that of Mars is a cold dry desert. Our knowledge of the composition and chemistry of the surfaces of Venus and Mars comes from various methods: (i) elemental analyses by α-particle X-ray spectroscopy (APXS), γ-ray, and X-ray fluorescence spectroscopy (XRFS), (ii) analyses of Fe-bearing minerals by Mössbauer spectroscopy, (iii) Earth-based and spacecraft radar observations, (iv) various types of imagers and imaging spectrometers, (v) thermal analysis instruments, (vi) physical property measurements of the surfaces, (vii) laboratory studies of chemical equilibria and reaction rates and (viii) models of chemical equilibria or photochemical reactions between atmospheric gases and presumed surface minerals. Not all methods have been used on each planet or on every spacecraft mission.

For example, the US *Mars Pathfinder* rover and *Mars Exploration* rovers used APXS to analyze the surface of Mars. The US *Viking* landers and the Russian *Venera* and *Vega* landers used XRFS to analyze the surfaces of Mars and Venus. The US *Mars Odyssey* orbiter used γ-ray spectroscopy to analyze the elemental composition of the surface of Mars. The US *Mars Global Surveyor* and *Mars Odyssey* orbiters used IR spectroscopy to analyze the mineralogical composition of the surface of Mars. We do not have the room to discuss all the results from the spacecraft armada that have been exploring Mars and Venus, but we discuss key results about their surface composition and chemistry below. We give recommended reading at the end of this chapter that lists special issues of scientific journals that contain papers describing results from the recent Mars and Venus missions.

The elemental composition of the surfaces of Venus and Mars are summarized in Tables 4.12 (Venus) and 4.13 (Mars). The results for Venus are from XRFS of soil samples at the *Venera 13*, *Venera 14* and *Vega 2* landing sites. The results for Mars are from XRF and APX spectroscopy of soil samples at the *Viking, Mars Pathfinder* and *Mars Exploration Rover* landing sites. Elements lighter than Mg could not be detected by the *Venera* and *Vega* XRF spectrometers and the Na content was estimated by the instrument team using geochemical

Table 4.12 Chemical composition (wt%) of Venus' surface (modified from ref. 51).

Oxide	Venera 13	Venera 14	Vega 2
Al_2O_3	15.8	17.9	16
CaO	7.1	10.3	7.5
Cl	<0.3	<0.4	<0.3
FeO[a]	9.3	8.8	7.7
K_2O	4.0	0.2	0.1
MgO	11.4	8.1	11.5
MnO	0.2	0.16	0.14
Na_2O[b]	2	2.4	2.0
SO_3	1.62	0.88	4.7
SiO_2	45.1	48.7	45.6
TiO_2	1.59	1.25	0.2
Total	98.1	98.7	95.4

[a]All Fe reported as FeO for all analyses.
[b]Na_2O content estimated from geochemical correlations.

Table 4.13 Chemical composition (wt%) of the Martian surface.

Oxide	MER-A Gusev	MER-B Meridiani	MPF Ares Vallis	Viking 1 Chryse[a]	Viking 2 Utopia[a]
Al_2O_3	10.2	9.39	8.0	7.3	7[b]
CaO	6.37	7.15	6.5	5.7	5.7
Cl	0.76	0.67	0.6	0.8	0.4
Cr_2O_3	0.29	0.41	0.3	—	—
FeO	15.6	18.0	20.1	17.5[c]	17.3[c]
K_2O	0.46	0.48	0.6	<0.5	<0.5
MgO	8.49	7.64	8.7	6	6[b]
MnO	0.31	0.37	0.5	—	—
Na_2O	3.02	2.17	1.1	—	—
NiO	0.06	0.05	<0.24	—	—
P_2O_5	1.01	0.85	1.0	—	—
SO_3	6.32	5.57	6.8	6.7	7.9
SiO_2	46.2	46.2	42.3	44	43
TiO_2	0.95	0.97	1.0	0.62	0.54
Total	100.04	99.92	97.5	91[d]	90[d]

[a] – Indicates value not determined.
[b]Estimated as equal to Viking 1.
[c]Fe_2O_3.
[d]Includes 2% other.

correlations (such as discussed earlier in this chapter). The XRF analyses done by the *Viking* landers did not determine several elements (Cr, Mn, Na, Ni, P) that were analyzed by the APX spectrometers on the *Mars Pathfinder* and *Mars Exploration* rovers. Comparisons with elemental analyses of terrestrial rocks, normative computations of mineralogy, geochemical correlations and geological interpretations of visible, IR and radar images of the landing sites on Mars and Venus have been used to suggest plausible types of rocks at the landing sites. In addition, Mössbauer spectra of Fe-bearing minerals were obtained by the *Mars Exploration* rovers and were used to constrain the rock types analyzed by the rovers. Gamma-ray data for K, U and Th are available for several landing sites on Venus and for K and Th over most of the Martian surface.

The analytical data show that Mars and Venus are differentiated planets (like the Earth) and that their surfaces are predominantly basaltic, unlike Earth's continental crust but like its oceanic crust. At least part of the surfaces of both planets has been weathered by Cl- and S-bearing fluids (Mars) and gases (Venus). Analyses of Martian soils at the *Viking*, *Mars Pathfinder* and *Mars Exploration* rover landing sites, up to 7000 km apart, are remarkably similar. Apparently the global dust storms have homogenized soils on the Martian surface down to depths of several centimeters (or less likely all the soils formed by weathering of similar precursor rocks). More detailed analyses of the data at the landing sites on Venus and Mars, and along the rover traverses on Mars, are given in the Further Reading cited at the end of this chapter, but we mention a few important points here.

The *Venera* 13 XRF data suggest weathered high-potassium alkaline basalts at its landing site on Venus. The *Venera* 14 XRF results look like terrestrial mid-ocean ridge basalt (MORB) that has been weathered by atmospheric SO_2. The Vega 2 γ-ray and XRF data also suggest a weathered MORB at that landing site. The analyzed samples at all three landing sites contained fairly large amounts of sulfur that correspond to 2.8%, 1.5% or 8.2% by mass anhydrite ($CaSO_4$) at the *Venera* 13, 14 and *Vega* 2 landing sites, respectively. This interpretation agrees with laboratory studies and thermodynamic models showing that SO_2 reacts with Ca-bearing minerals at Venus surface conditions to form anhydrite. In general the *Venera* and *Vega* XRF analyses show the presence of rocks containing the minerals thought to be responsible for regulating the partial pressures of CO_2, HCl, HF and other gases *via* gas–solid phase equilibria (discussed below).

Mars Exploration rover data revealed various types of unweathered basaltic rocks at Gusev Crater, weathered basalts enriched in S, Br, Cl and P in the Columbia Hills and sedimentary rocks cemented by sulfates at Meridiani Planum. Hematite spherules (called blueberries in press reports) were also found at Meridiani. The hematite spherules and the high Br, Cl and S concentrations in rocks and soil are evidence of a hydrosphere and aqueous weathering on early Mars.

On Earth, most of the chlorine and bromine are concentrated in sea-water (see Figure 4.5), but there are no oceans on Mars today. The halogens, which were presumably concentrated in ancient Martian sea-water, are now found in rocks and soil on Mars. The halogens possibly occur as evaporate minerals in some locations or as the result of chemical weathering by volcanic volatiles and/or aqueous solutions in other lo-cations. Terrestrial seawater contains significant amounts of sulfur as dissolved sulfate, which is the second most abundant anion in seawater after chloride. On Mars the sulfate that would be found in seawater is now found as magnesium and/or calcium sulfates in rocks and soil. Phosphate minerals are the biggest phosphorus reservoir on Earth and phosphorus in oceanic and terrestrial biomass is the second biggest reservoir. The bio-genic phosphorus reservoir is absent on Mars and all phosphorus occurs in rocks and soils instead (presumably as phosphate minerals).

The high temperatures and pressures at the surface of Venus (740 K, 95.6 bar) drive chemical reactions of CO_2, SO_2, H_2O, OCS, H_2S, HCl and HF with rocks and minerals. In 1963, microwave observations by the *Mariner II* spacecraft showed high surface temperatures on Venus. Several geochemists realized that the high temperatures on Venus were like those reached during metamorphism on Earth. Under these con-ditions it is plausible that the lower atmosphere of Venus is at least partially equilibrated with the surface rocks and that the atmospheric composition provides information about the surface mineralogy.

For example, reactions of CO_2 with pure silicate minerals are uni-variant phase equilibria and in principle could control the CO_2 pressure on Venus, or another hot planet such as a Venus-like exoplanet. The calcite–quartz–wollastonite (CQW) equilibrium occurs in contact metamorphic rocks on Earth and was studied by the Norwegian geo-chemist Victor Goldschmidt (1888–1947) in a pioneering application of chemical thermodynamics to petrology. The CQW equilibrium is:

$$CaCO_3 \text{ (calcite)} + SiO_2 \text{ (quartz)} = CaSiO_3 \text{ (wollastonite)} + CO_2(g) \quad (4.64)$$

The equilibrium constant K_{eq} for this reaction is the CO_2 equilibrium partial pressure:

$$K_{eq} = P_{CO_2} \quad (4.65)$$

The temperature variation of the CO_2 equilibrium partial pressure is:

$$\log_{10} P_{CO_2}(\text{bar}) = 7.97 - \frac{4456}{T} \quad (4.66)$$

The CO_2 pressure of ~ 92 bar on Venus is virtually identical to that from Reaction (4.64) at $740\,K$, the average surface temperature on Venus.[ii]

Although carbonates are expected to be thermodynamically stable on the Martian surface, CO_2 in the Martian atmosphere is definitely not in equilibrium with carbonate–silicate mineral assemblages on Mars. At $220\,K$, the average surface temperature on Mars, the CO_2 equilibrium partial pressure for Reaction (4.64) is $\sim 5 \times 10^{-13}$ bar, which is about ten billion times smaller than the observed CO_2 pressure. The CO_2 equilibrium partial pressure is still very small even if pure wollastonite is not present and $CaSiO_3$ is dissolved in pyroxenes.

Enstatite ($MgSiO_3$) is a common pyroxene mineral present on Earth, Mars and Venus. The reaction of enstatite with CO_2 is analogous to the CQW reaction and is:

$$MgCO_3 \text{ (magnesite)} + SiO_2 \text{ (quartz)} = MgSiO_3 \text{ (enstatite)} + CO_2(g) \quad (4.67)$$

If all minerals are pure phases, the equilibrium constant (K_{eq}) for this reaction is the CO_2 equilibrium partial pressure:

$$K_{eq} = P_{CO_2} \quad (4.65)$$

The temperature variation of the CO_2 equilibrium partial pressure for reaction (4.67) is:

$$\log_{10} P_{CO_2}(\text{bar}) = 8.62 - \frac{4054}{T} \quad (4.68)$$

At the average Martian surface temperature of $220\,K$, the CO_2 equilibrium partial pressure for the reaction of magnetite ($MgCO_3$) plus quartz is very small ($\sim 2 \times 10^{-10}$ bar) in comparison to the observed partial pressure.

If Reactions (4.64) and (4.67) reached chemical equilibrium in geologically short times, CO_2 would be totally absent from the Martian atmosphere, which would be dominantly N_2 plus Ar. This is clearly not the case and thermochemical weathering reactions that occur on Venus do not occur on Mars today. However, as originally suggested by O'Connor in 1968, "fossil" products from chemical weathering

---

[ii] Physical chemists may be interested to know that the temperature and pressure at Venus' surface correspond to 2.4 times the critical temperature ($304\,K$) and 1.3 times the critical pressure (72 bar) of CO_2, which behaves as an ideal gas, or very nearly so, under these conditions. Thus the partial pressure can be used instead of the fugacity in the equations above.

reactions in a warmer, wetter Martian paleo-environment may be preserved on Mars today. Interestingly, the sequestration of CO_2 by
reaction with silicates is currently being studied to reduce the amount
of CO_2 in Earth's atmosphere.

As noted above there is about 30 ppmv water vapor in Venus' lower
atmosphere below the clouds. Several water-bearing amphibole and
mica minerals take part in metamorphic reactions on Earth, and some
of these minerals are thermodynamically stable on Venus' surface.
For example, the dehydration and rehydration of eastonite mica is a
possible reaction to control the water vapor partial pressure in Venus'
atmosphere:

$$KMg_2Al_3Si_2O_{10}(OH)_2 \text{ (eastonite)} = MgAl_2O_4 \text{ (spinel)} + MgSiO_3 \text{ (enstatite)}$$
$$+ KAlSiO_4 \text{ (kalsilite)} + H_2O \text{ (g)}$$

$$(4.69)$$

Reaction (4.69), like the two carbonate–silicate reactions above, is an
univariant equilibrium. If all minerals are pure phases the equilibrium
constant for Reaction (4.69) is equal to the H_2O equilibrium partial
pressure, which is 3.5×10^{-3} bar at 740 K. This is equivalent to 37 ppmv
water vapor, which is identical within uncertainties to the H_2O mixing
ratio of 30 ppmv in Venus' lower atmosphere.

As discussed earlier, the average water vapor abundance in the
Martian atmosphere is about 140 ppmv. The Martian surface is too
cold for dehydration–hydration reactions involving amphibole and
mica minerals, which are thermally stable at Martian surface conditions.
Instead, the hydration–dehydration equilibria of hydrated salts such
as Ca and Mg chlorides [antarcticite ($CaCl_2 \cdot 6H_2O$), bischofite
($MgCl_2 \cdot 6H_2O$)] and sulfates [*e.g.*, gypsum ($CaSO_4 \cdot 2H_2O$), epsomite
($MgSO_4 \cdot 7H_2O$)] plausibly occur as a function of latitude, season, and
burial depth below the Martian surface.

Table 4.10 shows that HCl is present at Venus' atmosphere at about
0.5 ppmv (a column density of $\sim 7 \times 10^{20}$ molecules cm^{-2}). This does
not seem like much, but it is 35 000 times larger than the amount of
HCl in Earth's atmosphere and at least 1.4 million times larger than HCl
on Mars, where it is not observed and the upper limit is <2 ppbv
($< 5 \times 10^{14}$ molecules cm^{-2}). Some of the HCl in the terrestrial troposphere arises from anthropogenic sources and not from volcanic
emissions or from sea salt. By analogy with Earth, where HCl is observed in volcanic gases, volcanic out gassing is probably the ultimate
source of HCl on Venus. Hydrogen chloride vapor is corrosive and its

atmospheric abundance is probably regulated by reaction with Cl-bearing minerals on Venus' surface, for example:

$$2Na_4[AlSi_3O_8]_3Cl \text{ (marialite)} + Al_2SiO_5 \text{ (andalusite)}$$
$$+ 5SiO_2 \text{ (quartz)} + H_2O \text{ (gas)} = 8NaAlSi_3O_8 \text{ (albite)} + 2HCl \text{ (gas)} \quad (4.70)$$

Reaction (4.70) is a divariant equilibrium because the HCl equilibrium partial pressure depends upon temperature and the H_2O partial pressure. If all minerals are pure phases, the equilibrium constant for this reaction is:

$$K_{eq} = \frac{P_{HCl}^2}{P_{H_2O}} \quad (4.71)$$

Using the average water vapor mixing ratio of 30 ppmv in the lower atmosphere of Venus gives Equation (4.72) for the equilibrium mole fraction of HCl (X_{HCl}):

$$\log_{10} X_{HCl} = \log_{10}\left(\frac{P_{HCl}}{P_T}\right) = 4.22 - \frac{7860}{T} \quad (4.72)$$

At 740 K, the average surface temperature on Venus, the equilibrium HCl mole fraction is 0.4 ppmv, which is identical within uncertainties to the observed value of 0.5 ppmv. The P_T in this equation is the total pressure at a given temperature in Venus' atmosphere.

At present the Martian surface and lower atmosphere are too cold for HCl vapor to be present except transiently during active volcanic eruptions. Any gaseous HCl will be removed rapidly from the atmosphere by reaction with atmospheric dust and the Martian surface, e.g., by forming halite NaCl or other Cl-bearing minerals.

Hydrogen fluoride is present in Venus' atmosphere at a mixing ratio of 4.5 ppbv (Table 4.10). The corresponding column density is 7×10^{18} HF molecules cm^{-2}, which is 14 000 times larger than the HF column density of 5×10^{14} in Earth's troposphere (an HF mixing ratio of 25 parts per trillion by volume, pptv). Most of the HF in Earth's atmosphere is due to anthropogenic emissions. One source is emissions of HF from firing F-bearing clays in kilns to make bricks, pipes, whiteware and other ceramics. Silicon tetrafluoride (SiF_4) vapor is emitted at the same time and reacts with atmospheric water vapor to form HF and silica. The second source is HF mixed downward from the stratosphere where it is produced from photolysis of chlorofluorocarbon (CFC) gases. Only a small amount of the HF in the terrestrial troposphere is due to volcanic emissions.

The HF in Venus' atmosphere probably originated from volcanic outgassing, as HF is the major F-bearing gas in terrestrial volcanic gases. Hydrogen fluoride is more corrosive than HCl vapor and is used for etching glass. Thus, the HF partial pressure in Venus' atmosphere is probably regulated by reactions between HF and F-bearing minerals, for example:

$$2HF + NaAlSiO_4 \text{ (nepheline)} + 2CaMgSi_2O_6 \text{ (diopside)}$$
$$+ Mg_2SiO_4 \text{ (forsterite)} + MgSiO_3 \text{ (enstatite)} \tag{4.73}$$
$$= NaCa_2Mg_5Si_7AlO_{22}F_2 \text{ (fluoredenite)} + H_2O$$

Reaction (4.73), like reaction (4.70) involving HCl, is a divariant equilibrium because the HF equilibrium partial pressure depends upon temperature and the H_2O partial pressure. Equation (4.74) gives the equilibrium mole fraction of HF for the average water vapor mixing ratio (30 ppmv) in Venus' lower atmosphere:

$$\log_{10} X_{HF} = \log_{10}\left(\frac{P_{HF}}{P_T}\right) = 0.22 - \frac{6426}{T} \tag{4.74}$$

The HF equilibrium mole fraction for Reaction (4.73) at 740 K is 4.5 ppbv, which is identical within uncertainties to the observed amount of HF (about 5 ppbv). The minerals involved in the equilibria for controlling the partial pressures of CO_2, H_2O, HCl and HF are common in alkaline rocks on Earth, which indicates the presence of these rocks on Venus. As mentioned earlier, the X-ray fluorescence analysis at the *Venera 13* landing site is consistent with the presence of alkaline rocks on Venus.

Hydrogen fluoride gas is not observed in the Martian atmosphere, but it should not be present unless there is active volcanism. Any HF erupted into the atmosphere of Mars will be removed rapidly by reaction with atmospheric dust and the Martian surface. Thus, fluorine-bearing minerals are expected to be the major F reservoir on Mars.

4.5 ORIGIN AND EVOLUTION OF TERRESTRIAL PLANET ATMOSPHERES

Three (Venus, Earth and Mars) of the four terrestrial planets have significant, long-lived atmospheres. Mercury has a rarified, variable atmosphere composed of monatomic H, He, O, alkali metals and alkaline earths with a surface pressure $< 10^{-12}$ bar. Its atmosphere originates from several sources, including solar wind, sputtering of atoms from Mercury's surface, outgassing from Mercury's surface, and vaporization of impacting objects. Mercury's atmosphere is very different to those on the other three terrestrial planets, and we do not discuss it further in this section.

As shown in Figure 4.22, the atmospheres of Venus, Earth and Mars differ significantly. To a first approximation, Earth's atmosphere is a binary gas mixture of N_2 (78.08%) + O_2 (20.95%) with small and variable amounts of water vapor (0–4%) and trace Ar (0.93%). The large O_2 abundance reflects the importance of biological processes for Earth's atmospheric composition. The atmosphere of Venus is a hot, dense (740 K, 95 bar) mixture of CO_2 (96.5%) + N_2 (3.5%) with much smaller amounts of SO_2 (0.015%), H_2O (0.003%), and corrosive gases such as HCl and HF. The HF, HCl, SO_2 and other sulfur gases in Venus' atmosphere reflect the importance of thermochemical processes for Venus' atmospheric composition. The atmosphere of Mars is a cold, thin (220 K, 0.006 bar) ternary gas mixture of CO_2 (95.3%) + N_2 (2.7%) + Ar (1.6%) with trace amounts of O_2 (0.13%), CO (0.08%), and variable trace amounts of water vapor. The small amounts of O_2, CO,

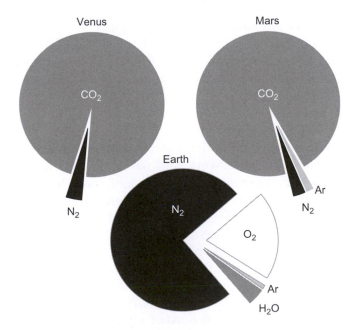

Figure 4.22   Major gases in the atmospheres of Venus, Earth and Mars.

O_3, H_2O_2 (hydrogen peroxide) and other gases reflect the importance of photochemical processes for Mars' atmospheric composition. The details of atmospheric composition and chemistry (discussed earlier in this chapter for Venus and Mars and in Chapters 5 and 6 for the Earth) describe more differences between these three planets. However, all of their atmospheres probably share a common origin.

The Earth's oceans and the atmospheres of Venus, Earth and Mars were produced by chemical reactions that released water and gases from volatile-bearing solids during and/or after planetary accretion. This secondary origin is in contrast to a primary origin by the capture of solar nebula gas by the terrestrial planets during their formation. The basis for the secondary atmosphere model is the low abundances of the chemically inert noble gases (Ne, Ar, Kr, Xe) relative to higher abundances of chemically reactive elements such as H, C, N and O at the surface of the Earth. Neither helium nor radon are important for this comparison, but for different reasons. Both ^{3}He and ^{4}He are light enough that they continually escape from the Earth's atmosphere. All radon isotopes are radioactive and decay on short time-scales. Thus, the He and Rn abundances do not constrain atmospheric formation models.

This disparity between the abundances of the noble gases and other volatile elements was first realized 87 years ago by the British scientist F. W. Aston (1877–1945), who received the 1922 Nobel Prize in Chemistry for his pioneering work in mass spectroscopy. In 1924 Aston published a graph showing the abundance of different atomic species on the Earth. He found that the noble gases (He, Ne, Kr, Xe) are abnormally scarce on the Earth in comparison to other elements. Aston argued that the most plausible explanation for the rarity of the noble gases is that Earth is depleted in these elements. Aston's conclusion was reinforced in 1933 by the American astronomers Henry Norris Russell (1877–1957) and Donald H. Menzel (1901–1976). They noted that astronomical spectra showed neon was abundant in the cosmos yet geochemists found it was scarce on Earth. This prior work and new data on the solar system abundances of the elements were synthesized into the secondary atmosphere model in the late 1940s by two pioneering cosmochemists – Harrison Brown (1917–1986) and Hans Suess (1909–1993). Brown and Suess independently showed that *all* volatile elements are depleted on Earth relative to their solar abundances. However, the noble gases are much more depleted on Earth than the chemically reactive volatiles (H, C, N, O gases). Spacecraft analyses of the atmospheres and surfaces of Venus and Mars show similar disparities in elemental abundances on these planets – much larger depletions of noble gases than for CO_2, N_2, H_2O, F, Cl and S.

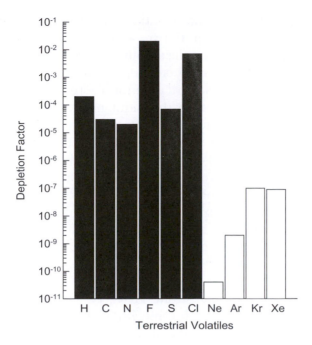

Figure 4.23 Depletions of the noble gases (Ne, Ar, Kr, Xe) and chemically reactive volatiles (H, C, N, F, S, Cl) on the Earth.

Figure 4.23 shows a modern version of the argument made by Brown and Suess.

The depletion factors for the terrestrial abundances of the noble gases and the chemically reactive elements H, C, N, F, S and Cl are plotted on a logarithmic scale. The depletion factors are the terrestrial elemental abundance relative to silicon divided by the solar elemental abundance relative to silicon (*i.e.*, the cosmochemical scale discussed in Chapter 1). For example, the depletion factor for Ne (DF_{Ne}) is:

$$DF_{Ne} = \frac{(Ne/Si)_{Earth}}{(Ne/Si)_{solar}} \qquad (4.75)$$

The numerator is the Ne/Si mass ratio in the Earth and the denominator is the Ne/Si mass ratio in solar composition material. Similar equations give the depletion factors for Ar, Kr and Xe. In each case the volume fraction of the noble gas in Earth's atmosphere (given in Table 5.1) is converted into the equivalent noble gas mass using the average molecular weight (28.97 g mol^{-1}) and total mass (5.137×10^{18} kg) of Earth's atmosphere.

These computations assume that the atmosphere contains the total inventory of noble gases on Earth, which is strictly a lower limit. However, it is implausible that the amount of Ne, Ar, Kr, or Xe trapped inside the Earth is sufficient to compensate for their very low abundances.[iii] Mass ratios are used because most terrestrial abundances are given in mass units (%, parts per thousand, micrograms per gram, *etc.*). The terrestrial Si abundance is essentially the Si abundance in the Earth's mantle because the Si content of the core is unknown, and the mantle is 99.4% of the mass of the Earth's lithosphere. The amount of Si in the atmosphere, biosphere and hydrosphere is negligible compared to the Si content of the lithosphere ($\sim 21.6\%$). Of course the Si content of Earth's lithosphere varies from one compilation to another, but these uncertainties are small and affect all element ratios to the same extent. Likewise, neglecting the Si content of Earth's core has the same effect on all element ratios.

The depletion factors for chemically reactive volatiles are computed using the same equation (substitute H, C, N, F, S or Cl for Ne) but computation of the terrestrial inventories of these elements is more involved because they occur in the atmosphere, hydrosphere and lithosphere. For example, the terrestrial carbon inventory includes atmospheric CO_2, dissolved inorganic and organic carbon in the hydrosphere and carbon-bearing materials in the lithosphere (carbonates, coal, oil, natural gas, *etc.*). Carbon could be dissolved in the Earth's core, but the amount of carbon in this reservoir is completely unknown. Strictly speaking, the value for the terrestrial carbon inventory is a lower limit and its depletion factor an upper limit, *i.e.*, carbon may be less depleted. The terrestrial inventories of the other chemically reactive volatiles are computed analogously and the same reasoning applies. Chapter 6 provides information on the terrestrial inventories of many of the chemically reactive volatiles and gives data on the major reservoirs for the volatiles shown in Figure 4.23.

Figure 4.23 shows that all volatiles – chemically reactive and inert – are depleted on Earth relative to their abundances in the solar nebula. The depletions are smallest for F and Cl, larger for H, S, C, and N, and largest for Ne, Ar, Kr and Xe. The reason for this is simple – chemically reactive volatiles were incorporated into the solid grains accreted by the Earth during its formation but the noble gases were not. The chondritic meteorites are relatively unaltered samples of solid material from the

---

[iii] Some Xe may be trapped in xenon clathrate hydrates $Xe \cdot 6H_2O$, which may be found as pure phases or a Xe-CH_4 clathrate hydrate solid solution $(CH_4,Xe) \cdot 6H_2O$. The importance of these phases for the terrestrial Xe inventory remains to be seen.

solar nebula and their composition is a guide to that of the solid grains accreted by the Earth.

The solar elemental abundances of F, Cl and S are small enough that these three elements can be completely incorporated into various minerals such as halite (NaCl), apatite [$Ca_5(PO_4)_3(F,Cl,OH)$] and troilite (FeS) found in chondritic meteorites. Mass balance prevents complete incorporation of H, C and N into minerals formed with Mg, Si, Fe, *etc.* because the elemental abundances of H, C and N are much larger than those of the rock-forming metals. (For example, the Si/C atomic ratio is only 0.5 so at most only 50% of the total carbon abundance could be retained as silicon carbide.) However, a small fraction of these elements is found in meteorites as organic matter, carbides, carbonates, diamonds, nitrides, hydrous minerals and dissolved in Fe alloy. In contrast, the noble gases do not form minerals and occur only in very small amounts in meteorites.

The depletions shown in Figure 4.23 reveal that Earth's atmosphere did not form by capture of solar nebula gas. If this had happened the Ne/N_2 abundance ratio in Earth's atmosphere would be 91 000 times larger than actually observed. Neon and nitrogen have similar atomic weights and solar abundances so a physical process such as diffusion would not alter their abundance ratio significantly. Yet, Earth is much richer in nitrogen than neon. Likewise, argon and sulfur have similar solar abundances and capture of solar nebula gas would give a terrestrial Ar/S abundance ratio of ~ 0.2 instead of 0.000 03 as observed. On the other hand, isotopic analyses of He and Ne emitted from Earth's mantle suggest that Earth contains some noble gases that may be remnant volatiles from the solar nebula. These observations do not change the conclusion drawn from Figure 4.23.

Figure 4.24 and Table 4.14 show that the carbonaceous, enstatite, and ordinary chondrites contain more H, C and N than the Earth. As discussed in Chapter 2, these are the three major groups of chondritic meteorites and are representative of the solids existing in the solar nebula. Accretion of chondritic material provides more than enough carbon, hydrogen and nitrogen for the terrestrial inventories of these elements. Table 4.14 lists important volatile-bearing host phases observed in meteorites and the potential outgassed volatiles from each phase.

The nature of the outgassed volatiles depends upon the temperature, pressure and the oxidation state of the solid materials that are outgassing. The elements carbon, nitrogen and sulfur may be outgassed predominantly as CH_4, CO or CO_2, NH_3 or N_2, and H_2S, SO_2 or SO_3 as conditions become progressively more oxidizing at constant temperature

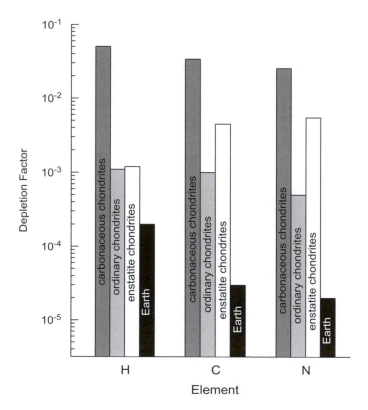

Figure 4.24   Comparison of the abundances of H, C and N in the Earth and in the major groups of chondrites (carbonaceous, ordinary and enstatite).

and pressure. Chemical equilibrium calculations of the volatiles released during outgassing give the following picture of Earth's early atmosphere. Similar results apply to Venus, Mars and extrasolar terrestrial planets (see the next section).

As discussed earlier in this chapter, accretion of the Earth took about 100 million years (from about 4.6–4.5 billion years ago). The average accretion rate during this period was about 60 000 megatons per year, and plausibly decreased from higher to lower values as Earth grew. Two exceptions to this decrease are probably during the moon-forming impact near the end of this time, and the late heavy bombardment at about 4.0–3.9 billion years ago. During the most vigorous phase of its accretion (and the moon-forming impact) the Earth was heated to high temperatures above the melting point of metal and silicates (Table 4.8). At this time, the major gases in Earth's atmosphere were volatiles released by vaporization of molten rock. For example, monatomic Na, O, O_2 and SiO were the major gases in the Earth's atmosphere at 3000 K.

Table 4.14 Volatile-bearing phases in chondrites and potential outgassed volatiles.

Name	Ideal formula	Occurs in[a]	Potential outgassed volatiles[b]
Anhydrite	$CaSO_4$	C, OC	SO_2, H_2S, OCS, S_x
Apatite	$Ca_5(PO_4)_3(F,Cl,Br,OH)$	Many	HF, HCl, Cl_2, HBr, Br_2, H_2O, H_2, O_2, P_x, PO_x, PH_3
Carbon	C dissolved in metal	Many	CH_4, CO, CO_2
Cohenite	$(Fe,Ni)_3C$	Many	CH_4, CO, CO_2
Gypsum	$CaSO_4 \cdot 2H_2O$	C, OC	SO_2, H_2S, OCS, S_x
Halite	NaCl	C, OC	HCl, Cl_2
Insoluble organic matter	$C_{100}H_{72}N_3O_{22}S_{4.5}$	C	CH_4, CO, CO_2, H_2O, H_2, N_2, NH_3, S_x, H_2S, OCS, SO_2
Nitrogen	N dissolved in metal	Many	N_2, NH_3
Osbornite	TiN	E, CH	N_2, NH_3
Sinoite	Si_2N_2O	E	N_2, NH_3
Schreibersite	$(Fe,Ni)_3P$	Many	P_x, PO_x, PH_3
Serpentine	$(Mg,Fe)_3Si_2O_5(OH)_4$	C	H_2O, H_2, O_2
Talc	$(Mg,Fe)_3Si_4O_{10}(OH)_2$	C	H_2O, H_2, O_2
Troilite	FeS	Many	S_x, H_2S, OCS, SO_2

[a]Abbreviations denote the following types of chondrites: C = carbonaceous chondrites, CH = CH chondrites, E = enstatite (EH, EL) chondrites, OC = ordinary (H, L, LL) chondrites, or many for many types of chondrites.
[b]The nature of the potential outgassed volatiles depends on several factors, including the temperature, pressure and oxygen fugacity during outgassing. Elemental fluorine does not form because it is too reactive. Hydrogen and oxygen are generated *via* equilibria of water vapor with Fe-bearing phases such as metal, magnetite and FeO-bearing silicates.

As accretion of the Earth neared completion and impacts became less frequent, the rock vapor atmosphere condensed onto the surface (as a global magma ocean or local magma lakes and seas) and it was replaced by a "steam" atmosphere. The typical P and T assumed for Earth's steam atmosphere are about 1500 K and 100 bar, but these parameters are not well constrained. Although H_2O was a significant constituent of the Earth's "steam" atmosphere, it was probably not the major gas. Instead, H_2 and CO were the two major volatiles released by the outgassing of ordinary (H, L, LL) and enstatite (EH, EL) chondritic material, which made up most of the material accreted by the Earth. Water was the major gas in the "steam" atmosphere only during the very last stage of accretion when more oxidizing CI or CM2 chondritic material was impacting the Earth.

As the Earth cooled, liquid water condensed out of the massive "steam" atmosphere. The total atmospheric pressure decreased, and the atmospheric composition changed to that of a reducing atmosphere with CH_4, N_2, NH_3 and H_2, which are the major volatiles produced by outgassing ordinary and enstatite chondritic material. This type of reducing atmosphere is favorable for the production of organic compounds *via* lightning [as first demonstrated by the American chemists Stanley L. Miller (1930–2007) and Harold C. Urey (1893–1981) in the 1950s], UV sunlight and heat.

The favorable production of organic compounds in Earth's reducing atmosphere meant that it was ultimately unstable against destruction by lightning and UV sunlight, which converted CH_4 into H_2 plus CO_2 and NH_3 into N_2 plus H_2 in geologically short periods. The time required for conversion from a reducing into a neutral or more oxidizing atmosphere depended upon the hydrogen escape rate (Φ_H), which depends on the total mole fraction (X_i) of H atoms in Earth's atmosphere:

$$\Phi_H = 10^{13.4}\left(X_H + 2X_{H_2} + 2X_{H_2O} + 3X_{NH_3} + 4X_{CH_4} + \cdots\right) \text{atoms cm}^{-2}\,\text{s}^{-1}$$

(4.76)

At present $\Phi_H \approx 3.6\times10^{10}$ H-atoms cm^{-2} s^{-1} from Earth's atmosphere and the H escape rate in Earth's early reducing atmosphere may have been higher.

The existence of carbonates in 3.8 billion year old (early Archean) sedimentary rocks at Isua, West Greenland shows that carbonate formation took place at that time, which implies CO_2 in the atmosphere and dissolved inorganic carbon (aqueous CO_2, H_2CO_3, HCO_3^-, CO_3^{2-}) in the oceans. Atmospheric CO_2 levels during the Archean plausibly dropped due to continued carbonate precipitation and N_2 was probably the major gas in Earth's atmosphere at that time, as it is today.

Controversial arguments interpret the carbon isotopic fractionation between graphitic carbon and carbonate carbon in the Isua sedimentary rocks as evidence of photosynthetic fixation of carbon (see discussion in Chapter 6). If correct, this means that organisms such as cyanobacteria (also known as blue-green algae) existed by 3.8 billion years ago and were consuming CO_2 and producing O_2. Microfossils interpreted as cyanobacteria exist in the 3.46 billion year old Apex chert in Western Australia, and in many younger Archean rocks, *e.g.*, the 2.6 billion year old Campbell Group (in Cape Province, South Africa) and the 2.0 billion year old Gunflint Chert (in Canada). Oxygen production by these simple photosynthetic organisms slowly increased the amount of O_2 in the oceans and in Earth's atmosphere over time until a dramatic increase occurred about 2.3–2.4 billion years ago. The subsequent evolution of Earth's atmospheric composition was closely tied to the rise of more complex living organisms and can be modeled using various biological, chemical and geological indicators. We do not have space to discuss this topic in detail here, but provide recommended reading about current ideas at the end of this chapter.

4.6 EXTRASOLAR TERRESTRIAL PLANETS

Several hundred extrasolar planets have been detected. Most of these objects are so massive that they are probably gas giant planets, but a few of the extrasolar planets are less massive and are thought to be terrestrial, or rocky, planets. Table 4.15 summarizes some of the putative extrasolar terrestrial planets (ETPs), which have masses up to 15 times that of the Earth. Mass balance considerations and models of chemistry in protoplanetary accretion disks provide guidance about the bulk composition of ETPs. With decreasing temperature, the major solids accreted into ETPs are refractory (Ca, Al, Ti) oxides and silicates, Fe-rich metal, anhydrous ferromagnesian silicates, troilite FeS, magnetite Fe_3O_4, hydrous silicates and possibly C-bearing organic materials.

Table 4.15 Possible extrasolar terrestrial planets.[a]

Planet	M_{Earth}[b]	a (AU)[b]	P $(days)$[b]
PSR 1257 + 12 b	0.02	0.19	25.2
Gl 581 e	1.9	0.03	3.1
MOA-2007-BLG-192-L b	3.2	0.62	
PSR 1257 + 12 d	3.8	0.46	98.2
PSR 1257 + 12 c	4.1	0.36	66.5
HD 40307 b	4.2	0.05	4.3
Gl 581 c	5.4	0.07	12.9
OGLE-05-390L b	5.4	2.1	3500
Gliese 876 d	5.7	0.02	1.9
HD 40307 c	6.9	0.08	9.6
Gl 581 d	7.1	0.22	66.8
HD 181433 b	7.6	0.08	9.4
HD 285968 b	8.4	0.07	8.8
HD 40307 d	9.2	0.13	20.5
HD 7924 b	9.2	0.06	5.4
HD 69830 b	10.5	0.08	8.7
HD 160691 c	10.6	0.09	9.6
55 Cnc e	10.8	0.04	2.8
CoRoT-7 b	11.1	0.02	0.85
GJ 674 b	11.8	0.04	4.7
HD 69830 c	12.1	0.19	21.6
OGLE-05-169L b	12.7	2.8	3300
HD 4308 b	14.9	0.11	15.6

[a]Data from the *Extrasolar Planets Encyclopaedia*: http://exoplanet.eu/. (accessed August 2009).
[b]Approximate values for planetary mass in units of Earth masses; a = semimajor axis. P = orbital period in days.

Objects formed at lower temperatures also accreted significant amounts of water ice and would be more like Uranus and Neptune than like the Earth and other terrestrial planets.

The atmospheres of ETPs will depend upon several variables, including the planetary bulk composition, proximity to its parent star (and its spectral type), stage of planetary evolution and so on. The large number of possible variables suggests that ETP atmospheres will be more varied than encountered in our solar system.

The presence of ETPs in the habitable zones around other stars is of particular interest and space missions such as the NASA *Kepler* mission and the ESA *COROT* mission are looking for ETPs by observing their transits across the disks of their primary stars. These missions cannot make spectroscopic observations of the atmospheres and/or surfaces of ETPs. Future space missions to make such observations were proposed by NASA (*Terrestrial Planet Finder, TPF*) and ESA (*Darwin*), but *TPF* is unfunded and *Darwin* will not launch prior to 2016.

FURTHER READING

R. Canup and K. Righter (eds), *Origin of the Earth and Moon*, University of Arizona Press, Tucson, AZ, 2000, 555 pp.

R. A. Daly, *Igneous Rocks and the Depths of the Earth*, McGraw Hill Book Co., New York, 1933.

V. M. Goldschmidt, *Geochemistry*, Clarendon Press, Oxford, 1954.

G. Heiken, D. Vaniman and B. M French (eds), *Lunar Sourcebook: A User's Guide to the Moon*, Cambridge University Press, New York, 1991, 736 pp.

I. Jackson (ed.), *The Earth's Mantle: Composition, Structure, and Evolution*, Cambridge University Press, Cambridge, 1998.

H. H. Kieffer, B. M. Jakosky, C. W. Snyder and M. S. Matthews (eds), *Mars*, University of Arizona Press, Tucson, AZ, 1992.

B. Mason and C. B. Moore, *Principles of Geochemistry*, 4th edn, John Wiley & Sons, 1962, New York.

H. E. Newsom and J. H. Jones (eds), *Origin of the Earth*, Oxford University Press, New York, 1990, 378 pp.

F. Nimmo and K. Tanaka, *Annu. Rev. Earth Planet. Sci.*, 2005, **33**, 133.

K. Rankama and T. G. Sahama, *Geochemistry*, Chicago University Press, 1950.

K. H. Wedepohl (ed.), *Handbook of Geochemistry*, Springer Verlag, 1969.

Results from spacecraft missions to Venus and Mars are given in serveral special issues of the Journal of Geophysical Research (Planets).

Special issues for Venus include volume 112, Number E4 (2007), and volume 114, Numbers E5, E9 (2009). Special issues for Mars include volumes 111 Numbers E2, E9, E12 (2006), 112 Numbers E3, E5, E8 (2007), and volume 113 Numbers E6, E12 (2008).

REFERENCES

1. R. S. Woodward, *Science*, 1889, **14**, 167.
2. I. Newton, *Principia; the Mathematical Principles of Natural Philosophy*, translated by Andrew Motte 1848, published by Daniel Adee, New York, 1687.
3. C. Hutton, *Phil. Trans. R. Soc., London*, 1776, **68**, 689.
4. W. Thomson, *Br. Assoc. Rep., Sections*, 1867, p. 7.
5. W. Thomson, *Phil. Trans. R. Soc. London*, 1863, **153**, 573.
6. E. Roche, *Science*, 1881, **2**, 458.
7. S. G. Brush, Discovery of the earth's core, *Am. J. Phys.*, 1980, **48**, 705.
8. K. E. Bullen, *The Earth's Density*, Chapman & Hall, London, 1975.
9. G. R. Helffrich and B. Wood, *Nature*, 2001, **412**, 501.
10. F. W. Clarke, Data on Geochemistry, *U.S. Geol. Survey Bull.* 1924, 770, continued by M. Fleischer (ed.), *U.S. Geol. Survey Prof. Papers*, 1962, 440
11. A. W. Hofmann, *Earth Planet. Sci. Lett.*, 1988, **90**, 297.
12. H. Wänke, G. Dreibus and E. Jagoutz, in *Archean Geochemistry: The Origin and Evolution of the Archean Continental Crust*, A. Kroner, *et al.*, (ed.), Springer Verlag, Berlin, 1984, p. 1.
13. K. Wedepohl, *Geochim. Cosmochim. Acta*, 1995, **59**, 1217.
14. S. R. Taylor and S. M. McLennan, *The Continental Crust: Its Origin and Evolution*, Oxford, Blackwell Science Publishers, 1985, p. 312.
15. H. T. de la Beche, *A Geological Manual*, 2nd edn, Carey & Lea, London, 1832.
16. S. A. Wilde, J. W. Valley, W. H. Peck and C. M. Graham, *Nature*, 2001, **409**, 175.
17. S. A. Bowring and I. S. Williams, *Contrib. Mineral. Petrol.*, 1999, **134**, 3.
18. L. O'Neil, R. W. Carlson, D. Francis and R. K. Stevenson, *Science*, 2008, **321**, 1828.
19. E. Jagoutz, H. Palme, H. Baddenhausen, K. Blum, M. Cendales, G. Dreibus, B. Spettel, V. Lorenz and H. Wänke, *Proc. Lunar Planet. Sci. Conf. 10th*, 1979, 2031.
20. S. R. Hart and A. Zindler, *Chem. Geol.*, 1986, **57**, 247–267.

21. J. S. Kargel and J. S. Lewis, *Icarus*, 1993, **105**, 1.
22. C. J. Allegre, J. P. Poirier, E. Humler and A. W. Hofmann, *Earth Planet. Sci. Lett.*, 1995, **134**, 515.
23. H. Palme and H. St. C. O'Neill, Cosmochemical estimates of mantle composition, in *Treatise on Geochemistry, Volume 2*, R. W. Carlson (ed.) H. D. Holland and K. K. Turekian (exec. ed.), 2003, Elsevier, p. 1.
24. K. Lodders and B. Fegley, *The Planetary Scientist's Companion*, Oxford University Press, 1998.
25. M. J. Walter, H. E. Newsom, W. Ertel and A. Holzheid, Siderophile elements in the Earth and Moon: metal/silicate partitioning and implications for core formation, in *Origin of the Earth and Moon*, R. Canup and K. Righter, (ed.), University of Arizona Press, Tucson, 2000, pp. 235–289.
26. V. M. Goldschmidt, *Z. Elektrochem.*, 1922, **28**, 411.
27. G. Dreibus and H. Palme, *Geochim. Cosmochim. Acta*, 1996, **60**, 1125.
28. H. St. C. O'Neill and H. Palme, Composition of the silicate Earth: implications for accretion and core formation, in *The Earth's Mantle: Composition, Structure, and Evolution*, I. Jackson, (ed.), Cambridge University Press, Cambridge, 1998, pp. 3–126.
29. H. Y. McSween, *Meteoritics*, 1994, **29**, 757.
30. D. S. McKay, E. K Gibson, K. L. Thomas-Keprta, H. Vali, C. S. Romanek, S. J. Clemett, X. D. F. Chillier, C. R. Maechling and R. N. Zare, *Science*, 1996, **273**, 924.
31. H. Wänke and G. Dreibus, *Phil. Trans. R. Soc. London A*, 1988, **325**, 545.
32. K. Lodders and B. Fegley, *Icarus*, 1997, **126**, 373.
33. T. B. McCord, J. B. Adams and T. V. Johnson, *Science*, 1970, **178**, 745.
34. G. Dreibus, J. Brückner and H. Wänke, *Meteoritics Planet. Sci.*, 1997, **32**, A36.
35. H. Wänke and G. Dreibus, in *Tidal Friction and the Earth's Rotation II*, P. Brosche and J. Sündermann, (ed.), Springer, Berlin, 1982, p. 322.
36. H. E. Newsom and S. R. Taylor, *Nature*, 1989, **338**, 29.
37. H. St. C. O'Neill, *Geochim. Cosmochim. Acta*, 1991, **55**, 1135; 1991, **55**, 1159.
38. J. M. D. Day, D. G. Pearson and L. A. Taylor, *Science*, 2007, **315**, 217.
39. R. Korotev, *Chem. Erde*, 2005, **65**, 297.
40. R. A. Daly, *Proc. Am. Phil. Soc.*, 1946, **90**, 104.

41. W. K. Hartmann and D. R. Davis, *Icarus*, 1975, **24**, 504.

42. A. G. W. Cameron and W. R. Ward, *Lunar Planet. Sci. Conf. Ser. VII*, 1976, 120.

43. R. B. Baldwin, *Popular Astron.*, 1942, **50**, 356.

44. R. B. Baldwin, *Popular Astron.*, 1943, **51**, 117.

45. G. K. Gilbert, *Bull. Phil. Washington*, 1892, **12**, 241.

46. G. K. Gilbert, *Sci. Am. Suppl.*, 1894, **37**, 15016.

47. E. T. Drake and P. D. Komar, *Geology*, 1984, **12**, 408.

48. D. Putnam Beard, *Popular Astron.*, 1917, **15**, 167.

49. A. E. Ringwood, *Proc. R. Soc. London A*, 1984, **395**, 1; H. Wänke, *Phil. Trans. R. Soc. London A*, 1981, **303**, 287.

50. U. Von Zahn, S. Kumar, H. Niemann and R. Prinn, in *Venus*, D. M. Hunten, L. Colin, T. M. Donahue and V. I. Moroz, (ed.), University of Arizona Press, Tucson, AZ, 1983, pp. 299–450.

51. B. Fegley, *Venus*, in *Meteorites, Comets, and Planets*, ed. A. M. Davis, vol. 1 *Treatise on Geochemistry*, H. D. Holland and K. K. Turekian, (ed.), Elsevier-Pergamon, Oxford, 2004, pp. 487–507.

52. H. Wänke and G. Dreibus, *Lunar Planet Sci. Conf*, 1997, **28**, 1495.

53. T. Kleine, C. Münker, K. Mezger and H. Palme, *Nature*, 2002, **417**, 952.

54. T. Kleine, K. Mezger, C. Münker, H. Palme and A. Bischoff, *Geochim. Cosmochim. Acta*, 2004, **68**, 2935.

55. Q. Yin, S. B. Jacobsen, K. Yamashita, H. Blichert-Toft, P. Telouk and F. Albarede, *Nature*, 2002, **418**, 949.

56. S. S. Barshay, *Combined condensation-accretion models of the terrestrial planets*, Ph.D. Thesis, MIT, Cambridge, MA, 1981.

57. J. S. Lewis, *Science*, 1974, **36**, 440.

58. J. S. Lewis, Origin and Composition of Mercury, in *Mercury*, F. Vilas, C. R. Chapman, M. S. Matthews, eds, University of Arizona Press, Tucson, AZ, 1988, pp. 651–666.

59. H. C. Urey, *Proc Natl, Acad. Sci., USA*, 1955, **41**, 127.

60. T. P. Kohman and M. S. Robinson, *Lunar and Planetary Science Conf.*, 1980, **XI**, 564.

61. W. F. McDonough and S. S. Sun, *Chem. Geol*, 1995, **120**, 223.

62. W. A. Hofmann, *Phil. Trans. R. Soc. Lond.*, 1989, **A328**, 425.

Terrestrial Atmospheric Chemistry

5.1 INTRODUCTION

We discuss chemistry of Earth's atmosphere in this chapter. Terrestrial atmospheric chemistry is important for many reasons, including global climate change and ozone loss. Many of the concepts and terms developed in terrestrial atmospheric chemistry are now also applied to other planets. Thus, we start by defining these important concepts and terms. Then we review the major features of Earth's atmosphere and its chemical composition. With this background, we describe important features of tropospheric, stratospheric and upper atmospheric chemistry. Many of the reactions that occur on Earth are also important on its neighboring planets Venus and Mars.

Chapter 6 continues the discussions of this chapter. It describes climate change, the major biogeochemical cycles, and marine chemistry. The origin and evolution of Earth's atmosphere is discussed in Chapter 4 as part of our discussion of the origin and evolution of the atmospheres of the terrestrial planets.

5.2 BASIC DEFINITIONS

We begin by defining concepts and terms used to discuss the composition, thermal structure and chemical models of the atmosphere of the Earth, of other planets in our solar system, and of planets in extrasolar planetary systems.

Chemistry of the Solar System
By Katharina Lodders and Bruce Fegley, Jr.
Published by the Royal Society of Chemistry, www.rsc.org

5.2.1 Mixing Ratio

Atmospheric chemists express the abundances of gases in planetary atmospheres in terms of the volume-mixing ratio (mixing ratio), number density and column density (or column abundance). The volume-mixing ratio (X_i) is a dimensionless quantity also called the mole (or volume) fraction of a gas. It is the gas partial pressure (P_i) divided by the total pressure (P_T):

$$X_i = \frac{P_i}{P_T} \tag{5.1}$$

Mixing ratios are given as percentages for major gases and as parts per million (10^6), per billion (10^9) or per trillion (10^{12}) by volume (ppmv, ppbv or pptv, respectively) for trace gases. However, water vapor is often reported in parts per million by mass (ppmm); 1 ppmv water vapor $= 1.607$ ppm by mass.

The number density of a gas "i" is denoted by square brackets $[i]$ and has the dimensions of particles (atoms plus molecules) per unit volume, *e.g.*, particles cm^{-3}. The number density is equal to:

$$[i] = \frac{P_i N_A}{RT} = \frac{P_i}{kT} \tag{5.2}$$

where N_A is the Avogadro constant (6.02214×10^{23} mol^{-1}), R is the ideal gas constant (8.31451 J mol^{-1} K^{-1} = 82.0578 cm^3 atm mol^{-1} K^{-1}), k is Boltzmann's constant ($k = R/N_A = 1.38066 \times 10^{-23}$ J K^{-1}) and T is temperature in kelvin. For reference, the number density of Earth's atmosphere is 2.55×10^{19} particles cm^{-3} at sea level where $T = 288.15$ K and $P =$ one atmosphere ($101\,325$ Pa $= 1.01325$ bar).

5.2.2 Column Abundance

The column abundance (or column density) of a gas is the number of gas particles throughout an atmospheric column and has the dimensions of particles per unit area, *e.g.*, particles cm^{-2}. The column abundance (σ_i) of gas "i" is equal to the integral:

$$\sigma_i = \int_{z=z_0}^{z=\infty} [i]\mathrm{d}z \tag{5.3}$$

The integration limits are a reference altitude z_0 (*e.g.*, sea level on Earth) and infinity (the top of the atmosphere). The column density is also equal to:

$$\sigma_i = \frac{P_i N_A}{g\mu} \tag{5.4}$$

where μ is the formula weight of the gas and g is the gravitational acceleration as a function of altitude and latitude. The mean formula weight of a gas mixture, such as dry air, is the mole fraction weighted sum of the individual formula weights. For example, using gas abundances from Table 5.1:

$$\mu_{air} = X_{N_2}\mu_{N_2} + X_{O_2}\mu_{O_2} + X_{Ar}\mu_{Ar} + X_{CO_2}\mu_{CO_2} + \cdots = 28.97\,g\,mol^{-1} \tag{5.5}$$

In practice, this summation does not have to extend beyond CO_2 because of the much smaller amounts of other gases in dry air. The secular increase in atmospheric CO_2 causes a small increase in the mean formula weight of dry air over time as fossil fuel carbon is burned and added to the atmosphere.

The total atmospheric mass per unit area m_T is total pressure (P_T) divided by gravity (g). The column mass of Earth's atmosphere at sea level and 45° latitude is:

$$m_T = \frac{P_T}{g} = \frac{1.01325 \times 10^6\,dyn\,cm^{-2}}{980.665\,cm\,s^{-2}} = 1033.2\,g\,cm^{-2} \tag{5.6}$$

The corresponding column densities in terms of moles per cm^2 and particles per cm^2 are:

$$\frac{1033.2\,g\,cm^{-2}}{28.97\,g\,mol^{-1}} = 35.66\,moles\,cm^{-2} \tag{5.7}$$

$$\frac{N_A \times 1,033.2\,g\,cm^{-2}}{28.97\,g\,mol^{-1}} = 2.148 \times 10^{25}\,particles\,cm^{-2} \tag{5.8}$$

On a global basis, the atmosphere contains about 1.77×10^{20} moles of gas, which corresponds to about 1.07×10^{44} gas atoms and molecules. These calculations neglect the spatially and temporally variable amount of water vapor in air. The average water vapor content of the atmosphere is $1.3 \times 10^{16}\,kg$, which is 0.25% of the total atmospheric mass of $5.14 \times 10^{18}\,kg$. We can express these masses in terms of petagrams ($10^{15}\,g = 1\,Pg$), which are commonly used to discuss biogeochemical

Table 5.1 Chemical composition of Earth's troposphere.

Gas	Abundance[a]	Source(s)	Sink(s)
N_2	78.084%	Denitrifying bacteria	Nitrogen fixing bacteria
O_2	20.946%	Photosynthesis	Respiration & decay
H_2O	<4%, varies	Evaporation, transpiration	Condensation
Ar	9340 ppm	Outgassing, ^{40}K decay	—
CO_2	387 ppm	Respiration, decay, combustion	Photosynthesis, oceanic dissolution, weathering
Ne	18.18 ppm	Outgassing	—
^4He	5.24 ppm	Outgassing (U, Th)	Atmospheric escape
CH_4	1.75 ppm	Biological, agricultural	Oxidation by OH
Kr	1.14 ppm	Outgassing	—
H_2	0.55 ppm	Photochemistry, biology, combustion	Uptake in soils, oxidation by OH
N_2O	~320 ppb	Anthropogenic, biological	Stratospheric photolysis
CO	125 ppb	Photochemistry	Photochemistry
Xe	87 ppb	Outgassing	—
O_3	~50 ppb	Photochemistry	Photochemistry
NMHCs[b]	≤80 ppb	Foliar emissions, combustion, anthropogenic	Photooxidation
HCl	~1 ppb	Derived from sea salt	Rainout
H_2O_2	~0.3–3 ppb	Photochemistry	Photochemistry
NH_3	0.1–3 ppb	Biology	Wet & dry deposition
HNO_3	~0.04–4 ppb	Photochemistry (NO_x)	Rainout
Reduced S-gases[c]	≤500 ppt	Biology, anthropogenic	Photodissociation, photooxidation

Table 5.1 (*continued*)

Gas	Abundance[a]	Source(s)	Sink(s)
NO_x[d]	~30–300 ppt	Combustion, biology	Photooxidation
CFCs[e]	1013 ppt	Anthropogenic	Stratospheric photolysis
HCCs[e]	689 ppt	Anthropogenic	Reaction with OH
HCFCs[f]	153 ppt	Anthropogenic	Reaction with OH
PFCs[g]	83 ppt	Anthropogenic	Photolysis (upper atm.)
CH_3Br	22 ppt	Ocean, marine biota	Reaction with OH
HFCs[h]	21.5 ppt	Anthropogenic	Reaction with OH
SO_2	20–90 ppt	Combustion	Photooxidation
Halons[i]	6.3 ppt	Anthropogenic	Stratospheric photolysis
CH_3I	~2 ppt	Ocean, marine biota	Photolysis (troposphere)

[a]Abundances by volume in dry air (non-urban troposphere).
[b]Non-methane hydrocarbons: alkanes, alkenes, alkynes, aromatics, sterols.
[c]OCS, H_2S, CS_2, $(CH_3)_2S$.
[d]CF_2Cl_2, $CFCl_3$, $C_2Cl_3F_3$, $C_2Cl_2F_4$, C_2ClF_5, $CClF_3$, CCl_4.
[e]CH_3Cl, CH_3CCl_3.
[f]$CHClF_2$, $C_2H_3Cl_2F$, $C_2H_3ClF_2$.
[g]CF_4, C_2F_6.
[h]CHF_3, $C_2H_2F_4$.
[i]CF_2ClBr, CF_3Br.

cycles. Thus, the atmospheric water inventory is 13 000 Pg and the total atmospheric mass is 5 140 000 Pg.

Spectroscopic observations of the Earth from space, or of other planets, give column abundances for the gases in a planet's atmosphere. Astronomers often give column abundances in units of cm-, m-, or km-amagats. An amagat is a number density unit for gas atoms and molecules (particles). It is named after the French physicist Emile Amagat (1841–1915) who pioneered studies of gases at high pressure. One amagat is about 2.687×10^{19} particles cm^{-3} at $0\,°C$ and 1 atm pressure. These conditions are standard (or normal) temperature and pressure (STP or NTP). A number density of one amagat is the same as Loschmidt's Number, named after the Austrian chemist Josef Loschmidt (1821–1895). Chemists use amagat units as a dimensionless measure of gas density at high pressures, *e.g.*, in studies of critical phenomena of gases. Astronomers use amagat units to measure column densities in planetary atmospheres. The total column abundance of air on Earth is thus equal to:

$$\frac{2.148 \times 10^{25}\,cm^{-2}}{2.687 \times 10^{19}\,cm^{-3}\,Amagat^{-1}} = 7.994 \times 10^{5}\,cm - amagat \sim 8.0\,km - amagat \tag{5.9}$$

In addition, astronomers sometimes use the terms cm-, m- and km-atmospheres or simply cm-, m-, km- at STP (or NTP) synonymously for cm-, m- and km-amagats. The column abundances are converted into mixing ratios by dividing the column abundance of an individual gas by the total column abundance of all gases in a planet's atmosphere. For example, in 1948 the Belgian astronomer Marcel Migeotte (1912–1992) discovered CH_4 in the Earth's atmosphere from its 3.3-μm infrared absorption band in solar spectra. The methane column abundance was measured as 1.2 cm NTP. This is equivalent to a CH_4 mixing ratio of:

$$X_{CH_4} = \frac{1.2\,cm - amagat}{7.994 \times 10^{5}\,cm - amagat} = 1.5 \times 10^{-6} \tag{5.10}$$

This value, which is for 1948, is $\sim 14\,\%$ lower than the current mixing ratio of 1.75×10^{-6} because the CH_4 abundance in Earth's troposphere is increasing over time. We return to CH_4 later when we discuss tropospheric chemistry.

5.2.3 Hydrostatic Equilibrium, Scale Height and Barometric Equation

With a few exceptions such as the tenuous atmosphere on Io, Jupiter's innermost Galilean satellite, planetary atmospheres are in hydrostatic equilibrium. Pressure decreases with increasing altitude z because of the decreasing mass of the atmospheric column as altitude increases:

$$\frac{dP}{dz} = -\rho g \qquad (5.11)$$

The gravitational acceleration g is $9.80665 \, \text{m s}^{-2}$ at sea level and $45°$ latitude. It varies with latitude ϕ according to the equation:

$$g = 9.780356\left(1 + 5.28850 \times 10^{-3} \sin^2 \phi - 5.90 \times 10^{-6} \sin^2 2\phi\right) \text{m s}^{-2} \quad (5.12)$$

However, the variation in g from equator to poles is only about 0.5% and can be neglected to first approximation. The gravity (g) and mass density (ρ) of the atmosphere also vary with altitude. These variations are more significant, especially for the density. On Earth, g and ρ from sea level ($z = 0 \, \text{km}$) to the tropopause ($z = 12 \, \text{km}$) are:

$$g = 9.80665 - 3.0747 \times 10^{-3}z - 4.1392 \times 10^{-7}z^2 \, \text{m s}^{-2} \qquad (5.13)$$

$$\rho = 1.2225 - 0.1140z + 3.340 \times 10^{-3}z^2 \, \text{kg m}^{-3} \qquad (5.14)$$

Experimental observations of the composition of air as a function of altitude show that air behaves as a single, well-mixed gas throughout most of the atmosphere, even though it is a mixture of gases with very different formula weights. Thus there is no need to apply the barometric equation and associated equations to the different constituents of air until very high in the atmosphere, above about 100 km altitude.

The decrease of pressure with altitude varies from planet to planet because of their different gravities and atmospheric densities. The pressure scale height H gives a convenient comparison of different planetary atmospheres. It is the altitude over which pressure decreases by the same amount on each planet, namely by a factor of 1/e. The letter e in this fraction is the base of natural logarithms and is a transcendental number approximately equal to 2.718. The pressure scale height H is:

$$H = \frac{RT}{\mu g} \qquad (5.15)$$

At sea level on Earth, the pressure scale height is 8.43 km. The scale height (H) is proportional to temperature. On Earth H decreases with decreasing temperature, *i.e.*, with increasing altitude, until it reaches a value of 6.34 km at 12 km altitude, the average position of the tropopause – the top of the troposphere.

Substituting the scale height H into the equation of hydrostatic equilibrium, rearranging and integrating gives the barometric equation:

$$P_z = P_0 \exp\left(-\frac{z - z_0}{H}\right) \qquad (5.16)$$

The pressure P_z corresponds to an altitude z and P_0 is the pressure at the reference altitude (z_0). Hydrostatic equilibrium is valid up to the top of the atmosphere. In atmospheric regions where temperature varies with altitude (such as the troposphere), the barometric equation is rewritten as:

$$P_z = P_0 \left(\frac{T}{T_0}\right)^{-\beta} \qquad (5.17)$$

The exponent β is:

$$\beta = \frac{\mu g}{R(\mathrm{d}T/\mathrm{d}z)} \qquad (5.18)$$

The temperature gradient ($\mathrm{d}T/\mathrm{d}z$) is in $K\,km^{-1}$ and the other terms are the same as we defined them to be earlier.

5.2.4 Top of the Atmosphere

The top of the atmosphere is defined as the critical level at which the horizontal mean free path (λ_m) of a gas atom or molecule with diameter d and partial pressure P:

$$\lambda_m = \frac{kT}{\sqrt{2}\pi d^2 P} \qquad (5.19)$$

is equal to the pressure scale height H. The mean free path is the average distance a gas atom or molecule travels before colliding with another atom or molecule. This definition applies to neutral species as long as the change in the number density of the gas is small over the mean free path length. This is true in all directions at altitudes below ~ 130 km, but not at higher altitudes. Thus, the horizontal mean free path is used to define the critical level. The mean free path and scale height become equal at about 500 km altitude on Earth. In general the top of a planetary atmosphere is defined by the product of the total number density $[n_T]$, the collisional cross section Q and λ_m:

$$1 = [n_T]Q\lambda_m \qquad (5.20)$$

At the critical level, gas atoms and molecules with sufficient velocity away from Earth have a probability of $1/e$ ($\sim 37\%$) of escaping to space. Above the critical level, gas atoms and molecules have ballistic trajectories.

5.2.5 Eddy Diffusion Coefficient

One-dimensional models of chemistry in planetary atmospheres use a quantity called the eddy diffusion coefficient. The eddy diffusion coefficient (K_{eddy}) parameterizes mixing by turbulent atmospheric motions from small to large scale and is the same for all gases and aerosol particles in air. It has units of $m^2 s^{-1}$, the same as a molecular diffusion coefficient. Large K_{eddy} values correspond to rapid mixing in turbulent, convective regions of planetary atmospheres and small K_{eddy} values correspond to slow mixing in quiescent, isothermal regions of planetary atmospheres. Typical values of K_{eddy} are $\sim 20\,m^2 s^{-1}$ for vertical transport in the troposphere (rapid mixing), and $\sim 0.3\,m^2 s^{-1}$ for vertical transport in the stratosphere (slow mixing). In contrast, typical values for molecular diffusion coefficients at ground level in the troposphere are $D \approx 2 \times 10^{-5}\,m^2 s^{-1}$. Thus eddy diffusion is much more important than molecular diffusion for mixing gases in the ground level atmosphere. This remains true until molecular diffusion coefficients become comparable to eddy diffusion coefficients ($D \approx K_{eddy} \approx 100\,m^2 s^{-1}$) at about 100–110 km in Earth's atmosphere. Diffusive separation of the gases in air occurs at higher altitudes.

It is useful to define a characteristic time for vertical mixing in planetary atmospheres (t_{mix}). Typically, t_{mix} is the time (in seconds) to mix gases over an altitude equal to the pressure scale height H and is given as:

$$t_{mix} \approx \frac{H^2}{K_{eddy}} \qquad (5.21)$$

Taking average values of 7.4 km for the scale height (H) and $20\,m^2 s^{-1}$ for K_{eddy} in Earth's troposphere yields a characteristic vertical mixing time of:

$$\text{Tropospheric } t_{mix} \approx \frac{(7.4 \times 10^3\,m)^2}{20\,m^2 s^{-1}} = 2.74 \times 10^6 s \sim 32\,\text{days} \qquad (5.22)$$

This is rapid mixing and is much shorter than the corresponding t_{mix} in the stratosphere just above the tropopause where $H \approx 6.34$ km and $K_{eddy} \approx 0.3$ m^2 s^{-1}:

$$\text{Stratospheric } t_{mix} \approx \frac{(6.34 \times 10^3 \text{ m})^2}{0.3 \text{ m}^2 \text{ s}^{-1}} = 1.34 \times 10^8 \text{ s} \sim 4.25 \text{ years} \quad (5.23)$$

The stratospheres of other planets also have slow mixing times.

5.3 MAJOR FEATURES OF EARTH'S ATMOSPHERE

Figure 5.1 shows the average thermal structure of the terrestrial atmosphere as a function of altitude up to 120 km. The different regions of the atmosphere correspond to different chemical and physical processes that control the thermal structure. In order of increasing altitude, these regions are the troposphere (0–12 km), the stratosphere (12–48 km), the mesosphere (48–85 km), the thermosphere (85–500 km) and the exosphere (> 500 km). The ionosphere overlaps the mesosphere and thermosphere (in the 60–400 km range) and is the atmospheric region where

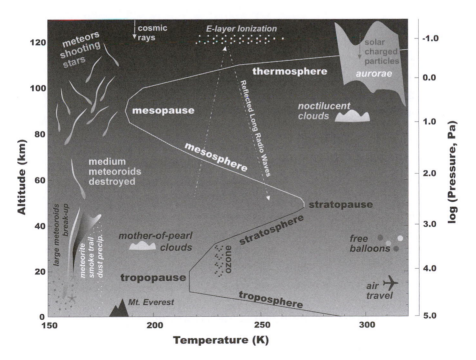

Figure 5.1 Schematic diagram of the major features of Earth's atmosphere.

the number of electrons is large enough to affect propagation of radio waves. Similar regions in the atmospheres of other planets are given the same names although different processes may be controlling the thermal structure of the atmosphere. In addition, there is no clear distinction between the stratosphere and mesosphere on several planets and we call the middle atmosphere of these planets the stratomesosphere.

5.3.1 Troposphere

The troposphere is the lowermost region of Earth's atmosphere and ranges from 0 to 12 km altitude on average, with its upper boundary varying from 8 to 18 km depending on latitude, season and weather. The troposphere contains about 82% of the mass of Earth's atmosphere. It is convective, which means that the temperature gradient (or lapse rate) is adiabatic. However, dry (cloud free) and moist (cloudy) regions of the troposphere have different lapse rates. The temperature gradient in cloud-free regions of Earth's troposphere is given by the dry adiabatic lapse rate:

$$\frac{dT}{dz} = -\frac{\mu g}{C_P} = -\frac{g}{c_p} \tag{5.24}$$

where C_P and c_p are the average molar heat capacity $(28.94\,\text{J}\,\text{mol}^{-1}\,\text{K}^{-1})$ and specific heat $(0.999\,\text{J}\,\text{g}^{-1}\,\text{K}^{-1})$, respectively, of dry air at sea level where the global average temperature is $15\,^\circ\text{C}$. The dry adiabatic lapse rate is about $-9.83\,\text{K}\,\text{km}^{-1}$ at sea level. The wet adiabatic lapse rate gives the temperature gradient in cloudy regions:

$$\frac{dT}{dz} = -\frac{\mu g}{C_P + \lambda_W X_W \left(\frac{\lambda_W}{RT^2} - \frac{1}{T}\right)} \tag{5.25}$$

The terms X_W and λ_W are the mole fraction and latent heat of condensation $(2499\,\text{J}\,\text{g}^{-1}$ at $0\,^\circ\text{C})$ of water vapor in the moist gas. The wet adiabatic lapse rate depends upon the temperature and moisture content of the air. It is smaller than the temperature gradient in dry air (at the same temperature) because water cloud condensation releases the latent heat of condensation to the atmosphere and warms the surrounding air. At sea level, the wet adiabatic lapse rate is about $-3.76\,\text{K}\,\text{km}^{-1}$ for air saturated with water vapor $(X_W \approx 0.017)$. The observed lapse rate in Earth's troposphere is about $-6.5\,\text{K}\,\text{km}^{-1}$ due to the presence of cloud free and cloudy regions (global average cloud cover of about 50%).

The absorption of solar radiation drives atmospheric circulation in the troposphere. The atmosphere absorbs about one-third and the Earth's surface about two-thirds of all solar energy absorbed. The equatorial regions absorb about four times more energy than the polar regions. Consequently, heat is transmitted from equator to poles *via* direct (Hadley and polar) and indirect (Ferrel) cells in each hemisphere. The two tropical Hadley cells are separated by a region of ascending air at the equator that is called the interhemispheric tropical convergence zone (ITCZ). The alternating seasonal dominance of one or the other tropical Hadley cells causes an oscillation of the ITCZ about its average position. This movement is primarily responsible for air exchange between the northern and southern hemispheres. Within the northern hemisphere (and mirrored in the southern hemisphere) zonal circulation is dominantly from east to west in the tropics and from west to east at mid-latitudes. The subtropical jet stream at about $30°$ latitude and about the same altitude as the tropopause flows west to east at $25–50 \, \text{m} \, \text{s}^{-1}$.

Atmospheric transport times in the troposphere and between the troposphere and stratosphere have been measured from the distribution of inert tracers such as SF_6, of environmental pollutants and of radioactive fallout from atmospheric nuclear testing, which primarily took place from 1945 to 1963. The characteristic magnitudes of these transport times are as follows:

- one hour for vertical mixing through the planetary boundary layer (the lowest 1–2 km next to Earth's surface),
- four weeks for vertical mixing from Earth's surface to the tropopause,
- two weeks for zonal transport (*i.e.*, around latitude circles) in the troposphere,
- three months for north–south transport in the troposphere,
- one year for interhemispheric exchange in the troposphere,
- three years for interhemispheric exchange in the stratosphere,
- one to two years for stratospheric air exchange with the troposphere,
- 50 years for tropospheric air exchange with the stratosphere.

The exchange time from the troposphere upward to stratosphere is longer than the exchange time from the stratosphere downward to troposphere because the mass of the troposphere (M_T) is larger than that of the stratosphere (M_S). The upward and downward air flux (F) is the same between the two regions. The ratio of the exchange times, which to a first approximation are the mass (M) divided by the flux (F), is:

$$\frac{t_T}{t_S} = \frac{M_T/F}{M_S/F} = \frac{M_T}{M_S} \sim \frac{1013 \text{mbar}}{120 \text{mbar}} \sim 8.4 \tag{5.26}$$

Recall that atmospheric mass is P/g. Hence, the mass ratio for the troposphere and stratosphere is simply the ratio of the pressures at sea level (1013 mbar) and at the tropical tropopause (120 mbar). The latter is where most of the air mass exchange occurs between the troposphere and stratosphere. The important implication of the long upward mixing time from troposphere to stratosphere is that gases released into the troposphere by anthropogenic or natural sources must have long life-times to survive mixing into the stratosphere. A gas has a long lifetime if it is inert to chemical or physical destruction at Earth's surface or in the troposphere. The CFC, or chlorofluorocarbon, gases that destroy ozone meet this requirement, and hence survive unscathed into the strato-sphere. In contrast, the hydrochlorofluorocarbon (or HCFC) gases, which are replacements for CFC gases, have shorter lifetimes and are destroyed in the troposphere.

5.3.2 Tropopause and Stratosphere

Figure 5.1 shows that on average the troposphere extends to about 12 km altitude where there is a temperature inversion called the tropo-pause. The French meteorologist Leon Philippe Teisserence de Bort (1855–1913) discovered the tropopause and the stratosphere around 1900 by making temperature measurements from balloons and kites. These showed the lapse rate decreased almost to zero from its average value of $\sim 6.5 \text{ K km}^{-1}$. The tropopause is the radiative–convective boundary for the thermal structure of Earth's atmosphere. Convective heat transport controls the thermal structure of the troposphere and radiative equilibrium with solar radiation controls the thermal structure of the stratosphere. The tropopause altitude varies from 8 to 18 km depending on the latitude, season and weather. The tropopause is at 8–10 km in the polar regions during winter time and is at 15–18 km in equatorial and tropical regions. A typical value for the tropical tropo-pause is 15 km with a temperature of 195 K and a pressure of 120 mbar.

The stratosphere lies directly above the tropopause. The temperature is nearly constant from the tropopause up to 35 km altitude in the lower stratosphere, and then increases with increasing altitude up to the stratopause at ~ 48 km. The temperature inversion at the tropopause and the thermal structure of the stratosphere are due to the absorption of solar ultraviolet (UV) sunlight by ozone O_3. The ozone layer extends from about 15 to 50 km and the thickest part is near 37 km. The stratosphere is very dry and contains about 5 ppmv water vapor in contrast to 1–4% water vapor in the troposphere. Most air mass exchange between the troposphere and stratosphere occurs in the ascending branches of the tropical Hadley cells and the tropical tropopause is a cold trap for water, which has a vapor pressure of 5×10^{-7} bar at 195 K.

Stratospheric chemistry is different from tropospheric chemistry because of the cold trap at the tropopause, the different thermal structure of the two regions, the shorter wavelength (and thus more energetic) ultraviolet sunlight in the stratosphere, the anthropogenic inputs into the troposphere and the sluggish mixing from troposphere to stratosphere. The formation and loss of ozone, which we discuss later, dominates chemistry of the stratosphere.

5.3.3 Stratopause and Mesosphere

The stratopause is a temperature inversion at about 48 km altitude ($T \approx 271$ K, $P \approx 1$ mbar). It marks the top of the stratosphere and the base of the mesosphere. The mesosphere extends to 85 km ($T \approx 190$ K, $P \approx 0.4$ mbar) where the mesopause is located. The summertime mesopause at high latitudes is the coldest part of Earth's atmosphere with temperatures reaching 120 K. The temperature decreases with increasing altitude throughout the mesosphere due to CO_2, O_3 and NO radiating infrared energy to space. The mesosphere is a poorly understood region of Earth's atmosphere because it lies between the highest operable altitudes for aircraft and the lowest operable altitudes for artificial satellites. The ongoing NASA AIM (Aeronomy of Ice in the Mesosphere) spacecraft mission is currently studying the formation of noctilucent clouds (also known as polar mesospheric clouds). The mesosphere is extremely dry like the stratosphere and contains about 1–5 ppm H_2O by mass. Water vapor condenses to form noctilucent clouds in the summer polar mesosphere. Transport from the lower atmosphere, rocket exhaust plumes and infalling meteoroids are sources of water vapor in the mesosphere.

5.3.4 Thermosphere, Exosphere and Ionosphere

The thermosphere is the region above the mesopause and extends from ~ 85 km to ~ 500–1000 km (the thermopause). The thermopause altitude varies with solar activity. The thermosphere is strongly heated by photolysis and ionization of molecular oxygen, which absorbs extreme UV radiation and X-rays ($\lambda < 180$ nm). Temperatures increase with altitude up to ~ 250 km, then become roughly isothermal, but vary widely (600–2000 K) depending on solar activity.

The aurora borealis (northern lights) and aurora australis (southern lights) occur at about 100 km altitude in the thermosphere. The auroras are produced by solar wind bombardment of the upper atmosphere and are more intense during periods of high solar activity. The green and red colors of the auroras arise from spectral lines of atomic oxygen. Red and blue colors arise from lines of N_2 and N ions.

The exosphere is the outermost region of Earth's atmosphere. The base of the exosphere is the critical level where the mean free path of a gas atom or molecule is equal to the pressure scale height. The critical level is at about the same altitude as the thermopause (500–1000 km).

The ionosphere, where the number density of electrons is high enough to affect transmission of radio waves, starts at about 60 km in the mesosphere. It is divided into several parts called the D layer (60–90 km), E layer (90–120 km), F-1 layer (a daytime layer at about 150 km) and the F-2 layer (200–400 km). The major positive and negative ions are different in these layers and are H_3O^+, $(H_2O)_nH^+$ ($n > 1$) positive ions and Cl^-, NO_3^- and HCO_3^- negative ions in the D layer, O_2^+ and NO^+ positive ions and electrons in the E layer, and O^+, N^+ and electrons in the F1 and F2 layers. At very high altitudes (500–1000 km) H^+ and He^+ ions become more important. Peak electron densities $> 10^6$ cm^{-3} occur in the 200–400 km region in the thermosphere. The homopause is the boundary between the well-mixed region of the atmosphere (the homosphere) and the diffusively unmixed region (the heterosphere). The homopause occurs at about 100–110 km in the thermosphere. The homopause (also called the turbopause) is different for different gases because it is the altitude where the molecular diffusion coefficient D_i of a gas is equal to the vertical eddy diffusion coefficient K_z (units of m^2 s^{-1} for D_i and K_z).

5.4 ATMOSPHERIC COMPOSITION

Table 5.1 lists the abundances, sources and sinks of major gases and some important trace gases in Earth's atmosphere. Some subsequent tables expand upon this one, for example, the listing of halogen-bearing gases in Table 5.2. There are literally hundreds of gases in Earth's atmosphere that have been detected and measured by various methods,

Table 5.2 Abundances of some halogen-bearing gases in Earth's troposphere.

Gas	Abundance[a]	Source(s)	Sink(s) & lifetimes
CH_3Cl (methyl chloride)	620	Ocean, biomass burning	Reaction with OH
CF_2Cl_2 (F12)	533	Anthropogenic ($+4.4$ pptv yr^{-1})	Photolysis (stratosphere), 100 years
$CFCl_3$ (F11)	268	Anthropogenic (-1.4 pptv yr^{-1})	Photolysis (stratosphere), 45 years
$CHClF_2$ (HCFC-22)	132	Anthropogenic ($+5$ pptv yr^{-1})	Reaction with OH, 11.9 years
CCl_4 (carbon tetrachloride)	102	Anthropogenic (-1.0 pptv yr^{-1})	Photolysis (stratosphere), 35 years
HCl (hydrogen chloride)	100	Volcanic, anthropogenic	Rainout, 10 days
$C_2Cl_3F_3$ (F113)	84	Anthropogenic (0 pptv yr^{-1})	Photolysis (stratosphere), 85 years
CF_4 (F14)	80	Anthropogenic ($+1.0$ pptv yr^{-1})	Reaction with O$^+$ (ionosphere); 330 000 years
CH_3CCl_3 (methyl chloroform)	69	Anthropogenic (-14 pptv yr^{-1})	Reaction with OH, 4.8 years
HF (hydrogen fluoride)	27	Volcanic, stratospheric	Rainout, 5.5 days
CH_3Br (methyl bromide)	22	Ocean, marine biota	Reaction with OH, 1.5 years
$C_2Cl_2F_4$ (F114)	15	Anthropogenic (<0.5 pptv yr^{-1})	Photolysis (stratosphere), 300 years
CHF_3 (HFC-23)	14	Anthropogenic ($+0.55$ pptv yr^{-1})	Reaction with OH, 260 years
CH_3CF_2Cl (HCFC-142b)	11	Anthropogenic ($+1$ pptv yr^{-1})	Reaction with OH, 19 years
CH_3CFCl_2 (HCFC-141b)	10	Anthropogenic ($+2$ pptv yr^{-1})	Reaction with OH, 9.3 years
$C_2H_2F_4$ (HFC-134a)	7.5	Anthropogenic ($+2.0$ pptv yr^{-1})	Reaction with OH, 13.8 years
C_2ClF_5 (F115)	7	Anthropogenic ($+0.4$ pptv yr^{-1})	Photolysis (stratosphere), 1700 years
SF_6 (sulfur hexafluoride)	4.2	Anthropogenic ($+0.24$ pptv yr^{-1})	Photolysis (upper atm), 3200 years
CF_3Cl (F13)	4	Anthropogenic ($+0.1$ pptv yr^{-1})	Photolysis (upper atm), 640 years
CF_2ClBr (Halon-1211)	3.8	Anthropogenic ($+0.2$ pptv yr^{-1})	Reaction with OH, 11 years
C_2F_6 (F116)	3	Anthropogenic ($+0.08$ pptv yr^{-1})	Reaction with O$^+$ (ionosphere); 420 000 years
CF_3Br (Halon-1301)	2.5	Anthropogenic ($+0.1$ pptv yr^{-1})	Reaction with OH, 65 years
HBr (hydrogen bromide)	2	Sea-salt	Rainout, reaction with OH, 5 days
CH_3I (methyl iodide)	0.8	Ocean, marine biota	Photolysis (troposphere), 5 days

[a]Abundances in parts per trillion by volume (pptv $= 10^{-12} =$ picomole) in dry air (non-urban troposphere). The mixing ratios for total F, Cl, Br, and I atoms are 2466, 3665, 30, and 0.8 pptv, respectively.

including but not limited to the following:

- chemical analysis (O_2, CO_2, NH_3, N_2, Ar),
- fractional absorption on activated charcoal (He, Ne),
- infrared spectroscopy (CO, CO_2, CH_4, HDO, HF, N_2O, NH_3, O_3, SO_2),
- mass spectroscopy (noble gases, isotopomers of O_2, O_3),
- gas chromatography (halocarbons, volatile hydrocarbons, H_2S, OCS, N_2O),
- measurements of radioactivity (Rn, $^{14}CO_2$, HTO),
- refractive index measurements (O_2/N_2 ratio).

Many gases can be measured by more than one technique, although one method may be preferable over others for various reasons. Most of these gases were detected and measured relatively recently; a 1937 review by the Austrian radiochemist Friedrich A. Paneth (1887–1958) lists eleven constituents of dry air (N_2, O_2, Ar, CO_2, Ne, He, Kr, H_2, Xe, O_3 and Rn). There are two reasons for the large increase in the number of gases found in air. The first is that the discovery of new gases occurs after advances in analytical methods are applied to atmospheric chemistry. Improved astronomical observations of the solar spectrum led to the discovery of CO, CH_4, HDO, HF, N_2O and O_3 in the atmosphere. Likewise, development of the electron conductivity detector for gas chromatographs led to the discovery of chlorofluorocarbon (CFC) gases in the atmosphere. The second reason is that some gases were not present in air in detectable amounts in the past. The chlorofluorocarbons, which are anthropogenic, are the prime example. Infrared solar spectra (at the same resolution) taken at the Jungfraujoch Observatory in Switzerland show bands of CF_2Cl_2 (F12) in spectra from 2000 but not in spectra from 1951.

Table 5.1 gives the composition of dry air in the non-urban troposphere (*i.e.*, the composition of the unpolluted atmosphere). In contrast to the major constituents, many of the trace gases are temporally and spatially variable. Moist air is a mixture of water vapor plus dry air, so the gaseous concentrations in moist air are simply:

$$C_{\text{moist}} = X_{\text{dry}} C_{\text{dry}} = (1 - X_{H_2O}) C_{\text{dry}} \qquad (5.27)$$

However, the abundance of water vapor is temporally and spatially variable, so it makes sense to tabulate the composition of dry air. One key point in Table 5.1 is that the large O_2 abundance shows that biological processes, with important anthropogenic and photochemical influences, control Earth's atmospheric composition. The atmospheres of Venus and Mars contain much smaller amounts of O_2, which are produced by solar UV photolysis of CO_2, the major gas in their atmospheres. Only photosynthesis can produce the large amounts of O_2 seen in Earth's atmosphere. Another significant point is that minor and trace gases, such as methane, nitrous oxide, ozone and halocarbons, are important for the greenhouse effect and ozone loss. For example, the halogen-bearing gases listed in Table 5.2 are either greenhouse gases or destroy ozone or sometimes do both.

5.5 TROPOSPHERIC CHEMISTRY

5.5.1 OH Radical and Oxidation Reactions

The hydroxyl OH radical is the cleansing agent of Earth's troposphere and plays a central role in tropospheric chemistry *via* oxidation of anthropogenic and natural air pollutants. It is unreactive with the major gases in air (N_2, O_2, H_2O, Ar, CO_2), but reactive with many of the trace gases in air such as CO, CH_4, H_2, H_2CO, hydrocarbons, O_3, NO_2, NH_3, SO_2, H_2S and other volatile sulfides. In general, the atmospheric lifetimes of most reactive gases in Earth's troposphere depend upon the OH concentration. Hydroxyl, and not O_2, which is about 520 trillion times more abundant, is the key species for cleansing the troposphere. In addition, the long atmospheric lifetimes of the chlorofluorocarbon (CFC) gases arise because they are not destroyed by reaction with OH radicals.

Hydroxyl radicals are produced by reaction of electronically excited oxygen atoms $O(^1D)$ with water vapor:

$$H_2O + O(^1D) \rightarrow OH + OH \quad k = 2.2 \times 10^{-10}\, cm^3\, molecule^{-1}\, s^{-1} \quad (5.28)$$

This is a rapid reaction at all temperatures and it proceeds at, or close to, the rate of molecular collisions, *i.e.*, at the gas kinetic rate. The arrow in this and subsequent reactions indicates that the elementary reaction takes place only in the direction written. The k value given next to the reaction is the reaction rate constant, or rate constant in units of cubic centimetre per molecule per second (henceforth given as $cm^3\, s^{-1}$ as done in the atmospheric chemistry literature). These are the typical units for bimolecular reactions, *i.e.*, elementary reactions with two reactants, and are necessary for the reaction rate to have the proper units of molecules per cubic centimetre per second (*i.e.*, concentration per unit time).

The OH abundance in Earth's troposphere is very small and its presence was first detected using UV absorption spectroscopy with a laser light source over a long path length in the atmosphere. Hydroxyl is difficult to measure, but careful measurements by several methods give an OH number density of about 10^6 OH cm^{-3}. This corresponds to a mixing ratio of 4×10^{-14} at sea level (total number density $= 2.55 \times 10^{19}$ cm^{-3}).

The principal removal reactions for OH are oxidation of CO and oxidation of methane. As discussed later, CO abundances vary in the troposphere, but 125 ppbv (a number density of $\sim 3.2 \times 10^{12}$ at sea level) is an average for the entire troposphere. Carbon monoxide oxidation is a fast reaction:

$$OH + CO \rightarrow H + CO_2 \ (k = 1.5 \times 10^{-13} \text{cm}^3 \text{ s}^{-1}) \qquad (5.29)$$

The chemical lifetime for OH loss by this reaction is:

$$t_{\text{chem}}(OH) = \frac{1}{k[CO]} = \frac{1}{(1.5 \times 10^{-13} \text{cm}^3 \text{ s}^{-1})(3.2 \times 10^{12} \text{ cm}^{-3})} \sim 2.1 \text{ s} \quad (5.30)$$

The chemical lifetime of a gas is the characteristic time for its loss by some reaction.

The oxidation of CO consumes OH but produces an H atom, which eventually produces more hydroxyl. Thus, CO oxidation is not a net loss of hydroxyl radicals.

The second major removal process is the oxidation of methane *via* the reaction:

$$OH + CH_4 \rightarrow CH_3 + H_2O \ [k = 4.6 \times 10^{-12} \exp(-1965/T) \text{cm}^3 \text{ s}^{-1}] \quad (5.31)$$

The rate constant for CH$_4$ oxidation is 5.0×10^{-15} cm^3 s^{-1} at 288 K, which is the average tropospheric temperature at sea level. As noted earlier, the CH$_4$ abundance is about 1.75 ppmv. This corresponds to a number density of 4.4×10^{13} cm^{-3} at sea level. The chemical lifetime for OH loss by CH$_4$ oxidation is thus:

$$t_{\text{chem}}(OH) = \frac{1}{k[CH_4]} = \frac{1}{(5.0 \times 10^{-15} \text{ cm}^3 \text{ s}^{-1})(4.4 \times 10^{13} \text{ cm}^{-3})} \sim 4.6 \text{ s} \quad (5.32)$$

This reaction produces water vapor and the methyl radical (CH$_3$). Methane oxidation is a net loss of OH radicals, and the increasing CH$_4$ abundance may lead to a decreasing OH radical abundance.

Overall, the OH lifetime due to its removal by reaction with CO and with CH_4 is:

$$t_{chem}(OH) = \left[\frac{1}{2.1} + \frac{1}{4.6}\right]^{-1} = 1.4\,s \qquad (5.33)$$

Thus, OH radicals exist for about 1.4 s before reactions with CO and CH_4 consume them. In general, the overall lifetime of a gas that is destroyed by several different chemical reactions is the inverse of the sum of the inverse lifetimes, *i.e.*:

$$t_{chem} = \left[\frac{1}{t_1} + \frac{1}{t_2} + \frac{1}{t_3} + \cdots\right]^{-1} \qquad (5.34)$$

The t_1, t_2, t_3 and so on in the equation above are the chemical lifetimes for reactions 1, 2, 3, *etc.* that destroy the gas. The same formula is used if the gas is destroyed by chemical reactions and lost by physical processes.

5.5.2 Tropospheric Ozone

Tropospheric ozone is a harmful pollutant and a greenhouse gas. The global average ozone abundance in the troposphere is 50 ppbv, corresponding to 370 teragrams (1 Tg = 1 megaton (Mt) = 10^{12} g = 10^9 kg). However, ozone is temporally and spatially variable and its abundance ranges from about 10 ppbv in the unpolluted troposphere to over 100 ppbv in polluted urban areas. The abundance of tropospheric ozone is increasing with time and is about 36% larger today than in the nineteenth century. Photolysis of tropospheric O_3 provides the electronically excited $O(^1D)$ atoms that react with water vapor to produce OH radicals. Ozone absorbs UV and visible light out to 360 nm in the visible end of the Huggins bands (300–360 nm). These are named for the English astronomer Sir William Huggins (1824–1910) who discovered these bands (due to terrestrial O_3) in spectra of the star Sirius. Photolysis of ozone by blue sunlight is the major source of electronically excited oxygen atoms in the troposphere:

$$O_3 + h\nu \rightarrow O(^1D) + O_2 \,[J = 1.2 \times 10^{-5}\,s^{-1}(1\,km\,altitude)] \qquad (5.35)$$

The $h\nu$ in this and subsequent equations represents a photon of UV or visible sunlight. Most of the electronically excited oxygen atoms collide with N_2 or O_2, and lose their excess energy to become oxygen atoms in their ground (^{3}P) electronic states, for example:

$$O(^1D) + N_2 \rightarrow O(^3P) + N_2 \tag{5.36}$$

Most of the ground state O atoms reform ozone *via* reaction with O_2 and a third body M:

$$O + O_2 + M \rightarrow O_3 + M \tag{5.37}$$

The M is a third body or collision partner that removes excess energy but does not take part in a reaction. For example, N_2, O_2 and Ar are the three most likely collision partners in Earth's atmosphere because they are the three most abundant gases. The ozone remade in this reaction can be photolyzed to give more electronically excited oxygen atoms. Some of these $O(^1D)$ atoms are quenched back to ground state oxygen, but others react with water vapor to form OH radicals as shown earlier.

The *J* value listed above is the photodissociation rate constant, which is a function of several factors, including the wavelengths at which a molecule absorbs light, the light flux as a function of wavelength, the zenith angle of the Sun, latitude and the presence of other gases, which absorb and/or scatter the incident light. The photochemical lifetime t_{chem} for O_3 is its concentration divided by its photolysis rate:

$$t_{chem}(O_3) = \frac{[O_3]}{d[O_3]/dt} = \frac{[O_3]}{J[O_3]} = \frac{1}{J} = 83333\,\text{s} \sim 23\,\text{h} \tag{5.38}$$

This is a short time and a source of tropospheric ozone is needed to replenish that lost by photolysis. What is this source?

Two major sources for tropospheric O_3 are transport downward from the stratosphere and photolysis of tropospheric nitrogen dioxide (NO_2), an anthropogenic air pollutant due to vehicles. This is why ozone alerts in cities often occur during high traffic conditions in summertime. The following reaction sequence forms O_3 and also oxidizes CO to CO_2:

$$OH + CO \rightarrow H + CO_2 \tag{5.29}$$

$$H + O_2 + M \rightarrow HO_2(\text{hydroperoxyl}) + M \tag{5.39}$$

$$HO_2 + NO \rightarrow OH + NO_2 \tag{5.40}$$

$$NO_2 + h\nu \rightarrow NO + O\,[J = 4.6 \times 10^{-3}\,\text{s}^{-1}(1\,\text{km altitude})] \tag{5.41}$$

$$O + O_2 + M \rightarrow O_3 + M \qquad (5.37)$$

$$\textit{Net reaction: } CO + 2O_2 \rightarrow CO_2 + O_3 \qquad (5.42)$$

This reaction scheme is important in areas with air pollution because CO and NO are higher than in the unpolluted troposphere.

As mentioned earlier, the abundance of tropospheric ozone is increasing at present. The increase is taking place because the sources for tropospheric ozone are larger than the sinks removing it. The three major sources are photolysis of nitrogen dioxide, oxidation of methane (discussed later) and transport downward from the stratosphere. The three major ozone sinks are photolysis, HO_x-catalyzed loss and deposition on the surface of the Earth. The imbalance is apparently larger in the northern hemisphere than in the southern hemisphere, probably because most anthropogenic emissions of nitrogen oxides and methane are concentrated in the northern hemisphere.

5.5.3 Carbon Monoxide

In 1949, Migeotte discovered telluric absorption lines (4.67 µm) of CO in solar spectra. Subsequent IR spectra by Migeotte and others confirmed the presence of CO in Earth's atmosphere and showed its abundance was variable. Today several methods are used to measure CO, including gas chromatography, IR spectroscopy and atomic absorption spectroscopy. The latter method relies on the reduction of mercuric oxide (HgO) to elemental mercury (Hg) by CO.

Carbon monoxide abundances are 1–10 ppmv in urban areas and drop to 200 ppbv (0.2 ppmv) outside urban areas in the northern hemisphere and 50 ppbv in the southern hemisphere. The major sources for CO are oxidation of CH_4 and other hydrocarbons (45%), biomass burning (30%), fossil fuel combustion and industrial operations (19%), and emissions from the oceans and vegetation (6%). Human population is concentrated in the northern hemisphere. Thus, the anthropogenic sources for CO (such as fossil fuel combustion, industrial operations, biomass burning) are also concentrated in the northern hemisphere. The major sinks for CO are oxidation by OH (80%), biologically mediated consumption by soils (16%) and transport into the stratosphere (4%). Most of Earth's landmass is in the northern hemisphere, thus, CO consumption by soils is more important in the northern than in the southern hemisphere.

Carbon monoxide oxidation is a fast reaction:

$$OH + CO \rightarrow H + CO_2 \quad (k = 1.5 \times 10^{-13}\, cm^3\, s^{-1}) \qquad (5.29)$$

320   *Chapter 5*

The chemical lifetime for CO loss by this reaction is:

$$t_{\text{chem}}(\text{CO}) = \frac{1}{k[\text{OH}]}$$

$$= \frac{1}{(1.5 \times 10^{-13}\,\text{cm}^3\,\text{s}^{-1})(10^6\,\text{cm}^{-3})} \sim 6.7 \times 10^6\,\text{s} \sim 77\,\text{days} \quad (5.43)$$

The overall lifetime for CO in the troposphere is shorter and is about 67 days (2 months) because bacterial consumption in soils also removes CO.

In the southern hemisphere, the anthropogenic sources and biologically mediated consumption in soils are less important than in the northern hemisphere. To a first approximation, the CO abundance of 50 ppbv results from a photochemical steady state where the production of CO by methane oxidation balances its loss by reaction with OH:

$$\text{Production} = \text{Loss} \quad (5.44)$$

$$k_{31}[\text{OH}][\text{CH}_4] = k_{29}[\text{OH}][\text{CO}] \quad (5.45)$$

$$[\text{CO}] = \frac{k_{31}}{k_{29}}[\text{CH}_4] = \frac{5.0 \times 10^{-15}\,\text{cm}^3\,\text{s}^{-1}}{1.5 \times 10^{-13}\,\text{cm}^3\,\text{s}^{-1}}\,1750\,\text{ppbv} = 58\,\text{ppbv} \quad (5.46)$$

This estimate depends upon the assumed temperature (taken as the global average of 288 K at sea level) and CH_4 concentration (taken as 1750 ppbv). The nominal values used above give good agreement with the CO abundance in the southern hemisphere.

5.5.4 Methane

As mentioned earlier, in 1948 the Belgian astronomer Migeotte discovered 1.5 ppmv CH_4 in Earth's atmosphere from its 3.3-μm infrared absorption band in solar spectra. The strong IR absorption of CH_4 at this and other wavelengths also makes it an important greenhouse gas, contributing about 15% to total greenhouse warming. Methane is the major hydrocarbon in the atmosphere and has natural and anthropogenic sources. Its abundance is about 1.75 ppmv (equivalent to 4.97 Pg CH_4) and is about 5% higher in the northern than in the southern hemisphere. The CH_4 abundance in Earth's atmosphere is increasing at about 1% per year. This increase is probably due to the increase in cattle and rice paddies and in slash and burn land clearing because of population growth. Unlike water vapor, CH_4 does not condense out of the atmosphere at the tropopause and CH_4 is an important source of water in the stratosphere. Finally, CH_4 is a key gas in tropospheric chemical cycles involving the OH radical, ozone and carbon monoxide, and in stratospheric chemical cycles involving OH and ozone.

Biogenic sources produce most of the methane in Earth's atmosphere. These sources and their estimated contribution to global CH_4 production are wetlands and landfills (25%), rice paddies (23%), cattle and other animals (20%), fossil methane (21%), biomass burning (10%) and all other sources (1%). Fossil methane is CH_4 released during coal, gas, oil exploration, production and transmission. Biomass burning produces CH_4 because of incomplete combustion. Methane production from all other sources includes emissions from the oceans and fresh water, decomposition of methane clathrate hydrate (ideally $CH_4 \cdot 7H_2O$) in permafrost and continental shelf areas, and geological processes such as serpentization. During serpentization, H_2 released by conversion of anhydrous rock into serpentine reacts with CO_2-bearing fluids to produce methane. This process is insignificant for CH_4 production on Earth because biogenic sources predominate. However, serpentization is plausibly the dominant source for CH_4 on Mars.

Oxidation by the OH radical is the major sink for methane in the troposphere:

$$OH + CH_4 \rightarrow CH_3 + H_2O \,[k = 4.6 \times 10^{-12}\exp(-1965/T)\,cm^3\,s^{-1}] \quad (5.31)$$

The rate constant for CH_4 oxidation is $5.0 \times 10^{-15}\,cm^3\,s^{-1}$ at 288 K, which is the average tropospheric temperature at sea level. As noted earlier, the OH number density is $10^6\,cm^{-3}$ at sea level. The chemical lifetime for CH_4 oxidation is thus:

$$t_{chem}(CH_4) = \frac{1}{k[OH]} = 2.0 \times 10^8\,s \sim 6.3\,\text{years} \quad (5.47)$$

This lifetime is long enough that CH_4, in contrast to CO, is well mixed in the troposphere.

As shown above, oxidation of methane by the OH radical produces a CH_3 radical, which undergoes further reactions that yield formaldehyde H_2CO:

$$OH + CH_4 \rightarrow CH_3 + H_2O \quad (5.31)$$

$$CH_3 + O_2 + M \rightarrow CH_3O_2 \,(\text{peroxymethyl}) + M \quad (5.48)$$

$$CH_3O_2 + NO \rightarrow CH_3O \text{ (methoxy)} + NO_2 \qquad (5.49)$$

$$CH_3O + O_2 \rightarrow H_2CO + HO_2 \qquad (5.50)$$

The formaldehyde abundance in the unpolluted troposphere is about 0.2 ppbv. However, its abundance is much higher in urban areas where it is produced in photochemical smog. Formaldehyde is only an intermediate and it is rapidly destroyed.

Photolysis destroys formaldehyde and proceeds by two pathways:

$$H_2CO + h\nu \rightarrow H_2 + CO \, [J = 2.6 \times 10^{-5} \, s^{-1} \, (1 \, km \, altitude)] \qquad (5.51)$$

$$H_2CO + h\nu \rightarrow HCO + H \, [J = 1.8 \times 10^{-5} \, s^{-1} \, (1 \, km \, altitude)] \qquad (5.52)$$

To first approximation, each reaction contributes 50% to H_2CO photolysis, although the relative importance of each is wavelength dependent. Taking the average J value for both reactions combined as $2.2 \times 10^{-5} \, s^{-1}$ gives a photochemical lifetime of about 12.5 h. Reaction with OH radicals is another important sink for H_2CO:

$$H_2CO + OH \rightarrow HCO \text{(formyl)} + H_2O \, [k = 2.0 \times 10^{-12} \exp(480/T)] \qquad (5.53)$$

At 288 K, the H_2CO lifetime for this reaction is about 26 h. The relatively long lifetime is due to the very small OH abundance. A third sink is rainfall, which washes H_2CO out of the atmosphere within 10 days (the global average lifetime for atmospheric water vapor). The formaldehyde lifetime from all three processes is about 8 h, and is mainly limited by H_2CO photolysis.

As shown above, CO is formed in one of the pathways for H_2CO photolysis. The formyl radical formed in the other pathway is converted into CO by reaction with oxygen:

$$HCO + O_2 \rightarrow CO + HO_2 \, [k = 3.5 \times 10^{-12} \exp(140/T)] \qquad (5.54)$$

This is a rapid reaction because of the large rate constant and large O_2 abundance. The CO formed in this reaction is rapidly converted into CO_2 by reaction with hydroxyl.

We can now summarize the methane oxidation chain, which is the sequence of reactions involved in oxidizing methane and ultimately forming carbon dioxide:

$$OH + CH_4 \rightarrow CH_3 + H_2O \qquad (5.31)$$

$$CH_3 + O_2 + M \rightarrow CH_3O_2 + M \qquad (5.48)$$

$$CH_3O_2 + NO \rightarrow CH_3O + NO_2 \qquad (5.49)$$

$$CH_3O + O_2 \rightarrow H_2CO + HO_2 \qquad (5.50)$$

$$H_2CO + h\nu \rightarrow H_2 + CO \qquad (5.51)$$

$$OH + CO \rightarrow H + CO_2 \qquad (5.29)$$

$$NO_2 + h\nu \rightarrow NO + O \qquad (5.41)$$

$$O + O_2 + M \rightarrow O_3 + M \qquad (5.37)$$

$$H + O_2 + M \rightarrow HO_2 + M \qquad (5.39)$$

Net reaction: $CH_4 + 2OH + 4O_2 \rightarrow CO_2 + H_2O + H_2 + O_3 + 2HO_2$ (5.55)

Methane oxidation consumes four O_2 and produces one O_3, CO_2, water vapor and hydrogen. Two OH radicals are converted into two HO_2 radicals, giving no net change in the oxidizing capacity of the troposphere. The ozone produced regenerates some OH *via* the reaction of electronically excited O (^1D) atoms with water vapor. Formaldehyde is photolyzed to H_2 and CO in this chain. The alternative photolysis pathway giving HCO and H is part of an analogous methane oxidation chain summarized by the net photochemical reaction:

$$CH_4 + 2OH + 6O_2 \rightarrow CO_2 + H_2O + O_3 + 4HO_2 \qquad (5.56)$$

This chain also produces one ozone per methane oxidized. Both chains show that the chemistry of CH_4, CO, NO_x and OH are linked together.

5.6 STRATOSPHERIC CHEMISTRY

5.6.1 Ozone

The formation and destruction of ozone O_3 is probably the most important aspect of stratospheric chemistry. Although ozone is a harmful pollutant in the troposphere, it is a beneficial constituent of the

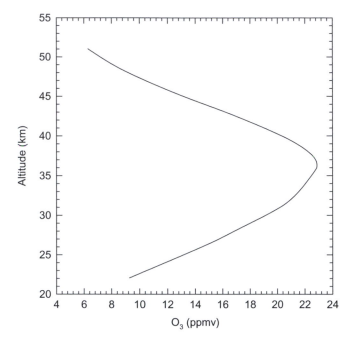

Figure 5.2   Ozone layer in Earth's stratosphere. The graph shows the annual average
ozone profile for 45° latitude from data tabulated by Klenk *et al.*,
J. Climate Appl. Meteorol. 1983, **22**, 2012.

stratosphere. The stratospheric ozone layer is at 15–50 km. The highest
O_3 partial pressure is at ~ 24 km and the highest concentrations near
37 km (Figure 5.2). The difference between the two maxima arises from
the exponential decrease of pressure with altitude. At the peak of the
ozone layer, which is about 37 km altitude, $T \approx 234$ K, $P \approx 5$ mbar and
the O_3 mixing ratio is ~ 23 ppmv. The O_3 partial pressure is $\sim 1.2 \times 10^{-4}$
mbar and the O_3 number density is $[O_3] \approx 3.6 \times 10^{12}$ cm^{-3}. However,
despite being a trace gas, O_3 shields the lower atmosphere and surface
from harmful UV sunlight ($\lambda < 290$ nm) and makes land-based life
possible on Earth. The story behind humanity's discovery of ozone,
the existence of the ozone layer and the reactions threatening to destroy
it is a fascinating account of the interplay between discoveries in
chemistry, meteorology and physics over 150 years.

5.6.2   History of the Discovery of the Ozone Layer

The English chemist Joseph Priestley (1733–1804) and the Swedish
chemist Carl Wilhelm Scheele (1742–1786) independently discovered

oxygen in the 1770s. Scheele prepared oxygen in 1771 and 1772, but did not publish his results until 1777. Priestley prepared oxygen in 1774 and published his work shortly thereafter. Prior to their work the element and gas oxygen were unknown. The Scottish physician and chemist Daniel Rutherford (1749–1819) discovered nitrogen in 1772. Thus, the chemical composition of air began to be unraveled.

Less than 70 years later, in 1840, the German chemist Christian Friedrich Schönbein (1799–1868) discovered ozone while working with electrical equipment in his poorly ventilated laboratory at the University of Basel, Switzerland. Several scientists dating back to 1785 had commented on the pungent odor produced by electrical discharges in air. You smell this same odor from a laser jet printer or photocopier in a poorly ventilated room. However, Schönbein proved that ozone was a distinct chemical compound, gave it the name ozone (from the Greek "ozo," smell), showed it was an allotrope of oxygen and detected it in rainwater after thunderstorms. Schönbein also used paper impregnated with starch and iodide to detect ozone in air – the ozone oxidizes the iodide to violet-colored iodine. Schönbein's interest in environmental chemistry was such that he coined the German word geochemie (geochemistry) in 1838.

In 1879, the French physicist Marie Alfred Cornu (1841–1902) proved that Earth's atmospheric absorption limited the solar spectrum, but he could not identify the cause of the absorption. Shortly thereafter in 1880–1881, the English chemist Sir Walter Hartley (1846–1913) discovered the strong UV absorption bands of O_3 at 200–300 nm, which are now named after him. These bands are so strong that they are as opaque to UV light as metals are to visible light (on an equal mass basis). It is worth emphasizing the strong UV absorption of ozone because this protects land-based life on Earth from harmful UV radiation and also makes possible sensitive measurements of the ozone column abundance from ground-based and/or satellite UV spectrometers.

The Beer–Lambert law gives the intensity of light (I) transmitted through a column of gas (or other material) with a column density (σ, in molecules cm^{-2}) and an absorption cross section (q, in cm^2). The light shining on the gas column initially has an intensity (I_0):

$$\frac{I}{I_0} = \exp(-q\sigma) \qquad (5.57)$$

The Hartley bands are strongest at ~ 255 nm where $q = 1.15 \times 10^{-17}$ cm^2. A typical O_3 column density in Earth's atmosphere is about

8.9×10^{18} molecules cm^{-2} (*i.e.*, this is the total number of O_3 molecules from ground level up to outer space). Substituting these numbers into the Beer–Lambert law, we find that:

$$\frac{I}{I_0} = \exp\left[-\left(8.9 \times 10^{18}\right)\left(1.15 \times 10^{-17}\right)\right] = \exp(-102.4) = 3.5 \times 10^{-45} \quad (5.58)$$

In other words, Earth's atmospheric O_3 column transmits virtually no UV light at this wavelength. This is a very good thing for us and all other land-based life!

Hartley also used experimental data on the UV absorption and abundance of O_3 and other atmospheric gases to explain Cornu's observations of the solar spectrum. Hartley proposed that the upper atmosphere contains more ozone than the lower atmosphere and that the amount of ozone in the upper atmosphere absorbs UV sunlight shortward of 293 nm. Over 30 years later, R. J. Strutt (1875–1947, the 4th Baron Rayleigh) verified Hartley's prescient ideas.

In 1917 the British astronomer Alfred Fowler (1868–1940) and Strutt showed that absorption bands near 330 nm in solar and stellar spectra are due to the Huggins bands of atmospheric ozone. Subsequently, in 1918 Strutt measured the UV transparency of the lower atmosphere over a 6.25 km path length across the Chelmer Valley in England. He showed that Rayleigh scattering, named for his father John William Strutt (1842–1919) the 3rd Baron Rayleigh, could explain the observed decrease in UV intensity at 254 nm. Alternatively, using absorption coefficients determined by Fabry and Buisson in 1913 (see below), and assuming that all of the UV absorption was due to O_3, Strutt calculated an upper limit of $\sim 4.2 \times 10^{-8}$ (42 ppbv) for O_3 in Earth's atmosphere. This is within the range of O_3 concentrations (10–100 ppbv) seen in the troposphere.

The French physicists Charles Fabry (1867–1945) and Henri Buisson (1873–1944) worked at the same time as Fowler and Strutt. Fabry was a co-inventor, with Alfred Pérot (1863–1925), of the Fabry–Pérot interferometer used throughout astronomy for high-resolution spectroscopic observations – some of which give us the photospheric elemental abundances discussed in Chapter 1. In 1913, Fabry and Buisson measured the wavelength dependent absorption coefficients of O_3 in the Hartley bands. The First World War interrupted their studies, and it took several years for them to apply their laboratory results to observations of Earth's atmosphere. In 1921, they made the first quantitative observations of O_3 absorption in the 290–315 nm region of the solar spectrum. It is worth quoting their results from their paper:

> "Ozone in the atmosphere is found to be equivalent to a layer about 3 mm thick at atmospheric pressure. The daily values for May and June, 1920, vary irregularly from 0.29 to 0.34 cm. It would be interesting to follow these variations through a solar period."

About 1920 the British physical chemist F. A. Lindemann (1886–1957), later known as Lord Cherwell, and the British physicist Gordon Dobson (1889–1976) used observations of meteors to deduce the structure of Earth's upper atmosphere. They found the upper atmosphere was much warmer than expected. Based on the work by Fabry and Perot and by Fowler and Strutt they suggested warming by ozone. Dobson then devised UV spectrometers to measure the ozone column abundance from ground-based measurements using the Sun as a light source and embarked on a campaign to do so.

Measurements by many scientists over the next 40 years provided fundamental information on the geographical distribution and temporal variations of O_3 column abundances and on the location of the stratospheric ozone layer. Starting in 1930, models were proposed to explain the production and loss of ozone and formation of the ozone layer. Development of the models and testing the models against observations required decades. The reasons for this are similar to the reasons that delayed understanding of the photochemistry of other planetary atmospheres, namely the lack of data for chemical reaction rates, O_3 abundance, solar UV flux and UV absorption coefficients. In addition, the ozone abundance varies temporally (days to months), geographically, spatially (altitude) and with weather so it was difficult to compare models with observations until the natural variations were understood. In some cases, the natural variations are large and rapid, *e.g.*, tens of percent changes in the O_3 column abundance over a few days time.

After World War II, scientists thought they understood the chemistry involved in ozone production, loss and the causes for the observed variations. However, as both measurements of ozone and reaction rates continued to improve it became clear that the accepted photochemical model did not match observations. The model predicted more ozone than actually observed, and the difference showed other chemical reactions were also destroying ozone. This led to the development of models for destruction of ozone by natural gases such as the OH radical.

The subsequent discovery of nitrogen oxides (collectively called NO_x) in the stratosphere (1968) and of chlorofluorocarbon (CFC) gases in the atmosphere (1970–1973) led to an explosion of activity in stratospheric chemistry models and observations and in laboratory measurement of reaction rates. Within a few years in the late 1960s and early 1970s scientists developed the foundation for modern understanding of ozone depletion and stratospheric chemistry. The work at this time by Paul Crutzen, Mario Molina and F. Sherwood Rowland on catalytic cycles for ozone loss eventually led to their sharing the 1995 Nobel Prize in Chemistry.

5.6.3   Ozone Abundance Measurements

The column abundance of ozone is measured in Dobson units (DU), which are named after Gordon Dobson. One Dobson unit is equivalent to a 0.01 mm thick layer of pure ozone at 0 °C and 1 atm pressure, *i.e.*, at STP. This is equivalent to an ozone column density of 2.687×10^{16} $O_3\,cm^{-2}$. For example, the results of Fabry and Buisson (0.29 to 0.34 cm) are equivalent to 290 to 340 Dobson units. The average O_3 column abundances in the stratosphere at low (15°), mid (45°) and high latitudes (75°) are about 260, 330 and 355 Dobson units, respectively. These are annual, zonal averages (around a latitude circle at all longitudes). For comparison, the global average ozone abundance (50 ppbv) in the troposphere corresponds to 34 Dobson units.

Today, ozone measurements are made from the ground with Dobson UV spectrometers and LIDAR instruments (light detection and ranging), *in situ* using aircraft, stratospheric balloons, rockets and from space on the Space Shuttle and on Earth-orbital satellites using ultraviolet, infrared and microwave spectroscopy. This listing is non-exhaustive but illustrates the variety of methods used to monitor ozone.

5.6.4   Chapman Cycle

A balance between production, destruction and transport between regions of net production and net destruction regulates the abundance of stratospheric ozone. The Chapman cycle describes the production and loss of O_3 in a pure oxygen atmosphere. The British geophysicist Sydney Chapman (1888–1970) proposed these reactions in 1930 shortly after Dobson's ozone measurements. The five reactions in the Chapman cycle are:

Photolysis of O_2 $O_2 + h\nu \rightarrow O + O$ $(\lambda = 180 - 240\,\text{nm})$ (5.59)

Ozone production $O + O_2 + M \rightarrow O_3 + M$ (5.37)

Ozone photolysis $O_3 + h\nu \rightarrow O(^1D) + O_2$ $(\lambda = 200 - 300\,\text{nm})$ (5.35)

Oxygen production $O + O + M \rightarrow O_2 + M$ (5.60)

Ozone destruction $O + O_3 \rightarrow O_2 + O_2$ (5.61)

Net reaction: $O + O_3 \rightarrow O_2 + O_2$ (5.61)

The rate of Reaction (5.60) is slow because the oxygen atom concentrations are small. Thus, Reaction (5.60) is often omitted from the Chapman cycle in many discussions.

Chapman's model also explains why the ozone layer forms. The top of the ozone layer is limited by the exponentially decreasing partial pressure of O_2 with higher altitude – solar UV photons are abundant but O_2 is not. Conversely, the bottom of the ozone layer is limited by the exponentially decreasing intensity of UV sunlight with decreasing altitude – O_2 is abundant but energetic UV photons to dissociate it are not. Convolution of the two exponential effects results in a layer with a limited altitude range.

5.6.5 Catalytic Cycles for Ozone Destruction

Natural and anthropogenic trace gases in the terrestrial stratosphere catalyze ozone destruction more rapidly than the Chapman cycle. The three important catalytic cycles involve oxides of hydrogen (HO_x), nitrogen (NO_x) and halogens (ClO, BrO, IO). The hydrogen and nitrogen oxides originate naturally and the halogen oxides are dominantly anthropogenic in origin. These oxides participate in different catalytic cycles, some of which are interconnected. The catalytic cycles are indirect routes for ozone loss that are faster than O_3 destruction by Reaction (5.61) in the Chapman cycle. The catalytic cycles are schematically written as:

$$X + O_3 \rightarrow XO + O_2 \quad (5.62)$$

$$XO + O \rightarrow X + O_2 \quad (5.63)$$

Net reaction: $\quad O + O_3 \rightarrow O_2 + O_2 \quad (5.61)$

The X and XO, which represent different hydrogen, nitrogen and halogen species such as H and HO, are interconverted but not irreversibly destroyed during the catalytic cycles and function as gas-phase catalysts. The catalytic cycles destroy ozone more efficiently than Reaction (5.61) because the number density of X is larger than that of monatomic oxygen, or the rate constant for Reaction (5.62) is larger than that for Reaction (5.61), or both. Furthermore, the X and XO species in the catalytic cycles go through the cycles many thousands of times before other reactions transform them into less reactive species.

5.6.6 Hydrogen Oxide (HO_x) Catalytic Cycles

In 1950, the Irish physicist Sir David Bates (1916–1994) and the Belgian physicist Baron Marcel Nicolet (1912–1996) proposed that H, OH and HO_2 destroyed O and O_3 in the mesosphere. About ten years later, Ronald Norrish (1897–1978), a co-recipient of the 1967 Nobel Prize in Chemistry, and his colleagues studied photochemistry of dry O_3 and $O_3 + H_2O$ mixtures ("wet" ozone). Norrish and colleagues confirmed prior reports that "wet" ozone photolyzed more rapidly than dry O_3 at UV wavelengths and showed that OH radicals formed during photolysis of "wet" ozone. They proposed that reaction of $O(^1D)$ plus water vapor generated OH and proposed a chain reaction mechanism, which is the third HO_x cycle below, to explain their observations. Their work and that of Bates and Nicolet led to models by Hunt of HO_x-catalyzed O_3 loss in 1966. His work used rate constants that were too large, but it led to other studies, including models by Crutzen that considered ozone loss *via* HO_x- and NO_x-catalyzed chemistry.

One cycle involves the H–HO pair and another cycle involves the HO–HO_2 pair. These cycles are:

$$HO_x \text{ Cycle I} \qquad H + O_3 \rightarrow HO + O_2 \qquad\qquad (5.64)$$

$$HO + O \rightarrow H + O_2 \qquad\qquad (5.65)$$

$$\textit{Net reaction:} \quad O + O_3 \rightarrow O_2 + O_2 \qquad\qquad (5.61)$$

$$HO_x \text{ Cycle II} \qquad HO + O_3 \rightarrow HO_2 + O_2 \qquad\qquad (5.66)$$

$$HO_2 + O \rightarrow HO + O_2 \qquad\qquad (5.67)$$

$$\textit{Net reaction:} \quad O + O_3 \rightarrow O_2 + O_2 \qquad\qquad (5.61)$$

These two cycles are important in the upper stratosphere. The third HO_x cycle also involves the hydroperoxyl HO_2 radical and is important in the lower stratosphere:

$$HO_x \text{ Cycle III} \quad HO + O_3 \rightarrow HO_2 + O_2 \quad (5.66)$$

$$HO_2 + O_3 \rightarrow HO + O_2 + O_2 \quad (5.68)$$

$$\textit{Net reaction}: \quad O_3 + O_3 \rightarrow 3O_2 \quad (5.69)$$

What is the source of the HO_x radicals in the stratosphere?

As stated earlier the stratosphere is extremely dry compared to the troposphere and contains about 5 ppmv water vapor. The two major sources of stratospheric water vapor are water transported upward through the tropical tropopause and methane oxidation. The former process accounts for about 60% (3 ppmv) of stratospheric water and the latter accounts for the rest. Methane oxidation occurs *via* the elementary reaction:

$$CH_4 + OH \rightarrow CH_3 + H_2O \quad (5.31)$$

The methane in Reaction (5.31) is transported upward from the troposphere and its concentration (~ 1.75 ppmv) is unaffected by condensation at the cold trap at the tropopause. The hydroxyl radicals are produced by reaction of electronically excited oxygen atoms $O(^1D)$ with water vapor:

$$H_2O + O(^1D) \rightarrow OH + OH \quad (5.28)$$

Photolysis of water vapor occurs in the mesosphere, but is unimportant in the stratosphere because O_2 absorbs the short wavelength UV sunlight ($\lambda < 190$ nm) that is absorbed by H_2O. Thus, photolysis of H_2O is not a source of OH radicals in the stratosphere.

What are the sinks for the HO_x radicals? In the upper stratosphere the OH and HO_2 radicals react with each other to reform O_2 and water vapor:

$$OH + HO_2 \rightarrow H_2O + O_2 \quad (5.70)$$

Reaction with nitric acid HNO_3 vapor removes OH from the lower stratosphere:

$$OH + HNO_3 \rightarrow H_2O + NO_3 \quad (5.71)$$

5.6.7 Nitrogen Oxide (NO_x) Catalytic Cycles

In 1968, nitric acid HNO_3 and nitrogen oxides NO_x were discovered in the stratosphere. Shortly thereafter, in 1970, the Dutch scientist Paul Crutzen proposed the NO_x cycle for ozone destruction. One year later, in 1971, the American physical chemist Harold Johnston independently proposed ozone destruction by the NO_x cycle. At this time, a fleet of several hundred supersonic transport (SST) aircraft was under consideration. The joint British–French SST, the Concorde, was built, but the hundreds of other proposed SST aircraft were not built. Initially scientists thought that the water vapor in the SST exhaust would destroy ozone *via* HO_x chemistry. However, the pioneering work by Crutzen and Johnston showed that the NO_x in SST exhaust posed a far greater problem.

The NO_x catalytic cycle involves the $NO–NO_2$ pair:

$$NO_x \text{ Cycle} \quad NO + O_3 \rightarrow NO_2 + O_2 \tag{5.72}$$

$$NO_2 + O \rightarrow NO + O_2 \tag{5.73}$$

$$Net\,reaction:\quad O + O_3 \rightarrow O_2 + O_2 \tag{5.61}$$

The major source of stratospheric NO_x is photolysis of nitrous oxide (N_2O, laughing gas) transported upward from the troposphere:

$$N_2O + h\nu \rightarrow NO + O(^1D) \ (\lambda < 240\,nm) \tag{5.74}$$

Some N_2O also decomposes to NO or to $N_2 + O_2$ *via* the reactions:

$$N_2O + O(^1D) \rightarrow NO + NO \tag{5.75a}$$

$$N_2O + O(^1D) \rightarrow N_2 + O_2 \tag{5.75b}$$

The branching ratio, *i.e.*, the ratio of the two sets of products, is about unity.

The sources for the upwardly mixed N_2O are soil bacterial denitrification of nitrate (NO_3^-) compounds and bacterial nitrification of NH_4^+ compounds. Consequently, artificial fertilization with NH_4NO_3 may increase N_2O production. Several reactions temporarily convert stratospheric NO_x into chemically less reactive reservoir species:

$$NO_2 + O_3 \rightarrow NO_3 + O_2 \tag{5.76}$$

$$NO_2 + NO_3 + M \rightarrow N_2O_5 + M \tag{5.77}$$

$$NO_2 + OH + M \rightarrow HNO_3 + M \tag{5.78}$$

However, solar radiation photolyzes NO_3 and nitric acid (HNO_3) and N_2O_5 is destroyed by collisions with a third body. All of these reactions regenerate NO_2:

$$NO_3 + h\nu \rightarrow NO_2 + O \quad (\lambda \sim 450 - 700\,nm) \tag{5.79}$$

$$HNO_3 + h\nu \rightarrow NO_2 + OH \quad (\lambda \leq 320\,nm) \tag{5.80}$$

$$N_2O_5 + M \rightarrow NO_2 + NO_3 + M \tag{5.81}$$

The only sinks for stratospheric NO_x are mixing NO_x and/or the reservoir species downward into the troposphere and destruction of NO by reaction with N atoms in the upper stratosphere where short wavelength UV light photolyzes NO:

$$NO + h\nu \rightarrow N + O \quad (\lambda < 190\,nm) \tag{5.82}$$

$$NO + N \rightarrow N_2 + O \tag{5.83}$$

5.6.8 Halogen Oxide (ClO_x, BrO_x, IO_x) Catalytic Cycles

The halogen oxide cycles involve oxides of Cl, Br and I, but not of fluorine for reasons discussed later. In 1973, the atmospheric chemists Richard Stolarski and Ralph Cicerone proposed chlorine oxide (ClO_x) catalyzed destruction of ozone. They discussed possible Cl sources, including industrial sources, volcanic eruptions and solid fuel rocket exhaust. At about the same time (1970–1973) the British scientist James Lovelock discovered $CFCl_3$, CCl_4 and SF_6 in Earth's atmosphere. His discovery immediately stimulated further measurements, *e.g.*, the American oceanographer Edward Goldberg discovered CF_2Cl_2, the most abundant CFC gas in Earth's atmosphere, shortly thereafter. We now know that a large number of halogen-bearing gases are present in Earth's atmosphere; Table 5.2 lists some of the more abundant ones that we discuss in this chapter and in Chapter 6.

Lovelock's discovery also led Mario Molina and F. Sherwood Rowland to propose that photolysis of CFC gases in the stratosphere initiated ClO_x-catalyzed ozone loss. Molina and Rowland recognized that the CFC gases are chemically inert in the troposphere but are destroyed by UV sunlight in the stratosphere. In turn, their work led to long-term monitoring of CFC gas abundances in the atmosphere, development of the BrO_x and IO_x catalytic cycles and, eventually, to the 1987 Montreal Protocol phasing out industrial production and use of CFC gases.

Anthropogenic emissions of chlorofluorocarbon (CFC) and halocarbon gases are predominantly responsible for the halogen oxide cycles, while natural emissions are predominantly responsible for the hydrogen and nitrogen oxide cycles. The major halogen oxide cycles for O_3 loss are:

$$\text{Chlorine cycle:} \quad Cl + O_3 \rightarrow ClO + O_2 \tag{5.84}$$

$$ClO + O \rightarrow Cl + O_2 \tag{5.85}$$

$$\textit{Net reaction:} \quad O + O_3 \rightarrow O_2 + O_2 \tag{5.61}$$

$$\text{Bromine cycle:} \quad Br + O_3 \rightarrow BrO + O_2 \tag{5.86}$$

$$Cl + O_3 \rightarrow ClO + O_2 \tag{5.84}$$

$$BrO + ClO \rightarrow Br + Cl + O_2 \tag{5.87}$$

$$\textit{Net reaction:} \quad O_3 + O_3 \rightarrow 3O_2 \tag{5.69}$$

$$\text{Iodine cycle} \quad I + O_3 \rightarrow IO + O_2 \tag{5.88}$$

$$Cl + O_3 \rightarrow ClO + O_2 \tag{5.84}$$

$$IO + ClO \rightarrow I + Cl + O_2 \tag{5.89}$$

$$\textit{Net reaction:} \quad O_3 + O_3 \rightarrow 3O_2 \tag{5.69}$$

Another iodine cycle with Br and BrO instead of Cl and ClO also occurs. Bromine also takes part in a catalytic cycle involving NO_x:

$$Br + O_3 \rightarrow BrO + O_2 \tag{5.86}$$

$$NO + O_3 \rightarrow NO_2 + O_2 \tag{5.72}$$

$$BrO + NO_2 + M \rightarrow BrNO_3 + M \tag{5.90}$$

$$BrNO_3 + h\nu \rightarrow Br + NO_3 \quad (\lambda \sim 200 - 500 \, nm) \tag{5.91}$$

$$NO_3 + h\nu \rightarrow NO + O_2 \tag{5.92}$$

$$\textit{Net reaction}: \quad O_3 + O_3 \rightarrow 3O_2 \tag{5.69}$$

We mentioned earlier that fluorine oxide catalytic cycles are unimportant for ozone destruction. There are two reasons for this. First, reactions with CH_4 and H_2O efficiently convert F atoms into HF, for example:

$$F + CH_4 \rightarrow CH_3 + HF \tag{5.93}$$

$$F + H_2O \rightarrow OH + HF \tag{5.94}$$

Photolysis of HF requires short wavelength UV light and only becomes significant at 80 km altitude in the mesosphere. Thus, no F atoms are available to react with O_3 in the same manner as Cl, Br and I atoms. Second, the reaction of FO with O_3 proceeds slower than the reaction of FO with NO, which removes FO from the catalytic cycle. Thus, fluorine is unimportant for O_3 loss. The HF produced by reactions such as (5.93) and (5.94) is effectively inert and is removed from the stratosphere by downward transport.

Table 5.2 lists the major chlorine gases in Earth's troposphere – these are the sources for stratospheric chlorine. The five most abundant Cl-bearing gases are methyl chloride (CH_3Cl), CF_2Cl_2 (F12), $CFCl_3$ (F11), $CHClF_2$ (HCFC-22) and CCl_4 (carbon tetrachloride). These account for about 83% of all Cl atoms entering the stratosphere.

Volcanic gases contain HCl, but with a few exceptions they are not an important source of stratospheric chlorine. The reason for this is simple. Most volcanoes erupt into the troposphere, which contains 1–4% water vapor, which is rained out within ten days. Hydrogen chloride is very soluble in water and rains out of the troposphere within the ten-day lifetime of atmospheric water vapor. The exceptions are volcanic gases erupted directly into the stratosphere by powerful volcanoes such as Mount Pinatubo in the Philippines. The stratosphere is dry and water-soluble gases are not rained out of it.

As Table 5.2 shows, methyl chloride is the most abundant halocarbon in Earth's troposphere. It has anthropogenic (biomass burning, incineration) and biogenic sources (biomass burning, marshes, wetlands, oceans). It has a lifetime of about 1.5 years in the troposphere because it is attacked by the hydroxyl OH radical:

$$CH_3Cl + OH \rightarrow HCl + CH_3 \tag{5.95}$$

In contrast, the anthropogenic chlorofluorocarbon (CFC) gases are chemically inert in Earth's troposphere. Hydroxyl radicals or other gases do not destroy them. They have small solubilities in water and are not lost by dissolution in the oceans. They do not react with dirt, rocks or vegetation on the surface of the Earth. The CFC gases have long atmospheric lifetimes (*e.g.*, CF_2Cl_2 – 100 years, $CFCl_3$ – 45 years, $C_2Cl_3F_3$ – 85 years) and survive unscathed into the stratosphere. The hydrochlorofluorocarbon (HCFC) gases, which are replacements for the CFC gases, are attacked by OH radicals and have shorter lifetimes, *e.g.*, 11.9 years for $CHClF_2$ (HCFC-22). However, HCFC gases with lifetimes comparable to 50 years are also transported from the troposphere into the stratosphere.

The CFC and HCFC gases are destroyed by UV sunlight in the stratosphere. For example, $CFCl_3$, the second most abundant CFC in the atmosphere, is destroyed *via*:

$$CFCl_3 + h\nu \rightarrow CFCl_2 + Cl \quad (\lambda < 265\,nm) \tag{5.96}$$

Photolysis occurs within the ozone layer at 25 km, but is much faster at 80 km in the mesosphere, as indicated by the photodissociation rate coefficients for Reaction (5.96):

$$J = 3.8 \times 10^{-8}\,s^{-1} \text{ at } 25\,km \tag{5.97}$$

$$J = 8.0 \times 10^{-6}\,s^{-1} \text{ at } 80\,km \tag{5.98}$$

The Cl atom produced by Reaction (5.96) enters into the chlorine catalytic cycle for O_3 loss. The $CFCl_2$ radical reacts with O_2 to form ClO, which also enters the chlorine catalytic cycle. Monatomic Cl is removed by reactions that convert it into HCl, which is a less reactive reservoir species:

$$Cl + CH_4 \rightarrow CH_3 + HCl \tag{5.99}$$

$$Cl + HO_2 \rightarrow O_2 + HCl \tag{5.100}$$

The H–Cl bond in HCl is weaker than the H–F bond in HF so HCl is only a temporary reservoir for chlorine with a lifetime of about one month in the lower stratosphere. It is transported downward into the troposphere and is destroyed by reaction with OH radicals:

$$HCl + OH \rightarrow H_2O + Cl \tag{5.101}$$

Chlorine monoxide is also removed by conversion into less reactive chlorine nitrate:

$$ClO + NO_2 + M \rightarrow ClNO_3 + M \qquad (5.102)$$

However, chlorine nitrate has a short lifetime of about 1 h in the lower stratosphere and is destroyed by UV sunlight, releasing Cl and ClO, which reenter the catalytic cycle:

$$ClNO_3 + h\nu \rightarrow NO_3 + Cl \quad (\lambda < 430\,\text{nm}) \qquad (5.103)$$

$$ClNO_3 + h\nu \rightarrow NO_2 + ClO \quad (\lambda < 430\,\text{nm}) \qquad (5.104)$$

Chlorine monoxide is also converted into Cl *via* hypochlorous acid (HOCl):

$$ClO + HO_2 \rightarrow HOCl + O_2 \qquad (5.105)$$

$$HOCl + h\nu \rightarrow OH + Cl \quad (\lambda < 420\,\text{nm}) \qquad (5.106)$$

However, this just exchanges one form of active chlorine for another and is not a sink.

The sources of stratospheric bromine and iodine are methyl bromide (CH_3Br), CH_3I and Br-bearing halon gases such as CF_3Br (halon 1301) and CF_2ClBr (halon 1211). Methyl bromide is the most abundant Br-bearing gas in the troposphere (22 pptv) and has anthropogenic and biogenic sources. It is a fumigant for grain and soil and is an herbicide. Marshes, oceanic biota and cruciferous plants emit CH_3Br. Anthropogenic and natural biomass burning is a source. Reaction with tropospheric OH destroys CH_3Br and it has a lifetime of 1.5 years, about the same as the transport time into the stratosphere.

Several inorganic bromine-bearing gases, including HBr, BrO, HOBr, Br and $BrNO_3$, occur in the troposphere. Their total abundance is 2–4 pptv, with HBr being dominant. The inorganic Br-bearing gases are produced from sea salt and contribute to the destruction of tropospheric ozone.

Methyl iodide (CH_3I) is the most abundant iodine-bearing gas in the atmosphere with an abundance of about 0.8 pptv. It has natural (oceanic biota, rice paddies) and anthropogenic (fumigant and herbicide use) sources. Photolysis rapidly destroys CH_3I, which has a lifetime of about five days. Oceanic biotas produce many other iodine-bearing organic gases such as CH_2I_2, ethyl iodide (C_2H_5I), CH_2ICl and CH_2IBr. Photolysis destroys these gases (within minutes to days) and produces iodine atoms, which react with tropospheric O_3 to form IO. The photolysis of IO regenerates I atoms and the O atoms react with O_2, reforming ozone.

Halons are used in gaseous fire suppression systems and are anthropogenic in origin. Ultraviolet sunlight photolyzes CH_3Br and halons in the stratosphere *via* reactions such as:

$$CH_3Br + h\nu \rightarrow CH_3 + Br \quad (\lambda < 290\,nm) \tag{5.107}$$

$$CF_3Br + h\nu \rightarrow CF_3 + Br \quad (\lambda < 300\,nm) \tag{5.108}$$

The Br atoms then enter into catalytic cycles for O_3 destruction.

The bromine catalytic cycles are terminated by conversion of active Br gases into less reactive reservoir species. The important reactions are:

$$BrNO_3 + H_2O \rightarrow HOBr + HNO_3 \tag{5.109}$$

$$HOBr + HCl \rightarrow BrCl + H_2O \tag{5.110}$$

The Br reservoir species are less stable than their Cl counterparts and are converted back into active Br gases on short timescales. The Br catalytic cycles are only effectively terminated when Br gases are mixed downward into the troposphere. Little is known about stratospheric iodine chemistry but its efficiency for O_3 destruction is 150–300 times greater than that of chlorine.

How important are the Chapman cycle and the HO_x, NO_x and halogen oxide catalytic cycles for O_3 loss? Models of O_3 chemistry show that the NO_x, HO_x, ClO_x and Chapman cycles account for 31–34%, 16–29%, 19–20% and 20–25%, respectively, of total O_3 destruction in Earth's stratosphere. The relative importance of each process varies with altitude and with concentrations of HO_x, NO_x and ClO_x gases. The NO_x cycle is most important at about 22–42 km in the middle of the O_3 layer and the HO_x cycle accounts for most ozone loss above ~ 42 km, and below ~ 22 km. The halogen oxide cycles do not dominate ozone loss at any altitude, but are the second most important loss process (after NO_x chemistry) over about 30–42 km. Chapman cycle chemistry is less important than other cycles throughout the ozone layer, but is the second most important loss process (after HO_x chemistry) for ozone in the upper stratosphere and mesosphere.

The importance of the catalytic cycles for ozone loss was recognized in 1995 when the Nobel Prize in Chemistry was awarded jointly to the Mexican chemist Mario Molina, the American chemist Sherwood Rowland and the Dutch chemist Paul Crutzen for their work on catalytic cycles for ozone depletion in Earth's atmosphere.

5.6.9 Antarctic Ozone Hole

Long-term observations by the British Antarctic Survey (starting in 1956) show a continued decrease in O_3 levels during springtime over Antarctica. Subsequently, satellite-based observations confirmed the ground-based observations. The data show ozone depletions in an area known as the Antarctic ozone hole. Ironically, the Antarctic ozone hole occurs at the end of the Earth, furthest away from the major production and use of chlorofluorocarbons in the northern hemisphere, yet its discovery finally validated the chlorine oxide catalytic cycles proposed in the early 1970s.

During the dark polar winter, the polar vortex isolates air over the Antarctic from the rest of the atmosphere. Consequently stratospheric temperatures drop to $\sim 190\,K$. Although the stratosphere is extremely dry, $190\,K$ is cold enough for polar stratospheric clouds composed of solid nitric acid trihydrate $HNO_3 \cdot 3H_2O$ to form. The surfaces of the nitric acid cloud particles are sites for heterogeneous reactions such as:

$$HCl + ClNO_3 \rightarrow HNO_3 + Cl_2 \tag{5.111}$$

$$ClNO_3 + H_2O \rightarrow HNO_3 + HOCl \tag{5.112}$$

$$HCl + HOCl \rightarrow H_2O + Cl_2 \tag{5.113}$$

These reactions convert reservoir species such as hydrochloric acid (HCl) and chlorine nitrate ($ClNO_3$) into more reactive compounds such as elemental chlorine (Cl_2) and hypochlorous acid (HOCl). Once polar winter is over and the Sun rises again, photolysis of $ClNO_3$, Cl_2 and HOCl releases chlorine atoms and chlorine monoxide ClO:

$$ClNO_3 + h\nu \rightarrow NO_3 + Cl \tag{5.103}$$

$$ClNO_3 + h\nu \rightarrow NO_2 + ClO \tag{5.104}$$

$$HOCl + h\nu \rightarrow OH + Cl \tag{5.106}$$

The Cl and ClO catalytically destroy O_3, leading to the observed O_3 depletion in Antarctic spring. In this case the catalytic cycle is slightly different than that discussed earlier and involves the following reactions:

$$Cl + O_3 \rightarrow ClO + O_2 \tag{5.84}$$

$$ClO + ClO + M \rightarrow Cl_2O_2 + M \tag{5.114}$$

$$Cl_2O_2 + h\nu \rightarrow Cl + ClOO \quad (\lambda \sim 190 - 450\,nm) \quad (5.115)$$

$$ClOO + M \rightarrow Cl + O_2 \quad (5.116)$$

$$\textit{Net reaction}: O_3 + O_3 \rightarrow 3O_2 \quad (5.69)$$

No oxygen atoms are involved in this cycle, instead chlorine peroxide (ClOOCl) and chlorine dioxide (ClOO) are involved. Chlorine dioxide is weakly bound and decomposes rapidly *via* Reaction (5.116). Thus, photolysis of chlorine peroxide effectively yields two Cl atoms and O_2 gas. In this cycle, ozone loss is proportional to the ClO concentration squared, thus giving a powerful nonlinear effect. Mixing of air over Antarctica with the rest of the atmosphere increases O_3 levels somewhat, but not enough to restore the annual loss and the O_3 depletions grow with time.

5.6.10   Oxygen Isotopic Fractionations in Ozone

Oxygen has three stable isotopes 16, 17 and 18, with abundances of about 99.762% (16), 0.038% (17) and 0.200% (18). Thus, most ozone is $^{48}O_3$ and other isotopomers, such as $^{49}O_3$ and $^{50}O_3$, are less abundant. In general, the relative abundances of the different isotopomers are calculable using statistics and can be expressed using the delta notation that was defined in Chapter 2. Starting in the early 1980s observations of stratospheric and tropospheric ozone showed that $^{49}O_3$ and $^{50}O_3$ are more abundant than expected by about 10%. The enrichments arise from isotope specific reaction rates for O_3 formation and for isotope exchange with monatomic oxygen. Ozone also transmits the isotope effects to other gases such as CO_2 and nitrogen oxides.

5.7   MESOSPHERIC AND THERMOSPHERIC CHEMISTRY

Important aspects of mesospheric and thermospheric chemistry include the production and loss of ozone, photolysis of water vapor and subsequent reactions of OH radicals, and the chemistry of metals deposited by infalling meteoroids. The first two topics involve reactions discussed in the last section and here we focus on the chemistry of meteoritic metals.

In 1929, the American astronomer V. M. Slipher (1875–1969) of Lowell Observatory discovered lines of monatomic sodium vapor at 589 nm in the airglow spectrum of the Earth. These lines arise from chemiluminescent reactions known as the Chapman airglow reactions:

$$Na + O_3 \rightarrow NaO + O_2 \qquad (5.117)$$

$$NaO + O \rightarrow Na + O_2 \qquad (5.118)$$

$$Net\ reaction: O + O_3 \rightarrow O_2 + O_2 \qquad (5.61)$$

Sodium dioxide forms by reaction of NaO with ozone:

$$NaO + O_3 \rightarrow NaO_2 + O_2 \qquad (5.119)$$

UV sunlight photolyzes NaO_2 during the daytime and regenerates Na and O_2:

$$NaO_2 + h\nu \rightarrow Na + O_2 \qquad (5.120)$$

Sodium hydroxide gas and sodium bicarbonate gas form by reactions of Na gases with mesospheric water vapor and CO_2, respectively:

$$NaO + H_2O \rightarrow NaOH + OH \qquad (5.121)$$

$$NaOH + CO_2 + M \rightarrow NaHCO_3 + M \qquad (5.122)$$

In 1958, rocket-borne mass spectrometry measurements by the Russian space scientist V. G. Istomin (1929–2000) discovered Ca, Fe, Mg and Si metal ions at about 95 km in the mesosphere. Subsequent measurements by mass spectrometry and other methods revealed that thin layers of Na and other metals, including Ca, Fe, K and Li, are present at 85–95 km in Earth's mesosphere. Infalling meteoroids deposit these and other metals in Earth's upper atmosphere at the rate of 130–240 tons per day (47–88 million kilograms per year). Stony meteoroids (mainly microscopic, a few larger) make up most of the infalling material and based on their average composition the ten most abundant metallic and rocky elements deposited into Earth's atmosphere should be Mg, Si, Fe, S, Al, Ca, Na, Ni, Cr and Mn. The metal ion abundances measured by *in situ* mass spectrometry do not agree with this sequence. The different measurements vary spatially and temporally, but typically show more Fe, Na and K, about the same amount of Ni and less Al (normalized to Mg). The different volatilities of rocky compounds of these elements is a possible explanation for the difference between the observed and expected elemental abundances.

5.8 IONOSPHERIC CHEMISTRY

As mentioned earlier, the ionosphere overlaps the mesosphere and thermosphere and extends from about 60 to 400 km. The ionosphere was discovered in 1901 when the Italian physicist Guglielmo Marconi (1874–1937) transmitted radio waves across the Atlantic Ocean from Poldhu, Cornwall to St. John's, Newfoundland. Radio waves, like light, travel along the line of sight and transatlantic transmission required a high altitude-reflecting layer. This is the E layer, originally known as the Kennelly–Heaviside layer.

The different conductive layers, in order of increasing altitude, are D, E, F1 and F2. Photo-ionization by solar radiation with $\lambda < 100$ nm is a source of ions and electrons in the E and F regions. Important reactions include:

$$N_2 + h\nu \rightarrow N_2^+ + e^- \tag{5.123}$$

$$N_2 + h\nu \rightarrow N + N^+ + e^- \tag{5.124}$$

$$O + h\nu \rightarrow O^+ + e^- \tag{5.125}$$

$$O_2 + h\nu \rightarrow O_2^+ + e^- \tag{5.126}$$

$$O_2 + h\nu \rightarrow O + O^+ + e^- \tag{5.127}$$

Ionized nitric oxide NO^+ is a major ion in the E layer. It forms *via* reactions such as:

$$O^+ + N_2 \rightarrow NO^+ + N \tag{5.128}$$

$$O + N_2^+ \rightarrow NO^+ + N \tag{5.129}$$

The dissociative recombination of O_2^+ and NO^+ are important loss processes for electrons:

$$O_2^+ + e^- \rightarrow O + O \tag{5.130}$$

$$NO^+ + e^- \rightarrow N + O \tag{5.131}$$

The D layer has positive cluster ions formed by water and negative ions in addition to electrons. Photo-ionization of NO by Lyman alpha

UV light at 121.6 nm gives NO^+:

$$NO + h\nu \rightarrow NO^+ + e^- \tag{5.132}$$

Negative oxygen ions form *via* the reaction:

$$O_2 + e^- + M \rightarrow O_2^- + M \tag{5.133}$$

The water cluster ions form *via* reactions initiated by oxygen ions:

$$O_2^+ + O_2 + M \rightarrow O_4^+ + M \tag{5.134}$$

$$O_4^+ + H_2O \rightarrow O_2 + O_2^+ \cdot H_2O \tag{5.135}$$

$$O_2^+ \cdot H_2O + H_2O \rightarrow H_3O^+ \cdot OH + O_2 \tag{5.136}$$

$$H_3O^+ \cdot OH + H_2O \rightarrow H^+ \cdot (H_2O)_2 + OH \tag{5.137}$$

The H_3O^+ ion is the hydronium ion and is the simplest water cluster ion.

FURTHER READING

S. S. Butcher, R. J. Charlson, G. H. Orians and G. V. Wolfe (eds), *Global Biogeochemical Cycles*, Academic Press, London, 1992.

J. W. Chamberlain and D. M. Hunten, *Theory of Planetary Atmospheres*, Academic Press, Orlando, FL, 1987.

Chemical Reviews, 2003, **103** (number 12) – atmospheric chemistry topical issue.

C. E. Junge, *Air Chemistry and Radioactivity*, Academic Press, New York, 1963.

A. Jursa (ed.), *Handbook of Geophysics and the Space Environment*, United States Air Force, 1985.

K. F. Klenk, P. K. Bhartia, E. Hilsenrath and A. J. Fleig, *J. Climate Appl. Meteorol.*, 1983, **22**, 2012.

J. S. Levine (ed.), *The Photochemistry of Atmospheres*, Academic Press, Orlando, FL, 1985.

K. Lodders and B. Fegley, Jr., *The Planetary Scientist's Companion*, Oxford University Press, New York, 1998.

P. Warneck, *Chemistry of the Natural Atmosphere*, Academic Press, San Diego, 1989.

R. P. Wayne, *Chemistry of Atmospheres*, 3rd edn, Oxford University Press, Oxford, UK, 2000.

Y. L. Yung and W. B. DeMore, *Photochemistry of Planetary Atmospheres*, Oxford University Press, New York, 1999.

CHAPTER 6

The Greenhouse Effect and Biogeochemical Cycles on Earth

6.1 INTRODUCTION

In this chapter we discuss the greenhouse effect and the major biogeo-chemical cycles. We start with radiative equilibrium, the greenhouse effect and greenhouse gases. This discussion complements material in Chapter 5 about tropospheric atmospheric chemistry of ozone, methane, CO and catalytic cycles for ozone loss in the stratosphere. Then we give an overview of the major biogeochemical cycles for water, carbon, nitrogen, sulfur and phosphorus.

6.2 GREENHOUSE EFFECT AND GREENHOUSE GASES

6.2.1 Solar Energy Received by the Earth

The Earth is in radiative equilibrium with the Sun, *i.e.*, the Earth re-emits to space all of the solar energy that it absorbs and no gain or loss of energy occurs. The annual average solar flux at Earth's mean distance from the Sun is the solar constant S_o, which is 1367 W per m^2 of Earth's total surface area. Earth receives this energy over its Sun-facing side, which has a circular area (πR^2), where R (6371.0 km) is Earth's average radius. The area of Earth's Sun-facing side is one quarter of its total surface area. Thus, the solar flux received by the Earth (F_{in}) is:

$$F_{in} = 0.25 \cdot S_o = 0.25 \cdot 1367 \, \text{W m}^{-2} = 342 \, \text{W m}^{-2} \quad (6.1)$$

Chemistry of the Solar System
By Katharina Lodders and Bruce Fegley, Jr.
© K. Lodders and B. Fegley, Jr. 2011
Published by the Royal Society of Chemistry, www.rsc.org

Most of the solar energy received by the Earth is visible light in the 400–700 nm range, peaking at about 450 nm.

6.2.2 Budget for the Incoming Solar Flux

Satellite observations show that the Earth reflects $107\,W\,m^{-2}$ ($\sim 31\%$) of the incoming solar energy back to space. Of this, $30\,W\,m^{-2}$ ($\sim 9\%$) comes from Earth's surface. The average reflectivity or albedo (A) of the Earth (surface, atmosphere and clouds) is 31%. The albedo of Earth's surface ranges from 4% for heavily forested areas, 5–6% for the oceans when the Sun is overhead, 28% for sand dunes, to 80–90% for the polar ice caps and fresh snow-covered regions.

The atmosphere (air molecules, suspended aerosols and dust) and clouds reflect $77\,W\,m^{-1}$ (22%) of the incoming solar radiation back to space. The annual average cloudiness, or cloud coverage, of Earth's atmosphere is about 52%. Cloud albedos range from 15% for thin cirrus clouds to 80% for thick cumulonimbus clouds. The remaining $235\,W\,m^{-2}$ of the incoming $342\,W\,m^{-2}$ of solar energy is absorbed.

The Earth's surface absorbs $168\,W\,m^{-2}$ (49%) of the incoming solar energy. The remaining $67\,W\,m^{-2}$ (20%) is absorbed by the atmosphere. Water vapor, O_3, O_2 and CO_2 are the major absorbers in the atmosphere. Ozone absorbs the incoming solar energy in the UV, visible and infrared in the Hartley (200–300 nm), Huggins (300–360 nm) and Chappius (450–900 nm) bands. Absorption of UV sunlight by ozone causes the temperature inversion at the tropopause and warms the stratosphere. Water vapor and H_2O dimers (>900 nm) and to a lesser extent CO_2 (>1 µm) are important absorbers in the infrared region. Oxygen weakly absorbs in the red (580, 630, 690, 760 nm) and infrared (1.06 µm, 1.27 µm) regions. In fact, a detailed analysis of the red bands of O_2 at 760 nm led to the discovery of ^{17}O and ^{18}O by Giauque and Johnston in 1929.

6.2.3 Budget for Earth's Outgoing Flux

The Earth re-emits the same amount of energy back to space ($F_{out} = 235\,W\,m^{-2}$) as it absorbs from the Sun. The emission occurs over its entire surface ($4\pi R^2$) at infrared wavelengths, with a broad peak around 12 µm. The energy re-radiated back to space by the Earth comes from the atmosphere ($165\,W\,m^{-2} = 70\%$), clouds ($30\,W\,m^{-2} = 13\%$) and the surface ($40\,W\,m^{-2} = 17\%$). The outgoing infrared radiation emitted by Earth's surface escapes at wavelengths known as atmospheric windows, where atmospheric IR absorption is low. The simplest approximation to the global energy balance equation is:

$$F_{in} = F_{out} \tag{6.2}$$

$$(1 - A)\pi R^2 S_o = (4\pi R^2)\varepsilon\sigma T_e^4 \tag{6.3}$$

where σ is the Stefan–Boltzmann constant $(5.67051\times10^{-8}\,W\,m^{-2}\,K^{-4})$ and ε is the emissivity of the Earth (ε = unity in this simplification). The emissivity corrects for the deviation from a perfect emitter (ε = 1). However, the emissivity is close to one for desert soil (0.96), liquid water (0.98) and ice (0.96), and is taken as unity in this simple calculation. The T_e is the equilibrium temperature of the Earth. This is the temperature in the absence of any greenhouse warming and is equal to:

$$T_e = \left[\frac{(1 - A)S_o}{4\sigma}\right]^{1/4} = 254\,K \tag{6.4}$$

This is –19 °C and is below the freezing point of water. The 34-degree increase of the Earth's average surface temperature (288 K) over its calculated equilibrium temperature (254 K) is due to warming by heat-absorbing gases in Earth's atmosphere. A heat-absorbing (or greenhouse) gas is a molecule that absorbs infrared radiation. Table 6.1 lists some of the important greenhouse gases in Earth's atmosphere. The natural greenhouse effect is primarily due to H_2O, O_3, CO_2, CH_4 and N_2O, all of which occur naturally in Earth's atmosphere.

The 254 K equilibrium temperature corresponds to an altitude of about 5 km in the troposphere. The average emission temperature T_o of the Earth is lower than the equilibrium temperature and is:

$$T_o = \frac{T_e}{\sqrt[4]{2}} \sim \frac{254}{1.2} = 212\,K \tag{6.5}$$

Table 6.1 Some important greenhouse gases in Earth's troposphere.

Gas	Abundance	Source(s)[a]	Lifetime	IR bands (μm)
H_2O	1–4%	N	10 days	2.66, 2.73, 6.27, ~80
CO_2	387 ppmv	N, A	14 years	1.2–2.5, 4.3, 15
CH_4	1.75 ppmv	N, A	8.4 years	3.31, 3.43, 6.52, 7.66
N_2O	320 ppbv	N, A	120 years	4.5, 7.8, 17.0
O_3	34 ppbv	N, A	4–18 days	4.7, 9.1, 9.6, 14.3
CF_2Cl_2	533 pptv	A	100 years	8.61, 9.1, 10.8, 15.0
$CFCl_3$	268 pptv	A	45 years	9.22, 11.8, 18.7

[a]N – natural, A – anthropogenic.

The average emission temperature corresponds approximately to the temperature of the lower stratosphere (217 K) in the 12–20 km region. However, emission temperatures vary with wavelength due to absorption of outgoing heat radiation by different greenhouse gases in Earth's atmosphere.

Nimbus weather satellite observations of infrared radiation emitted by the Earth to space show a complicated spectrum with prominent drops in emitted flux in three wavelength regions – (i) 12.5–16.7 μm due to the 15-μm CO_2 band, (ii) 9.6 μm due to the O_3 band at this wavelength and (iii) <8 μm due to water vapor bands. The lower flux regions correspond to emission from lower temperatures at higher altitudes in Earth's atmosphere. Greenhouse gas absorption is strong in these wavelength regions. Conversely, the higher flux regions correspond to emission from higher temperatures at lower altitudes in the atmosphere. Greenhouse gas absorption is weak in these wavelength regions, *e.g.*, the 10–12.5 μm region, which are the atmospheric windows mentioned earlier. For example, CO_2 absorbs strongly in the 12.5–16.7 μm range and (depending on humidity) is the fourth or fifth most abundant gas in the atmosphere and the first or second most abundant greenhouse gas. Thus, CO_2 absorbs and re-emits IR radiation until higher up in the atmosphere where the amount of CO_2 overhead is insufficient to keep reabsorbing the heat radiation, which escapes to space. For example, the 15-μm CO_2 band radiates IR energy to space and cools the upper stratosphere and lower mesosphere.

Finally, we consider the effect of radioactive heating on Earth's heat balance. Radioactive decay of ^{40}K, uranium, thorium and other radioactive elements in rocks generates 0.087 W m^{-2} of heat on Earth. However, radioactive heating is completely insignificant for the energy balance. The 0.087 W m^{-2} from radioactive decay is only 0.037% of the solar energy absorbed by the Earth (235 W m^{-2}). Heat flow in volcanic regions and hydrothermally active areas such as Yellowstone National Park in the US is higher, but still negligible compared to the absorbed solar flux. The other terrestrial planets Mercury, Venus and Mars are also in radiative equilibrium with the Sun and the amount of solar energy absorbed is identical to the amount re-emitted back to space. To first approximation, these planets contain the same amount of radioactive elements as the Earth and, thus, the amount of radioactive heating in them is insignificant relative to energy absorbed from sunlight.

However, the gas giant planets Jupiter, Saturn and Neptune do have internal heat sources and they re-radiate more energy to space than they receive from the Sun. Uranus is an exception and has a weak or nonexistent internal heat source that is only 14% (or less) of the absorbed solar flux.

6.2.4 Greenhouse Effect

The greenhouse effect of the Earth's atmosphere warms its surface above the temperature it would have otherwise. This basic idea dates back about 200 years. In the early 1800s, the French physicist Jean Joseph Fourier (1768–1830) first proposed that the Earth's atmosphere acts like the glass in a greenhouse because it lets through visible light but retains infrared radiation. The analogy is flawed because a greenhouse that transmits infrared radiation also remains warm. The famous American physicist Robert W. Wood (1868–1955), who received the Rumford Medal of the Royal Society in 1938, proved this in 1909 when he constructed a small greenhouse out of rock salt (NaCl), which is transparent to infrared radiation. A greenhouse remains warm because the glass reduces heat losses by convection. Nevertheless, the name greenhouse effect is entrenched and commonly used to describe warming of Earth's surface and lower atmosphere by heat-absorbing gases in air.

Mathematically, the greenhouse effect is the blackbody flux from the ground (B_g) due to the increased IR absorption (χ) of Earth's atmosphere from greenhouse gases:

$$B_g = \frac{F_{net}}{2\pi}(\chi + 2) \qquad (6.6)$$

The net flux is the difference between upward and downward IR radiation in Earth's atmosphere. For example, Earth's surface (288 K) emits about $390\,\mathrm{W\,m^{-2}}$ of IR energy upward, but downward emission of $324\,\mathrm{W\,m^{-2}}$ of IR energy from greenhouse gases in the troposphere and absorption by clouds partially cancels the surface emission. Only $40\,\mathrm{W\,m^{-2}}$ of the IR energy that is radiated by Earth's surface escapes to space *via* atmospheric windows.

6.2.5 Heat Absorbing or Greenhouse Gases

In principle, almost any molecular gas is a potential greenhouse gas. The importance of a gas as a greenhouse gas depends on its abundance (*i.e.*, column density), absorption band positions and band strengths. For example, SO_2 is an important greenhouse gas on Venus, where it is much more abundant than on Earth. Conversely, dichlorodifluoromethane

(CF$_2$Cl$_2$, F12) is an important greenhouse gas on Earth even though its abundance is only 533 parts per trillion by volume (pptv $= 10^{-12} =$ picomole/mole). The reason is that CF$_2$Cl$_2$ has strong IR bands in spectral regions where natural greenhouse gases such as H$_2$O, O$_3$, CO$_2$, CH$_4$ or N$_2$O do not absorb infrared radiation. In general, the halogen-bearing gases listed in Table 5.2 have strong IR bands in spectral regions where the natural greenhouse gases do not absorb IR radiation.

The wavelengths absorbed by greenhouse gases correspond to the energies of rotational and vibrational transitions of molecules. For example, H$_2$O is the most abundant heat-absorbing gas in Earth's atmosphere and is an important greenhouse gas on Venus, which is hotter than the planet Mercury (740 K on Venus *versus* 700 K at noontime on Mercury). Water vapor has vibrational infrared absorption bands centered at about 2.66 μm (3756 cm^{-1}, v_3), 2.73 μm (3657 cm^{-1}, v_1) and 6.27 μm (1595 cm^{-1}, v_2). It also has a strong rotational band centered at about 80 μm that is important for Earth's heat balance. In addition, H$_2$O has several weaker IR bands. There are many rotational states for each vibrational frequency and many separate rotational–vibrational transitions occur.

Homonuclear diatomic gases (N$_2$, O$_2$ and H$_2$) do not have permanent dipole moments but, even in these cases, collision-induced dipoles lead to weak IR absorption. The collision-induced absorptions are important for sufficiently large column densities of the gas. The collision-induced dipole IR absorption (1–2.5 μm vibrational bands, 10–100 μm rotational bands) of H$_2$ in the atmospheres of the gas giant planets Jupiter, Saturn, Uranus and Neptune is a good example of this behavior.

Without the presence of natural greenhouse gases such as water vapor, CO$_2$ and ozone in Earth's atmosphere much, if not all, of the Earth's surface would be too cold for humans to inhabit it comfortably and the average surface temperature would be below the freezing point of water. The concern about the greenhouse effect is that rapidly increasing concentrations of anthropogenic and natural greenhouse gases are increasing the Earth's average surface temperature by too much in too short a time. This is the enhanced greenhouse effect. Computer models predict an average temperature increase of 2–3 K for a doubling of atmospheric CO$_2$. These sophisticated models consider warming by greenhouse gases (including the overlap of their IR absorption bands), three-dimensional atmospheric circulation, heat transport by advection, convection and radiation, and the effects of greenhouse warming on evaporation and precipitation of water vapor. Finally, it is important to understand that greenhouse gases do not change the total amount of heat emitted by the Earth. Rather they change the thermal profile of the atmosphere with warming at the surface and cooling in the upper atmosphere where emission to space occurs.

6.3 BIOGEOCHEMICAL CYCLES

Biogeochemistry is the study of chemical reactions involving the atmosphere, biosphere (living organisms), hydrosphere (oceans and freshwater) and lithosphere (Earth's crust and mantle). It is concerned with the natural cycles of important biogenic elements (H, C, N, O, S, P, halogens and trace metals essential for animal and plant nutrition) and their compounds and with the effects of human (anthropogenic) activities on these cycles. The Russian geologist A. P. Pavlov (1854–1929) developed the concept of the anthropogenic era to emphasize the growing influence of human activities on the environment. The discipline of biogeochemistry evolved in the 1920s due to the wide ranging interests of several scientists, including the French biochemist Gabriel Bertrand (1867–1962) and the Russian geochemist Vladimir I. Vernadsky (1863–1945). The English zoologist G. Evelyn Hutchinson (1903–1991) pioneered applications of biogeochemistry in the western world in the 1940s and 1950s.

The large amounts of elements present in the different geochemical reservoirs on Earth are often expressed in terms of gigagrams $(1\,Gg = 10^9\,g = 10^6\,kg = 10^3$ metric tons), teragrams $(1\,Tg = 10^{12}\,g = 10^9\,kg = 10^6\,tons)$, petagrams $(1\,Pg = 10^{15}\,g = 10^{12}\,kg = 10^9\,tons)$, exagrams $(1\,Eg = 10^{18}\,g = 10^{15}\,kg = 10^{12}\,tons)$ and zettagrams $(1\,Zg = 10^{21}\,g = 10^{18}\,kg = 10^{15}\,tons)$.

6.3.1 Earth's Geochemical Reservoirs

The Earth is composed of the atmosphere, biosphere, hydrosphere, lithosphere and core. Biogeochemistry deals primarily with the reservoirs at the surface of the Earth and with the mantle to a lesser extent. As far as is known, the core does not take part in biogeochemical cycles, and we do not discuss it further. Table 6.2 summarizes the masses of the different geochemical reservoirs. Chapter 5 describes the atmosphere and we discuss the other geochemical reservoirs below.

6.3.1.1 Lithosphere. Most of Earth's mass is in the silicate lithosphere, which is made up of crust and mantle. The mantle dominates the lithosphere. The mass of the mantle is $4.007 \times 10^{24}\,kg$, while that of the crust is only $2.367 \times 10^{22}\,kg$. The mantle lies between the crust and

Table 6.2 Earth's geochemical reservoirs.[a]

Reservoir	Mass (kg)	% of Total
Atmosphere	5.14×10^{18}	8.6×10^{-5}
Biosphere	6.61×10^{15}	1.1×10^{-7}
Hydrosphere	1.70×10^{21}	2.85×10^{-2}
Lithosphere[b]	4.031×10^{24}	67.48
Core	1.941×10^{24}	32.49
Entire Earth	5.974×10^{24}	100.00

[a]Modified from K. Lodders and B. Fegley *The Planetary Scientist's Companion*, Oxford University Press, New York, NY, 1998.
[b]Lithosphere is crust plus mantle.

core and is itself divided into upper mantle, a transition zone and lower mantle by seismic discontinuities at 410 km and 670 km depth. The mantle takes part in biogeochemical cycles *via* eruption of volcanic gases (episodic with a timescale of a few years) and plate subduction (continuous with a timescale of 100 million years). Otherwise, the crust is the part of the lithosphere most involved in biogeochemical cycles.

The crust is made of continental $(1.522 \times 10^{22}$ kg) and oceanic $(8.450 \times 10^{21}$ kg) crust. The oceanic crust is 5–10 km thick, has basaltic composition, and on average is about 60 million years old. The oldest parts of oceanic crust are 200 million years old. In general, oceanic crust is much younger than continental crust, which has an average age of 960 million years and is up to 4.4 billion years old. Creation of new oceanic crust occurs continually *via* undersea volcanism and destruction of oceanic crust occurs continually *via* plate subduction back into the mantle.

Most of Earth's crust is continental crust. On average, the continental crust is about 35 km thick and is silica (SiO_2) rich, about 61.5% by mass. Table 6.3 lists the 17 most abundant elements in the continental crust. The other naturally occurring elements are less abundant than chlorine and they are not listed in the table; of these, only nitrogen, with an abundance of 59 $\mu g \, g^{-1}$, concerns us and we discuss the nitrogen biogeochemical cycle later.

6.3.1.2 Biosphere. The biosphere is the region of the Earth where life exists and it encompasses parts of the atmosphere, hydrosphere and lithosphere. The Austrian geologist Eduard Suess (1831–1914) introduced the term biosphere in his textbook *The Face of the Earth*. Incidentally, Suess's grandson Hans E. Suess (1909–1993) made important contributions to the study of the carbon cycle, which is probably the most important cycle in Earth's biosphere. Table 6.4

Table 6.3 Elemental composition of Earth's crust (mass%).[a]

Element	Crust
O	47.20
Si	28.80
Al	7.96
Fe	4.32
Ca	3.84
Na	2.35
Mg	2.20
K	2.14
Ti	0.40
H[b]	0.22
C[c]	0.20
P[c]	0.075 7
Mn	0.071 6
S[c]	0.069 7
Ba	0.058 4
F[c]	0.052 5
Cl[c]	0.047 2
Total	100.00

[a]Continental crust, K. H. Wedepohl, *Geochim. Cosmochim. Acta*, 1995, **59**, 1217, adjusted for pore water in sediments.
[b]Pore water in sediments is 1.97% of the crust.
[c]Also see the discussions for C, P, S, F, Cl and N. The latter is $5.9 \times 10^{-3}\%$ of the crust.

Table 6.4 Earth's biosphere.[a]

Reservoir	Pg (10^{12} kg) carbon	Total mass (Pg)[b]	% of Total
Terrestrial organic matter	1720	4410	66.7
Terrestrial organisms	830	2128	32.2
Marine organic matter	25	64	1.0
Marine organisms	3	8	0.1
Humans	0.09	0.46	0.007
Totals	2578	6610	100.01

[a]Modified from R. P. Wayne, *Chemistry of Atmospheres*, 3rd edn, Oxford University Press, New York, 1999.
[b]Average carbon content of 39% for all, except 19% for humans, see Table 6.5.

gives the mass of Earth's biosphere (total mass, and mass of carbon). Most biospheric mass is in decaying organic matter on land. This is a mixture of decaying animal and plant remains, microorganisms and substances made during decay of the animal and plant remains. Representative elemental abundances for this material on a dry weight basis range from 51–56% for carbon, 33–45% for oxygen, 3–6% for hydrogen, 0.7–4.6% for nitrogen and 0.3–1.1% for sulfur. Complete decay of the organic matter in soils takes several hundred to several thousand years (measured by carbon-14 dating). Terrestrial organisms are the next most important contribution to Earth's biosphere followed by decaying organic matter in the oceans and marine organisms. The approximately 6.5 billion human beings comprise an insignificant fraction (about 0.007%) of all biospheric mass on Earth.

We have tabulated the average elemental composition of Earth's biosphere in Table 6.5. This gives the average composition of phytomass (terrestrial land plants), of the human body and of alfalfa, one of the few plants that convert atmospheric N_2 into fixed, or chemically reactive, nitrogen. All of the compositions include water, which is most of an organism's total mass, *e.g.*, water comprises 60% of the human body and 75% of alfalfa by mass. Phytomass is rich in carbohydrates and poor in fats and proteins, which leads to lower nitrogen, phosphorus and sulfur than in the human body because N and S are abundant in proteins and P is abundant in fats. Average compositions for wood, peat, coal, oil and natural gas are given in Table 6.6. Tables 6.5 and 6.6 show that biomass (human body, plants, wood) is about 19–50% carbon by mass on a wet basis. Table 6.6 shows progressively increasing carbon, decreasing water and decreasing N contents in going from wood to fossil

Table 6.5 Composition of Earth's biosphere (mass%).

Element	Phytomass[a]	Human body[b]	Alfalfa[c]
O	52.43	61.34	77.90
C	39.35	22.82	11.34
H	6.59	9.99	8.72
N	0.50	2.57	0.825
Ca	0.38	1.43	0.58
K	0.23	0.20	0.23
Si	0.12	0.026	9.3×10^{-3}
Mg	0.098	0.027	0.082
S	0.071	0.20	0.11
Al	0.056	9×10^{-5}	2.5×10^{-3}
P	0.052	1.11	0.07
Cl	0.050	0.14	0.07
Fe	0.039	6×10^{-3}	2.7×10^{-3}
Mn	0.021	2×10^{-5}	3.6×10^{-4}
Na	0.019	0.14	0.04
Total	100.01	99.99	99.98

[a]Average composition of land plants [E. Deevey, *Sci. Am.*, 1970, **223**(3), 148]. To first approximation, this is the biospheric composition (see Table 6.4).
[b]W. S. Snyder, *Report of the Task Group on Reference Man*, Pergamon Press, Oxford, 1975.
[c]D. Bertrand, *Bull Am. Mus. Nat. History*, 1950, **94**, 403; K. Rankama and Th. G. Sahama, *Geochemistry*, University of Chicago Press, Chicago, IL, 1950.

Table 6.6   Chemical compositions of wood, peat and fossil fuels.[a]

Element	Wood	Peat	Coal[b] Brown	Coal[b] Soft	Coal[b] Hard	Crude oil[c]	Gas[d]
C	50	59	69	82	95	86.25	67.71
O	43	33	25	13	2.5	1.16	0.63
H	6	6	5.5	5	2.5	12.10	20.71
N	1	2	0.8	0.8	Trace	0.25	10.94
Totals	100	100	100.3	100.8	100	99.90	99.99

[a]From K. Rankama and Th. G. Sahama, *Geochemistry*, University of Chicago Press, Chicago, IL, 1950, unless otherwise noted.
[b]Brown coal – lignite, soft coal – bituminous, and hard coal – anthracite.
[c]Doba, Chad crude oil, also includes 0.14% sulfur, Dehkissia *et al.*, *Fuel*, 2004, **83**, 2157.
[d]Natural gas, Amarillo, TX, 72.94% CH_4, 18.96% C_2H_6, 7.71% N_2 and 0.39% CO_2 by volume. From M. Ruhemann, *The Separation of Gases*, 2nd edn, Clarendon Press, Oxford, UK, 1949.

Table 6.7   Earth's hydrosphere.[a]

Reservoir	Zg $(10^{18}$ kg$)$[b]	% of Total
Oceans	1370	80.67
Pore water in sediments	300	17.67
Icecaps & glaciers	20	1.18
Groundwater	8	0.47
Rivers & lakes	0.23	0.014
Atmospheric	0.013	0.00077
Totals	1698	100.00

[a]Modified from R. M. Garrels and F. T. Mackenzie, *Evolution of Sedimentary Rocks*, W. W. Norton, New York, 1971.
[b]One zettagram $= 1$ Zg $= 10^{21}$ g $= 10^{18}$ kg.

fuels. We include peat and fossil fuels in the crustal carbon inventory (as done by Wedepohl), the source of our crust composition.

6.3.1.3 Hydrosphere and Hydrologic Cycle. The hydrosphere is a small part of the entire Earth (Table 6.2), but it is the second most massive geochemical reservoir (after the crust) at the surface of the Earth. The hydrosphere covers most of Earth's surface [361.06×10^6 km^2 (70.8% of the total area) *versus* land area of 148.89×10^6 km^2 (29.2% of the total area)].

Table 6.7 shows that the total amount of water in the hydrosphere is about 1698×10^{18} kg (10^{18} kg $= 10^6$ km^3). Excluding pore water in sediments, the total exchangeable water reservoir is 1398×10^{18} kg, 98% of which is in the oceans. The remaining 2% of Earth's exchangeable water is primarily in the polar icecaps and glaciers (1.4%) and groundwater

(0.6%). The water in rivers, lakes and the atmosphere is less than 0.02% of the exchangeable water on Earth.

The hydrologic cycle is the largest biogeochemical cycle in terms of the annual flux of mass at the surface of the Earth. About 425 000 Pg water evaporates from the oceans each year. This evaporative loss is balanced by precipitation of 385 000 Pg and river flow of 40 000 Pg per year. The global average precipitation falling on the continents is 111 000 Pg per year. The river flow of 40 000 Pg yr^{-1} into the oceans and 71 000 Pg yr^{-1} of evaporation and transpiration balance the precipitation flux. The global average fluxes vary spatially and temporally. Zonal average precipitation is highest at the equator, has a minimum at 25° north and south of the equator, and drops off toward the poles. As a consequence, the lifetime of atmospheric water vapor varies from about 8 days in equatorial regions to 13–15 days near the poles, with an average of 10 days overall.

The average salinity, density, pH and alkalinity of seawater are 34.7‰, 1.025 g cm^{-3}, 8.1 and 2.4 milli-equivalents per kilogram, respectively. Salinity is defined as grams of NaCl per kilogram seawater and is expressed in per mille (parts per thousand) units. Oceanographers define salinity (S) on a mass basis because mass is conserved, while volume varies, as pressure and temperature change in the oceans. Salinity and chlorinity (Cl) are related by $S(‰) = 1.80655Cl(‰)$. As noted above, the average salinity of the oceans is 34.7‰. However, salinity variations of 33–38‰ occur in the oceans. The central gyre regions 25° north and south of the equator have high salinity, because the evaporation rate is larger than the rainfall rate. (The central gyre regions are cyclic patterns of surface currents.) The summer polar oceans have low salinity because melting of the ice sheets dilutes the oceans with freshwater. The alkalinity [Alk] is the acid neutralizing capacity of seawater in milli-equivalents (meq) per kilogram and is:

$$[\text{Alk}] = [\text{HCO}_3^-] + 2[\text{CO}_3^{2-}] + [\text{B(OH)}_4^-] + [\text{OH}^-] - [\text{H}^+] \qquad (6.7)$$

One equivalent is equal to one mole of H$^+$ ions per kilogram. The square brackets are concentration in millimoles (mmol) per kilogram. One mole of bicarbonate ions neutralizes one mole of H$^+$ ions *via* the formation of neutral carbonic acid:

$$\text{H}^+(\text{aq}) + \text{HCO}_3^-(\text{aq}) = \text{H}_2\text{CO}_3(\text{aq}) \qquad (6.8)$$

One mole of carbonate ions neutralizes two moles of H$^+$ ions *via* the sequential formation of bicarbonate and neutral carbonic acid. One mole of borate ions neutralizes H^+ ions *via* formation of neutral boric acid H_3BO_3 (aq).

The alkalinity of the oceans is mainly due to bicarbonate (2.2 mmol kg^{-1}), because it constitutes essentially 100% of all dissolved inorganic carbon at the average pH of the ocean. Dissolved CO_2 (g), neutral carbonic acid and carbonate ions are much less abundant than bicarbonate ions in the ocean. The next most important species for the alkalinity of the oceans are carbonate (0.27 mmol kg^{-1}) and borate (0.056 mmol kg^{-1}). At the average oceanic pH of 8.1, the hydroxide concentration is only $\sim 1.26 \times 10^{-3}$ mmol kg^{-1}. The average hydrogen ion concentration of $\sim 7.9 \times 10^{-6}$ mmol kg^{-1} reduces alkalinity, hence the minus sign in Equation (6.7). The alkalinity varies from ocean to ocean. It also varies with depth in the ocean due to the dissolution of the carbonate skeletons of dead organisms.

The carbonate compensation depth (CCD) is the depth at which carbonate completely dissolves *via* the net reaction:

$$CaCO_3(solid) + H_2CO_3(aq) = Ca^{2+}(aq) + 2HCO_3^-(aq) \qquad (6.9)$$

The CCD is 3700 m deep in the Atlantic and 1000 m deep in the Pacific. The CCD is shallower in the Pacific because water has a longer residence time in the deep Pacific Ocean than in the deep Atlantic Ocean due to circulation patterns. The longer residence time allows accumulation of dissolved CO_2 produced by the decay of dead organisms as they sink downward. In turn, the dissolved CO_2 forms more carbonic acid at a given depth in the Pacific Ocean *via* the net reaction:

$$CO_2(aq) + H_2O(aq) = H_2CO_3(aq) \qquad (6.10)$$

The larger amount of aqueous H_2CO_3 at a given depth in the Pacific (than in the Atlantic Ocean) dissolves carbonate at a shallower depth in the Pacific Ocean.

Tables 6.8 and 6.9 list the ten most abundant constituents in seawater and river water in units of milligrams (mg) per kilogram, which is the same as micrograms per gram ($\mu g \, g^{-1}$). River water is more dilute (exceptions are the concentrations of dissolved Al, Fe and silica, which are higher in river water than in the oceans), and the major cations and anions are different. River water chemistry varies seasonally and because different solutes are introduced from different rocks in different geological settings. River water is the major, but not the only, source of most elements in seawater. The inputs of dissolved solids and of suspended sediments both contribute to the oceanic input. Estimates for the

Table 6.8 Composition of seawater.[a]

Species	Abundance $(mg\,kg^{-1})$
Cl^-	19 353
Na^+	10 781
SO_4^{2-}	2 712
Mg^{2+}	1 280
Ca^{2+}	415
K^+	399
HCO_3^-	142
Br^-	67
Sr^{2+}	7.8
$B(OH)_4^-$	4.5

[a]Ten most abundant dissolved constituents.

Table 6.9 Composition of river water.[a]

Species	Abundance $(mg\,kg^{-1})$
HCO_3^-	58.4
Ca^{2+}	15
SiO_2	13.1
SO_4^{2-}	11.2
Cl^-	7.8
Na^+	6.3
Mg^{2+}	4.1
K^+	2.3
NO_3^-	1
Fe	0.37

[a]Global average, ten most abundant dissolved constituents.

global average flux of suspended sediments range from 18.3 to 32.5 Pg yr^{-1}. Globally, the average composition of the sediment is that of the continental crust. For example, the average composition of sediment carried by the Mississippi and Nile Rivers is dominated by SiO_2 (61.0%) and Al_2O_3 (14.3%) *versus* 61.5% SiO_2 and 15.1% Al_2O_3 in continental crust. Assuming steady state, the global average flow rate of rivers into the oceans (40 000 km^3 yr^{-1} as mentioned earlier) replaces all water in the oceans once every 34 250 years. (The residence time of ocean water is the mass of water in the ocean divided by the global average flow rate of rivers into the oceans. A similar definition gives the residence time of dissolved oceanic species.)

Six elements (Cl, Na, S, Mg, Ca and K) dominate the composition of seawater and have constant ratios relative to one another despite variations in salinity. These elements and others showing the same behavior (*e.g.*, B, Br and F) are conservative elements. The addition or removal of pure water to seawater explains variations in the concentrations of conservative elements. The conservative elements have constant ratios relative to one another because they have oceanic residence times longer than that of ocean water itself. For example, the residence times for Cl, Br, Na, Ca and F are 120, 100, 75, 1.1 and 0.5 million years, respectively. Calcium is conservative only to first approximation because $CaCO_3$-shell formation by oceanic organisms (*e.g.*, clams, forams, oysters) depletes Ca in surface water relative to its concentration in deep ocean water.

The other elements in seawater are non-conservative for various reasons. Oceanic organisms deplete nutrient elements such as C, N and P and shell-building elements such as Ca and Si (used for the silica shells of diatoms) in surface waters. Conversely, decomposition of sinking detritus and dissolution of shells enriches C, N, P, Ca and Si in deep ocean water. Photosynthetic marine organisms produce O_2 and its concentration is higher in the photic zone. Input from the atmosphere enriches surface water in Fe (from mineral dust), radionuclides such as tritium T, ^{14}C, ^{90}Sr, ^{137}Cs and Pu (from atmospheric nuclear explosions), and Pb (from anthropogenic pollution). Input from undersea hydrothermal vents enriches ^3He, ^{222}Rn and CH_4 at intermediate depths. The oceanic circulation pattern enriches tritiated water HTO at intermediate depths.

The total volume of ocean water is $1370 \times 10^6 \, km^3$, mainly in the Pacific ($707.6 \times 10^6 \, km^3$), Atlantic ($323.6 \times 10^6 \, km^3$) and Indian ($291.0 \times 10^6 \, km^3$) oceans. The global average depth of the oceans is 3794 m, where the pressure is 386 bar. In contrast, average continental elevation is 840 m. The continental margin is the transition region between the continents and oceans. Continental margins form the minimum at 1.8 km below sea level in the Earth's bimodal hypsometric curve and the average elevation of the continents is the maximum.

Several different regions comprise the continental margins. The continental shelf is the landward part of the margin. This is the submerged continuation of the continents and varies in width, with an average width of 70 km. The continents drop off into the ocean at the continental shelf break, which occurs at about 130 m depth. The gradient rapidly changes from 1 : 1000 to 1 : 40 in this continental slope region. The continental slope may grade into the continental rise. This is a feature formed by deposition of sediments carried down the continental slope by ocean currents. In other instances, the slope forms one side of an oceanic trench. The greatest ocean depths occur in ocean trenches (up to 11.04 km in the Challenger Deep in the Marianas Trench).

The surface mixed layer of the oceans interacts with the atmosphere. Air–sea exchange of CO_2 or another soluble gas occurs at a velocity of about five meters seawater per day (the piston velocity). The surface mixed layer is the uppermost 75–200 m, which has an average temperature of 18 °C. Vertical diffusivities (D) in the surface mixed layer are in the range of 1–10 cm^2 s^{-1}. For a mixed layer thickness $z = 75$ m and diffusivity $D = 10$ cm^2 s^{-1}, the mixing velocity v_{mix} is:

$$v_{mix} \sim \frac{D}{z} \sim \frac{10\,\text{cm}^2\,\text{s}^{-1}}{(75\,\text{m})(100\,\text{cm m}^{-1})} \sim 1.3 \times 10^{-3}\,\text{cm s}^{-1} \sim 1.15\,\text{m day}^{-1} \quad (6.11)$$

The corresponding time for mixing across a 75-m thick surface layer is 65 days, or two months. Sinking of cold water at the poles removes 50×10^6 m^3 s^{-1} water from the surface mixed layer. At this rate, water in the top 75–200 m overturns once every 17–46 years.

The thermocline occurs below the surface mixed layer and extends to 1 km depth. The temperature decreases from that of the surface mixed layer (18 °C) to that of the abyssal (or deep) ocean (3.5 °C) across the thermocline. The deep ocean contains 95% of all seawater. The oceans have warm water on top and cold water below and thus vertical mixing between these density stratified regions is very slow. For example, vertical mixing from bottom of the surface mixed layer at 200 m depth down to the base of the thermocline at 1 km depth takes about 125 years (diffusivity $D \approx 1.6$ cm^2 s^{-1}).

This is measured by tracing the concentration of ^{14}C-bearing dissolved inorganic carbon (DIC, a mixture of dissolved CO_2, bicarbonate and carbonate) and HTO. Large amounts of both radioactive species resulted from atmospheric nuclear testing, although cosmic rays continually produce ^{14}C in smaller amounts. The thermocline or thermohaline oceanic circulation occurs *via* downwelling at high latitudes in both hemispheres and transport along constant density (isopycnal) contours to equatorial regions where upwelling occurs. This circulation pattern removes CO_2 from the atmosphere at high latitudes and slowly transports it toward equatorial regions. Carbon-14 dating of dissolved inorganic carbon gives ages of 275 years and 510 years for bottom waters in the Atlantic and Pacific Oceans, respectively. This "old" seawater has less DIC, indicating lower concentrations of atmospheric CO_2 prior to the Industrial Revolution.

6.3.1.4 Distribution of Biogenic Elements. The continental crust (and the atmosphere, biosphere and hydrosphere) are enriched in incompatible elements. These are elements that partition into molten rock and/or volcanic gas and have melt/rock or gas/rock concentration ratios greater than unity. To first approximation, the atmosphere, hydrosphere and continental crust formed by partial melting and out-gassing of the primordial silicate Earth. Thus, the incompatible elements concentrated into the melt and volcanic gas that later formed the crust, oceans and atmosphere. In general, the biogenic elements and their compounds (water, C, N, S, P, halogens) are incompatible and are concentrated at the surface of the earth.

Nitrogen is the pre-eminent example. The nitrogen abundance in Earth's atmosphere (78.084% by volume) corresponds to 75.5% by mass. The nitrogen abundance in the mantle is 0.0014% by mass. Thus, the atmosphere/mantle partition coefficient for nitrogen is about 53 930. Likewise, the oceanic Cl abundance (1.9353%) is about 14 890 times larger than its abundance (0.00013%) in the mantle. To give a third example, the average carbon abundance in biomass (39%) is 3900 times larger than its abundance (0.01%) in the mantle. Some elements such as nitrogen are concentrated in the atmosphere, others such as carbon and iodine are concentrated in the biosphere, and still other elements and compounds such as chlorine, bromine and water are concentrated in the oceans.

6.3.2 Carbon Cycle

6.3.2.1 Overview. Carbon is one of the elements known since ancient times. It is the fourth most abundant element in the solar system (after H, He and O) with an abundance of 7.19 million C atoms per 10^6 Si atoms. Carbon is the second most abundant element in biomass after oxygen (Tables 6.5 and 6.6), and is a minor constituent of the crust (0.20%), oceans (142 mg bicarbonate per kg seawater) and atmosphere (387 ppmv CO_2). The carbon content of mid-ocean ridge basalts (MORBs) is about $500\,\mu g\,g^{-1}$. Carbon is an incompatible element and has a melt/rock partition coefficient of ~ 10. Thus, the $500\,\mu g\,g^{-1}$ carbon content in MORBs corresponds to about $50\,\mu g\,g^{-1}$ carbon in upper mantle rocks.

Carbon has two stable isotopes (^{12}C and ^{13}C) with approximate atomic abundances of 98.9% (^{12}C) and 1.1% (^{13}C). The atomic $^{13}C/^{12}C$ ratio is 0.0112375 in the PDB standard for carbon isotopes. The PDB standard is the Cretaceous belemnite formation at Pee Dee, South Carolina, USA. A belemnite is an extinct marine cephalopod similar to squid, which existed on Earth from 208–65 million years ago during the Carboniferous Period to the end of the Cretaceous Period.

Carbon-14 is a radioactive isotope of carbon with a half-life of 5715 years. Cosmic ray bombardment of Earth's atmosphere produces neutrons that react with ^{14}N, producing ^{14}C and a proton. Atmospheric testing of nuclear bombs (primarily prior to 1963) produced large increases in atmospheric ^{14}C.

The abundance of ^{14}C in living organisms is constant because photosynthesis (by plants, or more accurately by photoautotrophs) or consumption of organic matter (by animals, or more accurately by heterotrophs) balances the continual decay of ^{14}C. The steady state abundance of ^{14}C in living organisms is equivalent to 13.56 disintegrations per minute per gram. In contrast, the abundance of ^{14}C in animal and plant remains and in inorganic compounds decreases with time after the organism dies or after the inorganic compound stops equilibrating with the $^{14}CO_2$ in Earth's atmosphere. The radioactive decay law describing this is:

$$A_t = A_o \exp(-\lambda t) \tag{6.12}$$

The ^{14}C abundance at time t (in years) after uptake of fresh ^{14}C stopped is A_t, the initial abundance of ^{14}C at zero time (*i.e.*, in the living organism or during equilibration with $^{14}CO_2$ in Earth's atmosphere) is A_o, and λ is the radioactive decay constant for ^{14}C, which is $1.21 \times 10^{-4}\, yr^{-1}$. Equation (6.12) is the same rate law used for first-order kinetics. The absolute $^{14}C/^{12}C$ atomic ratio is 1.176×10^{-12} (on January 1, 1950) in an oxalic acid (HOOC-COOH) standard from the US National Institute of Standards and Technology (NIST, formerly NBS).

We earlier alluded to the work of Hans Suess on the carbon cycle. The Suess effect is the dilution of atmospheric $^{14}CO_2$ by CO_2 produced by combustion of fossil fuels such as coal, natural gas and oil. The CO_2 from fossil fuels does not contain ^{14}C, which has long since decayed away.

6.3.2.2 Major Carbon Reservoirs at Earth's Surface. The crust contains most of the carbon at the surface of the Earth. The crust contains about 0.20% carbon (Table 6.3), which corresponds to $4.2 \times 10^{19}\, kg$ ($4.2 \times 10^7\, Pg$) carbon. Carbonates (0.16% by weight of sediments) and organic carbon (0.04% by weight of sediments) make up 79.7% of crustal carbon, with the remaining 20.3% carbon in the different rock types found in the crust. The relative amounts of carbonate and organic carbon in the crust are fixed at a mass ratio of about 4-to-1 by their isotopic compositions of $\delta^{13}C = 0‰$ (carbonate) and $\delta^{13}C = -25‰$ (organic carbon).

This distinction between isotopically "heavy" carbonates and isotopically "light" organic matter was discovered in 1939 by the American physicist Alfred O. C. Nier (1911–1994). Organic carbon is depleted in ^{13}C because the enzyme ribulose-1,5-bisphosphate carboxylase (commonly known as RuBp carboxylase or RuBisCO) preferentially incorporates ^{12}C from CO_2 into organic material during the first step of carbon fixation in the Calvin photosynthetic cycle. The depletion in ^{13}C is retained in organic carbon formed from remains of plants and other photoautotrophs.

The total number of ^{12}C and ^{13}C atoms on Earth is constant. Thus, isotopic mass balance requires that the total number of ^{12}C atoms is equal to the mass weighted sum of ^{12}C atoms in carbonate (carb) and organic (org) carbon in the crust. An analogous constraint holds for the total number of ^{13}C atoms. The mass fractions of carbonate and organic carbon sum to unity. In mathematical terms these mass balance constraints are:

$$^{12}C_{Total} = {}^{12}C_{carb}X_{carb} + {}^{12}C_{org}X_{org} \tag{6.13}$$

$$^{13}C_{Total} = {}^{13}C_{carb}X_{carb} + {}^{13}C_{org}X_{org} \tag{6.14}$$

$$X_{carb} + X_{org} = 1 \tag{6.15}$$

The total masses of carbonate and reduced carbon in Earth's crust are thus about 2.68×10^{19} kg and 6.7×10^{18} kg, respectively. This calculation was first performed by the Swedish geochemist F. E. Wickman in 1941, shortly after Nier's discovery of the isotopic dichotomy of inorganic and organic carbon in Earth's crust. However, the economically recoverable reserves of coal (9.1×10^{14} kg), natural gas (7.2×10^{13} kg) and oil (1.9×10^{14} kg) are only about 0.017% of the total mass of reduced crustal carbon.

The oceans are the second largest reservoir of carbon at Earth's surface and contain about 38 930 Pg carbon. Dissolved inorganic carbon is about 97.3% (37 900 Pg) of the total with dissolved organic carbon making up 2.6% (1000 Pg), and particulate carbon about 0.1% (30 Pg). As mentioned earlier, most of the dissolved inorganic carbon is bicarbonate because it is the major species in the CO_2–H_2CO_3 system at oceanic pH (~ 8.1).

The biosphere (2578 Pg) and atmosphere (827 Pg) are the other two carbon reservoirs at Earth's surface. Terrestrial organic matter (1720 Pg) and terrestrial organisms (830 Pg) contain most of the biospheric carbon (Table 6.4). Most atmospheric carbon is in CO_2, which is 387 ppmv as of February 2008. One ppmv of CO_2 is equal to 2.13 Pg-C or 7.8 Pg-CO_2. Thus, the present atmospheric CO_2 inventory is 823 Pg-C. The remainder of carbon in Earth's atmosphere is in CH_4 (3.7 Pg-C), CO

Table 6.10 Major carbon reservoirs at Earth's surface.

Reservoir	Pg (10^{12} kg) carbon	Notes
Atmosphere	827	Dominantly CO_2, 1 ppmv $CO_2 = 2.13$ Pg-C
Biosphere	2 578	From Table 6.4
Hydrosphere	38 930	Oceanic carbon, see text
Lithosphere	4.2×10^7	Crustal carbon, see text

(0.3 Pg-C) and other carbon-bearing gases (CH_3Cl, CFCs, *etc.*, that sum up to 0.2 Pg-C). Although the crust is the largest carbon reservoir at Earth's surface, the atmosphere and biosphere are the two most important reservoirs in terms of carbon transport on an annual basis. Table 6.10 summarizes our discussion of the major carbon reservoirs at Earth's surface.

6.3.2.3 CO_2 – H_2CO_3 System. The dissolution of CO_2 in fresh water and seawater, the formation of carbonic acid H_2CO_3, and its subsequent ionization to bicarbonate and carbonate are very important for the chemistry of groundwater, lakes, rainfall, rivers and the oceans. Typical concentrations of dissolved inorganic carbon ($CO_2 + H_2CO_3 + HCO_3^- + CO_3^{2-}$) are 2.3 mmol kg^{-1} in seawater, 0.1–5 mmol kg^{-1} in lake and river water, 0.01–0.05 mmol kg^{-1} in rainwater and 0.5–8 mmol kg^{-1} in groundwater. In general, the dissolved inorganic carbon species are responsible for most of the alkalinity (acid neutralizing capacity) of natural waters. We thus describe the aqueous chemistry of this system in some detail.

The dissolution of CO_2 in water occurs *via* the net thermochemical reaction:

$$CO_2(g) = CO_2(aq) \qquad (6.16)$$

The equilibrium constant for Reaction (6.16) is the Henry's law constant (K_H):

$$K_H = \frac{c_{CO_2}}{P_{CO_2}} \text{ mol L}^{-1} \text{ atm}^{-1} \qquad (6.17)$$

The abbreviations c and P in Reaction (6.17) stand for concentration and pressure, respectively. The Henry's law constant is a small number and it is conveniently expressed using the same notation as pH, *i.e.*:

$$pK_H = -\log K_H \qquad (6.18)$$

For example, at 18 °C, the average temperature of the surface mixed layer (the uppermost 75–200 m) of the oceans, $K_H = 0.0413$ and $pK_H = 1.384$ in pure water, while $K_H = 0.0350$ and $pK_H = 1.456$ in seawater of average salinity. This comparison shows that, at constant temperature, CO_2 solubility in water decreases with increasing salinity (the salting-out effect). At constant salinity, the solubility of CO_2 decreases with increasing temperature. The pK_H values at 5, 15 and 25 °C are 1.20, 1.34 and 1.47, respectively.

At the average oceanic pH of 8.1, water reacts with the dissolved CO_2 gas to form carbonic acid:

$$CO_2(aq) + H_2O(aq) = H_2CO_3(aq) \tag{6.10}$$

This reaction is slow and has first order kinetics. Thus, a large fraction of "carbonic acid" is actually $CO_2(aq)$. For example, at 5, 15 and 25 °C the CO_2/H_2CO_3 concentration ratios in pure water are 513, 463 and 386, respectively. The equilibrium constant for Reaction (6.10) is:

$$K_{eq} = \frac{c_{H_2CO_3}}{c_{CO_2(aq)}} \tag{6.19}$$

The corresponding pK_{eq} values for pure water at 5, 15 and 25 °C are 2.710, 2.666 and 2.587, respectively.

Neutral carbonic acid ionizes to bicarbonate ions and protons *via*:

$$H_2CO_3(aq) = HCO_3^-(aq) + H^+(aq) \tag{6.20}$$

The equilibrium constant for Reaction (6.20), the first ionization of carbonic acid, is denoted K_1 and is:

$$K_1 = \frac{c_{HCO_3^-} c_{H^+}}{c_{H_2CO_3}} \tag{6.21}$$

The pK_1 values at 5, 15 and 25 °C are 3.807, 3.754 and 3.764, respectively.

However, as mentioned above, most of the "carbonic acid" in water is actually dissolved CO_2 gas. The first ionization of neutral "carbonic acid" can thus be expressed as the combination of Reactions (6.10) and (6.20):

$$CO_2(aq) + H_2O(aq) = HCO_3^-(aq) + H^+(aq) \tag{6.22}$$

The combined reaction has the equilibrium constant:

$$K_1^* = \frac{c_{HCO_3^-} c_{H^+}}{c_{CO_2(aq)}} \tag{6.23}$$

Reaction (6.22) is the sum of two other reactions, and the equilibrium constant for it is the product of the equilibrium constants of the two combined reactions:

$$K_1^* = K_{eq} K_1 \tag{6.24}$$

$$pK_1^* = pK_{eq} + pK_1 \tag{6.25}$$

The pK_1^* values are 6.517, 6.420 and 6.352 at 5, 15 and 25 °C, respectively.

The second ionization of carbonic acid is the ionization of the bicarbonate ion:

$$HCO_3^-(aq) = CO_3^{2-}(aq) + H^+(aq) \tag{6.26}$$

$$K_2 = \frac{c_{CO_3^{2-}} c_{H^+}}{c_{HCO_3^-}} \tag{6.27}$$

The pK_2 values are 10.557, 10.430 and 10.329 at 5, 15 and 25 °C, respectively.

As discussed earlier, the alkalinity, or acid neutralizing capacity, of natural waters depends upon the concentrations of bicarbonate and carbonate ions because the reaction of these ions with protons removes acid from the aqueous solution. The bicarbonate ion provides most of the carbonate alkalinity of the oceans because it is the dominant form of dissolved inorganic carbon at the average pH of the oceans. The alkalinity of the oceans also varies with depth because the ionization constants of carbonic (and boric) acid increase slightly with increasing pressure. Formation of aqueous ion pairs such as $MgHCO_3^-$ and $MgCO_3$ counteracts the effect of pressure because ion pair formation decreases the concentrations of bicarbonate and carbonate ions in seawater and thus decreases the alkalinity. Some of the books listed at the end of this chapter discuss these effects in more detail.

Many lakes and rivers have pH values lying between the first and second pK values of carbonic acid. Thus, bicarbonate is the major dissolved carbon species and the major source of alkalinity in freshwater as well as in the oceans.

Anthropogenic Emissions of Carbon Dioxide. Historical Background: We now know that the atmospheric CO_2 abundance is increasing rapidly due to CO_2 produced by fossil fuel combustion, by cement manufacturing and by deforestation. However, it has taken over 100 years for this fact to be established.

The Swedish physical chemist Svante Arrhenius (1859–1927), who won the 1903 Nobel Prize in Chemistry for his theory of ionization of electrolyte solutions, was the first to examine the effect of increased CO_2 on Earth's climate. In 1896, Arrhenius calculated heating due to CO_2 and H_2O in air and showed that increasing CO_2 abundances led to higher temperatures. At the time, Arrhenius was interested mainly in the cause of Ice Ages. In 1904, he discussed CO_2 release by burning coal, oil and other fossil fuels. Arrhenius concluded that atmospheric CO_2 was increasing due to fossil fuel combustion and that the accelerated use of fossil fuels would accelerate the buildup of atmospheric CO_2.

About 40 years later, the British steam engineer and amateur meteorologist Guy S. Callendar (1898–1964) reiterated Arrhenius' conclusion. In a series of papers published from 1938 to 1958 Callendar argued that atmospheric CO_2 was increasing due to fossil fuel combustion and that this was leading to warmer temperatures. Callendar critically assessed published measurements of atmospheric CO_2 and selected the most reliable data to determine the temporal trend, if any, in atmospheric CO_2 abundances. He concluded that the CO_2 abundance increased from 290 parts per million by volume (ppmv) in 1900 to 325 ppmv in 1956. Callendar's result for 1900 is still valid, although we know now that the data he selected for 1956 are too high and that the CO_2 abundance was closer to 313 ppmv at that time. Although Callendar relied on literature data, his study demonstrated an increase in atmospheric CO_2 of about 8% in 56 years due to fossil fuel combustion. Callendar's results led to further research on the atmospheric CO_2 abundance and its temporal changes.

However, the percentage of emitted CO_2 that remained in the atmosphere, as opposed to dissolving in the oceans, was a key question affecting Callendar's conclusions. In his first paper published in 1938, Callendar estimated that 25% of emitted CO_2 dissolved in the oceans and 75% remained in the atmosphere. In 1957, which was the International Geophysical Year (IGY), the American oceanographer Roger Revelle (1909–1991) and the Austrian physical chemist Hans Suess (1909–1993) studied the partitioning of CO_2 between the oceans and atmosphere. They concluded that most of the anthropogenic emissions had been absorbed by the oceans and very little remained in the atmosphere. In fact, at present about half of emitted CO_2 dissolves in the oceans and about half remains in the atmosphere – closer to Callendar's original estimate.

With support from Revelle and the US Weather Bureau (now part of NOAA, the National Oceanic and Atmospheric Administration), the American geochemist Charles D. Keeling (1928–2005) began monitoring atmospheric CO_2 in 1958. With a few interruptions these measurements continue to the present and provide an important record of atmospheric CO_2 abundances (see the Keeling curve in Figure 6.1). The data show an increase in atmospheric CO_2 levels from about 315 ppmv in 1958 to 387 ppmv in February 2008, an increase of 72 ppmv ($\sim$23%) in 50 years. In comparison, Callendar's corrected results show an increase of $\sim$23 ppmv (8%) in atmospheric CO_2 over the 56 years from 1900 to 1956. The data show an accelerating increase in atmospheric CO_2 over time. We now consider two interesting questions. First, how have CO_2 emissions varied over time? Second, what are the major sources and sinks for CO_2 at the present time?

CO_2 Emissions over Time. Several different methods are used to estimate CO_2 abundances in Earth's atmosphere at different times during the past. As described earlier, Callendar critically assessed published analyses of CO_2 in air from the nineteenth and twentieth century literature. He concluded that throughout the nineteenth century, up to the year 1900, the atmosphere contained 290 ppmv CO_2. A second

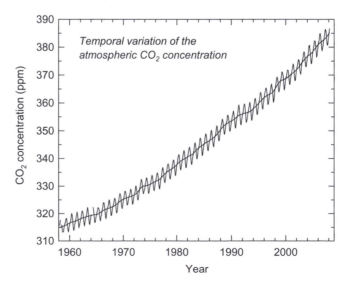

Figure 6.1   The Keeling curve shows atmospheric CO_2 concentrations (parts per million by volume) from air sampling at Mauna Loa Observatory, Hawaii. (Data are from the Scripps CO_2 Program http://scrippsco2.ucsd.edu)

method is to add up fossil fuel emissions and assume a constant partitioning of CO_2 between the atmosphere and oceans. For example, CO_2 emissions from 1850 to 1982 total 173 ± 10 Pg-C. At present the oceans take up about 55% of all CO_2 emitted. Assuming the same percentage over time gives 290 ppmv CO_2 in 1850. A third method involves measurements of the $^{12}C/^{13}C$ isotopic ratio in the wood of tree rings over time. The tree rings provide chronological dating (one tree ring per year). The carbon isotopic ratio increases with time because CO_2 emitted by fossil fuel combustion has a larger $^{12}C/^{13}C$ value than atmospheric CO_2 would have otherwise. This method gives a pre-industrial CO_2 abundance of 243 ppmv. A fourth method is analysis of total dissolved inorganic carbon (DIC) in "old" ocean water. The total DIC in ocean water is assumed to have equilibrated with the atmospheric partial pressure of CO_2 at the time. The ^{14}C content of the DIC gives the age of the ocean water. This method gives pre-industrial CO_2 abundances of 250–275 ppmv. Extraction and analyses of CO_2 in air trapped in ice cores is a fifth method, which gives a record of atmospheric CO_2 over the past 420 000 years. The ice core data show fluctuations of ~ 100 ppmv in atmospheric CO_2.

Major Sources and Sinks for CO_2. Neglecting the well-known diurnal variations, factors that affect the atmospheric CO_2 abundance on a short timescale include removal by photosynthetic organisms on land and in the oceans, respiration of CO_2 by animals and plants, fossil fuel combustion, cement manufacturing and land-use changes, for example, deforestation. Cement manufacturing is a source of CO_2 because (i) calcination of limestone (ideally $CaCO_3$) releases CO_2 as lime (CaO) is made for cement and (ii) CO_2 is released from fossil fuels burned to heat the cement kilns. Total CO_2 emissions from cement manufacturing were about 829 Tg-C yr^{-1} in the year 2000, about 3% of the CO_2 emitted by fossil fuel combustion. Deforestation (*c.* 1980) released 1.8–4.7 Pg-C yr^{-1}, comparable to fossil fuel emissions (~ 5 Pg-C yr^{-1}) at the same time. Carbon dioxide is generally the second most abundant gas (after steam) in volcanic gases, but the global average volcanic emissions of CO_2 (~ 6.6 Tg-C yr^{-1}) are negligible in comparison to the fluxes listed above. For example, over 1850–1982, the average CO_2 emissions from fossil fuel combustion are about 1.3 Pg-C yr^{-1}. This long-term annual average flux is about 200 times larger than the volcanic flux (calculated for the 1800–1969 period). The natural fluxes are in balance, and the increase in atmospheric CO_2 is due to fossil fuel combustion, cement manufacturing and land use changes. Factors that affect atmospheric CO_2 on much longer timescales include rock weathering, burial, uplift and erosion. We do not discuss these processes here and focus on shorter timescale processes.

Photosynthetic carbon fixation (assimilation) by photoautotrophs removes CO_2 from the atmosphere and is represented by the net reaction:

$$CO_2 + H_2O \rightarrow CH_2O + O_2 \qquad (6.28)$$

The CH_2O represents complex carbohydrates. The reverse of Reaction (6.28) represents the release of CO_2 by fossil fuel combustion, animal and plant respiration and by oxidation of decaying plant and animal remains. The global average rate of carbon fixation, or gross primary production (GPP), is $\sim 120\,Pg\text{-}C\,yr^{-1}$. About half of the fixed carbon is respired by plants during the night and global average net primary production (NPP) is about half of gross primary production ($60\,Pg\text{-}C\,yr^{-1}$). However, the net primary production varies widely from one ecosystem to another. For example, agricultural land has a high NPP of 3.6 kg carbon per square meter per year ($kg\text{-}C\,m^{-2}\,yr^{-1}$) and tropical rain forests have a NPP of $2.3\,kg\text{-}C\,m^{-2}\,yr^{-1}$. In contrast, shrubs in semi-desert areas have a NPP of $0.14\,kg\text{-}C\,m^{-2}\,yr^{-1}$ and tundra and alpine ecosystems have NPP of $0.22\,kg\text{-}C\,m^{-2}\,yr^{-1}$. Land use changes such as building in previously forested areas lower NPP and are thus a CO_2 source. Conversely, replanting trees in urban areas increases NPP and is a CO_2 sink.

The gross primary productivity for oceanic biota is $50\,Pg\text{-}C\,yr^{-1}$, less than half of that of land plants. The net primary productivity of oceanic biota is only $5\,Pg\text{-}C\,yr^{-1}$, and most of the fixed carbon is lost by respiration ($45\,Pg\text{-}C\,yr^{-1}$). The total CO_2 fluxes into and out of the oceans are $80\,Pg\text{-}C\,yr^{-1}$, and the $30\,Pg\text{-}C\,yr^{-1}$ not consumed by oceanic biota in the surface waters is exchanged between the surface and deep ocean water. As mentioned earlier, the ocean presently consumes about 55% of all CO_2 emissions. The size of the oceanic CO_2 sink can be constrained with measurements of annual and seasonal changes in the atmospheric O_2 content. The seasonal changes are about 20 ppmv O_2, several times larger than the annual changes. For comparison, the total O_2 abundance of 20.946% equals 209 460 ppmv, so the annual and seasonal changes are very small in comparison to the total amount of O_2 in the atmosphere.

The global average residence time (t_{res}) for atmospheric CO_2 is given by:

$$t_{res}(CO_2) = \frac{\text{atmospheric } CO_2 \text{ mass (Pg)}}{\text{NPP flux (Pg yr}^{-1})} = \frac{823\,Pg}{60\,Pg\,yr^{-1}} = 13.7\,yr \qquad (6.29)$$

In other words, the net carbon fixation (productivity) by photosynthetic organisms would consume all atmospheric CO_2 in less than 14 years, unless it was replaced by another process. Likewise, the global average residence time for carbon in plants (phytomass) is also about 13.7 years. However, the average residence times for carbon in different types of plants and plant remains vary widely. For example, residence times for carbon in flowers, fruit and leaves are a few years or less, carbon in wood has about a 50-year residence time and carbon in humic material in soils has a residence time of thousands of years.

6.3.3 Nitrogen Cycle

6.3.3.1 Overview. The Scottish physicist and chemist Daniel Rutherford (1749–1819) discovered nitrogen in 1772. Carl Wilhelm Scheele (1742–1786), Joseph Priestley (1733–1804) and Henry Cavendish (1731–1810) discovered nitrogen independently at about the same time. It is the sixth most abundant element after H, He, O, C and Ne in the solar system with an abundance of 2.12×10^6 N atoms per 10^6 Si atoms and is the fourth most abundant element (after O, C and H) in biomass (Table 6.5). However, nitrogen is a trace element in the Earth. It has an abundance of $59 \, \mu g \, g^{-1}$ in Earth's crust, almost equal to that of cerium $(60 \, \mu g \, g^{-1})$ and comparable to that of nickel $(56 \, \mu g \, g^{-1})$ and zinc $(65 \, \mu g \, g^{-1})$.

Earth's crust contains 1.4×10^9 Tg nitrogen, which is about 26.5% of Earth's nitrogen inventory. Fixed (*i.e.*, chemically combined) nitrogen and organic nitrogen in sediments $(7.5 \times 10^8$ Tg) and fixed nitrogen in igneous rocks $(5.8 \times 10^8$ Tg) constitute the bulk of crustal nitrogen. The rest of crustal nitrogen is elemental nitrogen in igneous rocks $(7.0 \times 10^7$ Tg), fixed nitrogen in coal (200 000 Tg) and nitrate minerals in caliche (100 Tg).

Sodium nitrate (soda niter, $NaNO_3$) is the major nitrogen-bearing mineral in the caliche deposits in the Atacama Desert of Chile. Similar deposits of smaller size occur in other arid regions, such as Death Valley in California. Potassium nitrate (niter, KNO_3) occurs in small amounts in association with soda niter in the caliche deposits of Chile, Death Valley and elsewhere. Niter also occurs as efflorescence on organic-rich soils. Less common nitrates include darapskite, $Na_3(NO_3)(SO_4) \cdot H_2O$, found abundantly in some of the Chilean caliche deposits, and nitrocalcite, $Ca(NO_3) \cdot 4H_2O$, in the Death Valley niter deposits. The latter is rare and reported from only a few localities. Buddingtonite, NH_4Al-Si_3O_8, is ammonium feldspar found in several hydrothermal areas. It is less abundant than soda niter. The nitride minerals carlsbergite (CrN), nierite (α-Si_3N_4), osbornite (TiN) and sinoite (Si_2N_2O) occur in meteorites, but apparently not on Earth, at least in any abundance.

Nitrogen has two stable isotopes (^{14}N and ^{15}N) with atomic abundances of 99.634% (^{14}N) and 0.366% (^{15}N), giving an average atomic ratio of 272 ($^{14}N/^{15}N$) in air, which is the standard for nitrogen isotopic measurements. However, unlike carbon there is no clear division between the nitrogen isotopic composition of organic and inorganic material. The two stable nitrogen isotopes are separated industrially using the exchange reaction:

$$^{15}NO(g) + {}^{14}NO_3^-(aq) = {}^{14}NO(g) + {}^{15}NO_3^-(aq) \qquad (6.30)$$

This reaction may also account for ^{15}N-rich sediments from the KT boundary 65 million years ago. The impact of a large bolide (an asteroid or a comet) at that time led to formation of large amounts of nitric acid HNO_3 *via* combustion of atmospheric N_2 to NO and subsequent oxidation of the NO to nitric acid.

Nitrogen is an essential constituent of many organic compounds important for life. Amino acids are the most important example. These have the general formula $RCH(NH_2)COOH$ where R is H (glycine), CH_3 (alanine) or other organic radicals (other amino acids). Amino acids are the building blocks of proteins, and nitrogen makes up about 17% (by mass) of proteins. Nitrogen is an essential part of many other biochemical compounds. Pyrroles (nitrogen heterocyclic rings) coordinate to Mg in chlorophyll and with Fe in hemoglobin. Nitrogen occurs in adenine (A), cytosine (C), guanine (G), thymine (T) and uracil (U), which are the base pairs in nucleic acids (DNA and RNA). Nitrogen also occurs in alkaloids (caffeine, cocaine, nicotine), amines [*e.g.*, methylamine (CH_3NH_2)] and vitamins (riboflavin, thiamin, vitamin B_{12}). The occurrence of nitrogen in urea (H_2NCONH_2) is historically significant because urea was the first organic compound synthesized from inorganic reagents. The German chemist Friedrich Wöhler (1800–1882) made urea from ammonium cyanate (NH_4CNO) in 1828. Wöhler's synthesis proved that organic compounds can be made abiotically and that life is not essential for their formation.

Over 180 years ago, the German chemist Justus von Liebig (1803–1873) emphasized that nitrogen, along with potassium and phosphorus, is one of the three primary nutrients for plants. The importance of nitrogen for plant nutrition is so great that industrial production of N-bearing fertilizers is the largest source of fixed nitrogen on Earth and exceeds the natural fixation of nitrogen by living organisms and by abiotic processes such as natural fires and lightning.

Molecular nitrogen is the major gas in Earth's atmosphere ($\sim 78\%$ of dry air, see Table 5.1). The atmospheric N_2 abundance is 3.87×10^9 Tg and the atmosphere is the major nitrogen reservoir (73.1% of all N) on Earth. Nitrogen, unlike CO_2, is not very soluble in seawater, about $13.5 \, \text{cm}^3$ N_2 dissolves per liter of seawater at $15\,^{\circ}\text{C}$, the global mean surface temperature. Oceanic nitrogen is mainly dissolved N_2 ($16.5 \, \text{mg kg}^{-1}$) with smaller contributions from dissolved nitrate ($1.86 \, \text{mg kg}^{-1}$), N_2O, ammonium and other compounds. The oceans contain 2.3×10^7 Tg nitrogen (0.4% of Earth's nitrogen inventory). Most terrestrial biomass is phytomass and using a mass ratio of ~ 0.013 for N/C in combination with 2578 Pg-C in biomass gives about 33 500 Tg nitrogen in biomass. Thus, the three major nitrogen reservoirs on Earth are the atmosphere (73.1%), crust (26.5%) and oceans (0.4%). The nitrogen content of Earth's mantle and core are not well constrained, and are not considered in the inventory. Table 6.11 summarizes the major nitrogen reservoirs at Earth's surface.

Nitrous oxide (N_2O) is the second most abundant nitrogen gas in Earth's atmosphere with an abundance of ~ 320 ppbv (~ 1590 Tg-N) and a lifetime of about 120 years. As mentioned earlier, N_2O is a greenhouse gas that is about 300 times as efficient as CO_2 for global warming. This is due to the location and strength of its IR absorption bands. Total N_2O sources amount to $16.4 \, \text{Tg yr}^{-1}$, with emissions due to denitrifying bacteria in the oceans and soil comprising about 58% of the total. Biomass burning, industrial sources and emissions from agricultural activities make up the remaining 42%. The major sink for N_2O is transport into the stratosphere where UV sunlight photolyzes it, forming nitric oxide NO and an electronically excited oxygen atom. The NO then enters into the NO_x catalytic cycle for ozone destruction (Chapter 5). The stratospheric sink for nitrous oxide is about 12.6 Tg yr^{-1}, which is less than the source strength. Consequently, the atmospheric abundance of N_2O is increasing by about 0.8 ppbv yr^{-1} (0.25% per year) now. Atmospheric monitoring of N_2O in air and analyses of air

Table 6.11   Major nitrogen reservoirs at Earth's surface.

Reservoir	Pg (10^{12} kg) nitrogen	Notes
Atmosphere	3.87×10^6	Dominantly N_2, see Table 5.1
Biosphere	33.5	Using N/C $= 0.013$ for phytomass, see Tables 6.5 and 6.10
Hydrosphere	23 000	Dissolved N_2 and nitrate in oceans, see text
Lithosphere	1.4×10^6	Total N in crust, see text

trapped in ice cores show increases in the N_2O abundance from about 270 ppbv in the period 1000–1800 to 320 ppbv now.

Ammonia, nitric acid vapor and nitrogen oxides also occur in the troposphere but have significantly smaller abundances (see Table 5.1) and shorter lifetimes than N_2O. About $54 \, Tg \, yr^{-1}$ of NH_3 is emitted from animal and human excrement (54%), decaying organic matter in soils (28%), and anthropogenic and natural combustion processes (18%). Rainout is the major sink for ammonia. Some NH_3 reacts with SO_2 and HNO_3 to form ammonium sulfate, $(NH_4)_2SO_4$, and ammonium nitrate (NH_4NO_3) aerosols in the atmosphere.

Volcanoes originally outgassed N_2 into Earth's atmosphere, and volcanic outgassing of N_2 continues today. The estimated volcanism rate on Earth is about $3\,000 \, Tg$ lava per year. Steam dominates most volcanic gases, and the average steam/lava mass ratio is about 1%. Volcanic gases with N_2/Ar ratios higher than those for air (83.60) and air-saturated water (38.5) contain indigenous N_2. The N_2/steam mass ratio for volcanic gases containing indigenous N_2 is in the range 0.1–1%. Thus, the volcanic N_2 flux is about $0.03–0.3 \, Tg \, yr^{-1}$. A long-term average value for the volcanic nitrogen flux is the total amount of nitrogen at the surface of the Earth ($4.6 \times 10^9 \, Tg$ in the atmosphere, oceans and crust) divided by the age of the Earth ($4.55 \times 10^9 \, yr$). This calculation gives $1.0 \, Tg \, yr^{-1}$, which is probably an upper limit because the rate of volcanism has decreased over time. The calculated volcanic N_2 flux of $<1 \, Tg \, yr^{-1}$ is small in comparison to many of the abiotic, anthropogenic and biogenic fluxes in the nitrogen cycle.

Volcanic gases contain insignificant amounts of other N-bearing gases because thermodynamic equilibrium in the hot, low oxygen fugacity, volcanic gases favors N_2 over NH_3 or nitrogen oxides. Thus, the net thermochemical reactions proceed strongly toward nitrogen and steam:

$$4NH_3 + 3O_2 \rightarrow 2N_2 + 6H_2O(\text{steam}) \qquad (6.31)$$

$$2NO + 2H_2 \rightarrow N_2 + 2H_2O(\text{steam}) \qquad (6.32)$$

However, the mixing of hot volcanic gases with air produces some NO, which subsequently oxidizes to NO_2 and nitric acid HNO_3 vapor. The global flux of oxidized nitrogen emitted directly from volcanoes must be less than that of N_2 itself because only part of the volcanic N_2 is oxidized upon contact with air *via* the net thermochemical reaction:

$$N_2 + O_2 = 2NO \qquad (6.33)$$

The standard Gibbs free energy for this reaction (298–2500 K) is:

$$\Delta_r G_T^\circ = 182\,578 - 0.5669T \log T - 23.3503T \; \mathrm{J\,mol^{-1}} \qquad (6.34)$$

Heating of air by hot lavas (1300–1500 K) may produce additional NO by the same net reaction. Formation of NO from hot air occurs during anthropogenic and natural combustion processes, impacts of large bolides, lightning and thermonuclear explosions. On average, these abiotic processes contribute a negligible flux of fixed nitrogen to the nitrogen biogeochemical cycle, but they are important under some circumstances such as the Tunguska impact in 1908 in Siberia and the KT impact 65 million years ago. These impacts deposited large amounts of energy into the atmosphere, and produced NO.

Molecular nitrogen is unreactive and most living organisms (except those responsible for biological nitrogen fixation) cannot utilize N_2 directly. The conversion of atmospheric N_2 into chemically reactive nitrogen (nitrogen fixation) and the conversion of chemically reactive nitrogen back into atmospheric N_2 (denitrification) are the major input and output, respectively, for biospheric nitrogen. The fixed (biospheric) nitrogen undergoes several biologically mediated conversions that transform it into one form or another of chemically reactive nitrogen utilized by different organisms. Ammonia assimilation converts NH_3 and NH_4^+ into organic nitrogen in organisms. This is a minor pathway because most organisms utilize nitrate (assimilatory nitrate reduction) instead of getting energy from NH_3 or ammonium compounds.

Mineralization is the reverse process, which occurs during decomposition of organic matter. It converts organic nitrogen, which is mainly nitrogen in amino acids and proteins, back into NH_3 or NH_4^+. A schematic reaction involving glycine, the simplest amino acid, illustrates the mineralization process:

$$2CH_2NH_2COOH \text{ (glycine)} + 3O_2 = 4CO_2 + 2H_2O(\text{liq}) + 2NH_3 \qquad (6.35)$$

The standard Gibbs energy ($\Delta_r G_{298}^\circ$) at 25 °C for this reaction is -1472 kilojoules (kJ). About $15\,\mathrm{Tg\text{-}NH_3\,yr^{-1}}$ are emitted by mineralization in soils. Ammonia is very soluble in water ($529\,\mathrm{g\text{-}NH_3\,l^{-1}}$ at 20 °C) and rains out of the atmosphere on a timescale comparable to the water vapor lifetime of ~ 10 days.

Nitrification is the oxidation of ammonia or ammonium to nitrite and the subsequent oxidation of nitrite to nitrate. In 1877, Müntz and Schlössing demonstrated that bacteria were responsible for nitrification. A few years later, in 1879, the English agricultural chemist Robert

Warrington (1838–1907) obtained liquid cultures of nitrifying bacteria that oxidized ammonia to nitrite, but did not produce nitrate. In 1881, Warrington obtained cultures that oxidized nitrite to nitrate, but which could not oxidize ammonia to nitrite. The pioneering Russian micro-biologist Sergei Winogradsky (1856–1953) isolated and identified the nitrifying bacteria *Nitrosomonas* and *Nitrosococcus* in the early 1890s. Bacteria of the genus *Nitrosomonas* oxidize NH_3 to nitrite NO_2^- *via* the net reaction:

$$2NH_3 + 3O_2 = 2NO_2^- (aq) + 2H_2O (liq) + 2H^+ (aq) \qquad (6.36)$$

The standard Gibbs energy change $(\Delta_r G^\circ)$ for this reaction at 298 K is -506 kJ. Bacteria of the genus *Nitrobacter* oxidize nitrite to nitrate *via*:

$$2NO_2^- (aq) + O_2 = 2NO_3^- (aq) \; \Delta_r G_{298}^\circ = -158 \, kJ \qquad (6.37)$$

Winogradsky proved that both types of bacteria are obligatory autotrophs that use the energy released from nitrification to drive their metabolisms.

6.3.3.2 Denitrification. At present, the major sources of N_2 are de-nitrifying bacteria in soils and oceans, which convert nitrate (NO_3^-) compounds back into N_2. The conversion proceeds *via* a series of intermediate nitrogen compounds – nitrite (NO_2^-), nitric oxide (NO) and nitrous oxide (N_2O). The conversion of N_2O into N_2 is incom-plete and denitrification emits some N_2O (molar N_2/N_2O ratios of ~ 15). Most denitrifying bacteria are chemoheterotrophs, *i.e.*, they use chemical energy to metabolize organic matter as a food source. Many bacteria are capable of denitrification and several common genera are *Pseudomonas*, *Micrococcus* and *Thiobacillus*. Denitrifying bacteria are facultative anaerobes and prefer O_2-poor conditions in waterlogged, organic-rich soils near neutral pH (6–8). A schematic, net denitrifying reaction, which produces N_2 and releases 2 385 kJ energy per mole of glucose consumed, is:

$$\begin{aligned} 5C_6H_{12}O_6 \, (glucose) &+ 24NO_3^- (aq) \\ = 30CO_2(g) &+ 18H_2O (liq) + 24OH^- (aq) + 12N_2(g) \end{aligned} \qquad (6.38)$$

A schematic denitrifying reaction producing N_2O and releasing 2280 kJ energy per mole glucose consumed is:

$$C_6H_{12}O_6 + 6NO_3^- = 6CO_2 + 3H_2O (liq) + 6OH^- (aq) + 3N_2O(g) \qquad (6.39)$$

Estimates of the total denitrification flux range from 90 to 240 Tg yr^{-1} and are dominated by denitrification in soils. Taking a mean value of 165 Tg yr^{-1} gives a replacement time of:

$$\frac{3.87 \times 10^9 \, \text{Tg N}_2}{165 \, \text{Tg yr}^{-1}} \sim 23.5 \, \text{million years} \tag{6.40}$$

for all N$_2$ in Earth's atmosphere. However, this ranges from 43 to 16 million years for the range of denitrification fluxes given above. A small part of the total flux (10%) is due to the use of artificial fertilizers. In the absence of any anthropogenic effects, the replacement time for atmospheric N$_2$ would thus be about 10% larger.

6.3.3.3 Nitrogen Fixation. The major sinks for N$_2$ are biological nitrogen fixation by bacteria in soils and oceans, industrial fixation of nitrogen by fertilizer production, natural fires, anthropogenic combustion processes and lightning. As mentioned earlier, industrial production of N-bearing fertilizers is the largest source of fixed nitrogen on Earth and exceeds the natural fixation of nitrogen by living organisms, natural fires and by lightning. According to industry statistics, in 2006 fertilizer production consumed 205 Tg of nitrogen. Ammonia fertilizer accounted for 122 Tg, urea fertilizer for 62 Tg, with the remaining 21 Tg nitrogen used in ammonium nitrate, ammonium sulfate and calcium ammonium nitrate fertilizers. Fertilizer production at this rate (205 Tg yr^{-1}) would consume all atmospheric N$_2$ in about 19 million years. Prior to development of ammonia synthesis around 1909 (the Bosch–Haber process), guano and the Chilean caliche deposits (the major source) were mined to produce nitrate fertilizers (about 0.360 Tg of fixed nitrogen in 2.25 Tg of caliche in 1910).

Humanity's understanding of biological nitrogen fixation dates back 170 years. In 1838, the French chemist Jean-Baptiste Boussingault (1802–1887) demonstrated that legumes restore nitrogen to the soil. He grew clover and peas in unfertilized, sterilized sand and showed the sand gained nitrogen as a result. However, the bacterial fixation of nitrogen remained unknown until 1887. At that time, Hermann Hellriegel (1831–1895) and Hermann Wilfarth (1853–1904) showed that bacteria in root nodules of legumes fixed N$_2$ from the atmosphere. One year later, the Dutch microbiologist Martinus Beijerinck (1851–1931) isolated *Rhizobium* from the nodules of pea plants.

The three major types of nitrogen-fixing organisms are symbiotic bacteria, asymbiotic bacteria and cyanobacteria. Symbiotic bacteria of genus *Rhizobium* occur in nodules on roots of leguminous plants (alfafa, chickpeas, clover, lupins and soybeans) and are responsible for much of the nitrogen fixation on land. Asymbiotic bacteria of genus *Azobacter* fix nitrogen under aerobic conditions and asymbiotic bacteria of genus *Clostridium* fix nitrogen under anaerobic conditions. The asymbiotic bacteria occur in organic-rich soils. The *Nostoc* cyanobacteria (sometimes called blue-green algae) fix nitrogen in aquatic systems. The nitrogenase enzyme carries out nitrogen fixation in these organisms. This generally contains two metal-bearing proteins, one a Mo-Fe protein that binds the N_2, the other an Fe protein that provides the electrons for reducing N_2 to ammonia. *Azobacter* uses Fe-, Mo- and V-bearing nitrogenase enzymes.

Anthropogenic and biological nitrogen fixation is energy intensive. Ammonia synthesis from the elements by the Bosch–Haber process requires high temperature (450 °C) and high pressure (100 bar). In contrast, biological fixation of 14 g nitrogen by nitrogenase occurs at ambient conditions but requires the energy equivalent to metabolizing 1 kg of glucose ($C_6H_{12}O_6$). Symbiotic bacteria such as *Rhizobium* get this energy from plant respiration, but asymbiotic bacteria such as *Azobacter* and *Clostridium* need to get this energy by metabolizing organic compounds in soils.

Liebig first suggested nitrogen fixation by lightning in 1827. Abiotic fixation of nitrogen by lightning and other high-temperature processes involves conversion of N_2 into nitric oxide NO *via* the net thermochemical reaction:

$$N_2 + O_2 = 2NO \tag{6.33}$$

We discussed this reaction earlier in connection with NO production by hot lavas in contact with air. Heating of air to high temperatures forms NO *via* a sequence of elementary reactions proposed by the Russian physicist Yakov B. Zel'dovich (1914–1987) and named after him:

$$O + N_2 \rightarrow N + NO \tag{6.41}$$

$$N + O_2 \rightarrow O + NO \tag{6.42}$$

Net reaction: $\quad N_2 + O_2 \rightarrow NO + NO \tag{6.33}$

Thermal dissociation of O_2 produces the oxygen atoms in Reaction (6.41):

$$O_2 + M \rightarrow O + O + M \tag{6.43}$$

Some of these O atoms recombine to regenerate O_2:

$$O + O + M \rightarrow O_2 + M \qquad (6.44)$$

The rest of the O atoms react with N_2 to form NO. Less thermal dissociation of N_2 occurs because the triple bond in N_2 is about twice as strong as the double bond in O_2 (the bond energies at 298 K are 945 kJ for N_2 *versus* 498 kJ for O_2).

The biological nitrogen fixation rate is $\sim 150\,\text{Tg yr}^{-1}$; estimates range from 99 to $170\,\text{Tg yr}^{-1}$. At the nominal rate of $150\,\text{Tg yr}^{-1}$, biological nitrogen fixation would consume all atmospheric N_2 in about 26 million years. The abiotic nitrogen fixation rates due to lightning and biomass burning are smaller and are about $0.5–10\,\text{Tg yr}^{-1}$ (lightning) and about $12–15\,\text{Tg yr}^{-1}$ for biomass burning (natural and anthropogenic). Taken together, industrial, biological and abiotic nitrogen fixation consume about $370\,\text{Tg yr}^{-1}$ nitrogen, giving a lifetime of about 10 million years for N_2 in Earth's atmosphere.

If the biological and industrial sinks were removed, while lightning and natural fires continued at their present rates ($\sim 5\,\text{Tg-NO yr}^{-1}$), the lifetime for N_2 would increase dramatically to ~ 770 million years. However, the Earth's atmosphere would contain much less O_2 in the absence of biological oxygen production. Nitric oxide production by lightning in a $N_2–CO_2–O_2$ atmosphere would be much less efficient, and natural fires may not occur in an O_2-poor atmosphere (*e.g.*, like that of Mars). The lifetime of N_2 on a lifeless Earth-like planet would be several billion years.

As mentioned earlier, industrial fixation of nitrogen exceeds natural fixation by biological and abiotic processes. This is causing an imbalance between fixation and denitrification and an increase in the runoff of nitrate and other nitrogen compounds into rivers and streams. This creates algal blooms that deplete the available dissolved oxygen, killing fish and other O_2-breathing organisms in the water (the process of eutrophication). The runoff of nitrate and phosphate caused eutrophication of Lake Erie in the late 1960s, which is now mitigated by limits on the runoff of these compounds. At present, nitrate and phosphate runoff from Midwest farmlands flows down the Mississippi River into the Gulf of Mexico, creating a dead zone at the river's mouth.

6.3.4 Sulfur Cycle

Sulfur (brimstone) is one of the elements know to the ancients and according to the Bible rained down upon Sodom and Gomorrah during their destruction. Sulfur is the tenth most abundant element (after H, He, O, C, Ne, N, Mg, Si, Fe) in the solar system (421 200 atoms per 10^6 Si atoms) and the 14th most abundant element in the Earth's crust ($697\,\mu g\,g^{-1}$, see Table 6.3).

Sulfur has four stable isotopes – ^{32}S (95.02%), ^{33}S (0.75%), ^{34}S (4.21%) and ^{36}S (0.02%). The troilite (FeS) in the Canyon Diablo iron meteorite, which made Meteor Crater in Arizona, is the standard for sulfur isotopic measurements. Sulfur, like oxygen, exhibits mass-in-dependent isotopic fractionations in some chemical reactions (*e.g.*, SO_2 photolysis). These fractionations provide evidence for the growth of O_2 in Earth's atmosphere over time.

Table 6.12 is an inventory of sulfur in the bulk silicate Earth. This includes the mantle, crust, oceans and atmosphere but not the Earth's core, which may be the largest sulfur reservoir on Earth. As discussed below, the amount of sulfur in Earth's core is inferred from seismic data and cosmochemical arguments and is very uncertain. The mantle is the largest sulfur reservoir in Table 6.12 with about 97% of the total. However, the sulfur content of Earth's mantle is uncertain ($124 \pm 90\,\mu g\,g^{-1}$) because of wide variations in the sulfur content of different types of mantle rocks. The next largest reservoir is the crust (2.3% of total S), followed by seawater (0.3%). The amounts of sulfur in the biosphere and atmosphere are negligible for Earth's inventory, but are important for atmospheric chemistry and biochemistry. The sulfur content of the atmosphere is about 690 pptv (dominantly in reduced sulfur gases), which is equivalent to 0.65 ng of sulfur per gram of atmosphere.

Most sulfur at the surface of the Earth is in sedimentary rocks in the crust and in seawater. The three major reservoirs are pyrite (FeS_2, the most abundant sulfide mineral), gypsum ($CaSO_4 \cdot 2H_2O$, the most abundant sulfate mineral) and seawater sulfate (2712 mg sulfate per kg seawater, equal to 898 mg sulfur per kg seawater). The amounts of pyrite (5.9×10^{18} kg-S), gypsum (4.7×10^{18} kg-S) and seawater sulfate

Table 6.12 Major sulfur reservoirs in the bulk silicate Earth.

Reservoir	*Sulfur content*	*Sulfur (g)*	*% of Total sulfur*
Atmosphere	$590\,\text{pptv} = 0.65\,\text{ng}\,g^{-1}$	3.35×10^{12}	6.57×10^{-10}
Biosphere	0.071%	4.69×10^{15}	9.20×10^{-7}
Continental crust	$697\,\mu g\,g^{-1}$	1.06×10^{22}	2.08
Oceanic crust	$123\,\mu g\,g^{-1}$	1.04×10^{21}	0.20
Seawater	$898\,\mu g\,g^{-1}$	1.49×10^{21}	0.29
Mantle	$124\,\mu g\,g^{-1}$	4.97×10^{23}	97.43
Total		5.10×10^{23}	100.00

$(1.5 \times 10^{18}\,\text{kg-S})$ are coupled by isotopic mass balance. Gypsum $(\delta^{34}S = 17‰)$ and seawater $(\delta^{34}S = 20‰)$ are isotopically heavy. Pyrite is isotopically light $(\delta^{34}S = -18‰)$ because sulfate-reducing bacteria preferentially use ^{32}S. The total number of ^{32}S and ^{34}S atoms on Earth is fixed, and the relative sizes and isotopic compositions of the gypsum, pyrite and seawater sulfate reservoirs are such that the average sulfur isotopic composition of Earth is slightly heavier $(\delta^{34}S = 0.3‰)$ than sulfur in troilite from the Canyon Diablo iron meteorite $(\delta^{34}S = 0‰$ by definition).

However, sulfur occurs in numerous compounds spanning six oxidation states (-2 in sulfides, -1 in disulfides, 0 in elemental sulfur, $+2$ in thiosulfate, $+4$ in SO_2 and sulfites and $+6$ in sulfate). Hydrogen sulfide occurs in volcanic gases, natural gas, petroleum and biogenic emissions; microorganisms emit dimethyl sulfide, $(CH_3)_2S$, and many metal sulfides [*e.g.*, chalcocite (Cu_2S), galena (PbS) and sphalerite (ZnS)] occur in the crust. Pyrite and its polymorph marcasite are the most common sulfides on Earth and occur in many environments. For example, coal contains 1–2% sulfur as pyrite impurities, and oil shale contains about 0.7% sulfur, mainly as pyrite. Elemental sulfur occurs around volcanic vents, in the cap rock of salt domes and in the sedimentary evaporite deposits of Southeast Poland. Volcanic gases emit considerable SO_2, which is generally the most abundant S-bearing species in volcanic gases.

In general, the composition of volcanic gases varies from one volcano to another and varies with time at a given volcano. However, with a few exceptions steam is the most abundant species (on a volumetric or molar basis) in volcanic gases followed by CO_2, SO_2, H_2, H_2S, CO, OCS, HCl, HF, S_2 and other less abundant gases. The global average annual emissions of sub-aerial volcanic gases are H_2O (30 Tg), CO_2 (6.6 Tg), SO_2 (7.5–10.5 Tg), H_2S (1.5–37.1 Tg) and OCS (0.094–320 Gg). Terrestrial volcanic gases are in chemical equilibrium and SO_2, H_2S, OCS and S_2 vapor are interconverted *via* net thermochemical reactions such as:

$$SO_2 + 3H_2 = H_2S + 2H_2O \qquad (6.45)$$

$$H_2O + OCS = H_2S + CO_2 \qquad (6.46)$$

$$2SO_2 + 2CO + 2H_2 = 2CO_2 + S_2 + 2H_2O \qquad (6.47)$$

The volcanic SO_2 flux is primarily emitted into Earth's troposphere where it is removed on short time scales by rainout. However, the volcanic fluxes of SO_2 and other sulfur gases are smaller than the anthropogenic and biogenic fluxes of the same gases. In general, S-bearing gases and aerosols are important for climate on Earth and Venus and possibly on Mars in its past.

Troilite (FeS) and Fe metal are common minerals in chondritic meteorites, which are thought to be the building blocks of the Earth and other terrestrial planets. Thus, formation of Earth's core plausibly started with melting of a eutectic mixture of iron and troilite (FeS) at ~ 1260 K, about 550 degrees below the melting point of pure iron. The fluidity of Earth's outer core and thus the existence of Earth's magnetic field depend, in part, upon the freezing point depression due to sulfur (and other light elements) dissolved in the core. Seismic data and laboratory measurements of the physical properties of Fe alloys at high temperatures and pressures suggest 9–12% S in Earth's core. Cosmochemical models suggest 1.7–4.1% sulfur in the core and imply that at least one other light element is needed in the core (see discussion in Chapter 4).

Sulfur is critically important for life on Earth. It is the seventh most abundant element in the average human body (0.20%) and is the ninth most abundant element in plants (0.07%). Sulfur is more abundant in humans (and other animals) than in plants because it mainly occurs in proteins, and plants contain less protein than animals. Sulfur occurs in structural proteins found in fingernails, fur and hair. The sulfur-bearing amino acids are methionine [$H_3CS(CH_2)_2CHNH_2COOH$], cysteine ($HSCH_2CHNH_2COOH$) and cystine ($SCH_2CHNH_2COOH)_2$. Their disulfide linkages provide three-dimensional structure and mechanical strength for the proteins in fingernails, fur and hair.

The most abundant sulfur gases in Earth's troposphere are reduced sulfur gases such as carbonyl sulfide (OCS, 500 pptv), H_2S (30–100 pptv), dimethyl sulfide [DMS, $(CH_3)_2S$, 5–60 pptv] and CS_2 (15 pptv). Other important reduced sulfur gases in Earth's atmosphere include methyl sulfide (CH_3SH), and dimethyl disulfide (DMDS, CH_3-S-S-CH_3). Sulfur dioxide (20–90 pptv), which has natural and anthropogenic sources, is the major oxidized sulfur gas.

In general, biogenic emissions are sources of reduced sulfur gases, and anthropogenic emissions are sources of oxidized sulfur gases (SO_2, SO_3). The exceptions to this are SO_2 emissions from biomass burning (7 Tg yr^{-1}) and volcanic eruptions (7.5–10.5 Tg yr^{-1}). However, these fluxes are much smaller than the anthropogenic sources (~ 100 Tg yr^{-1}). The annual mean volcanic SO_2 flux is highly variable because of the episodic nature of volcanism. For example, the 1815 eruption of the Tambora volcano in Indonesia is estimated to have produced 150 Tg SO_2. But, on average, the volcanic flux is about ten times smaller than the anthropogenic flux.

The major contributions to anthropogenic SO_2 emissions are coal combustion (60%), petroleum combustion (28%), metal smelting (10.4%) and the paper and pulp industry (1.6%). Anthropogenic emissions are increasing with time as industrial activity increases and are also concentrated in highly industrialized regions in the northern hemisphere ($>90\%$ of emissions).

Biogenic emissions from oceanic and terrestrial biota are the major sources of OCS and the other reduced sulfur gases. Volcanic emissions are negligible by comparison because the high temperature equilibria in volcanic gases favor SO_2. For example, one estimate gives the following major sources (in units of $Tg\,yr^{-1}$) for OCS: the oceans (0.4), oxidation of CS_2 (0.20), biomass burning (0.18), fossil fuel combustion (0.10) and soils (0.06).

Carbonyl sulfide has an atmospheric lifetime of about 2–17 years. Some of it is destroyed by reaction with OH (slow) and uptake by vegetation. However, transport into the stratosphere is the major sink. Once in the stratosphere OCS is destroyed by UV sunlight photo-oxidation to SO_2:

$$OCS + h\nu \rightarrow CO + S\,(\lambda < 250\,nm) \tag{6.48}$$

$$S + O_2 \rightarrow SO + O\,(k_{298} = 2 \times 10^{-12}\,cm^3\,s^{-1}) \tag{6.49}$$

$$SO + O_2 \rightarrow SO_2 + O\,(k_{298} = 9 \times 10^{-18}\,cm^3\,s^{-1}) \tag{6.50}$$

$$SO + O_3 \rightarrow SO_2 + O_2\,(k_{298} = 6 \times 10^{-14}\,cm^3\,s^{-1}) \tag{6.51}$$

Massive volcanic eruptions, which are much more energetic than normal, are the second major source of stratospheric sulfur dioxide. Three such massive eruptions are the 1991 eruption of Mount Pinatubo in the Philippines, the 1982 eruption of El Chichón in Mexico and the 1882 eruption of Krakatoa in Indonesia. Subsequently, the SO_2 is converted into sulfuric acid aerosols that form the Junge layer at 20 km in the stratosphere. The reactions involved are:

$$OH + SO_2 + M \rightarrow HSO_3 + M \tag{6.52}$$

$$HSO_3 + O_2 \rightarrow HO_2 + SO_3 \tag{6.53}$$

$$SO_3 + H_2O \rightarrow H_2SO_4 \tag{6.54}$$

$$H_2SO_4(gas) = H_2SO_4\,(liquid) \tag{6.55}$$

The M in Reaction (6.52) can be any other gas atom or molecule. The third body acts as a collision partner to remove excess energy and does not take part in the chemical reaction. The sulfuric acid aerosol droplets in the Junge layer are about 75 mass% sulfuric acid and 25 mass% water.

In contrast to the massive eruptions mentioned above, typical volcanic eruptions, such as Kilauea in Hawaii or Etna in Italy, emit SO_2 into Earth's troposphere. Anthropogenic processes, such as high-temperature combustion of coal, sulfur-bearing oil and industrial processes, also emit SO_2 into the troposphere. Subsequently, gas phase and gas–liquid reactions oxidize tropospheric SO_2 to sulfurous (H_2SO_3) and sulfuric acids, which are removed as acid rain within a short time.

6.3.5   Phosphorus and Halogen Cycles

6.3.5.1   Phosphorus. Phosphorus is a minor element in the solar system with an abundance of only 8300 P atoms per 10^6 Si atoms, but it is the sixth most abundant element in the human body (1.11%), the eleventh most abundant element in phytomass ($52 \mu g\,g^{-1}$) and the twelfth most abundant element in Earth's crust ($757 \mu g\,g^{-1}$).

Table 6.13 is an inventory for phosphorus in the bulk silicate Earth. The mantle contains about $60 \mu g\,g^{-1}$ P, and contains 95.4% of total phosphorus. The crust is the second largest reservoir with 4.6% of total P, and all other reservoirs are negligible in comparison. With the exception of P in soils (the top 60 cm of continental landmasses), phosphorus in the crust and mantle is effectively immobile (190 million year residence time). Most P cycling occurs between the smaller reservoirs immediately at Earth's surface, and we consider these next.

The atmosphere is the smallest phosphorus reservoir and mainly contains phosphate-bearing aerosols and dust. However, phosphine gas (PH_3) is present in the urban troposphere at levels up to $157 \, ng\,m^{-3}$ and at $1 \, ng\,m^{-3}$ or less in the nonurban troposphere. Anaerobic

Table 6.13   Major phosphorus reservoirs in the bulk silicate Earth.

Reservoir	*Tg (10^9 kg) Phosphorus*	*Notes*
Atmosphere	28	P in aerosols & dust and PH_3 gas
Biosphere	73	Oceanic biota, see text
	1 110	Terrestrial biota, see text
Hydrosphere	2 710	Surface layer, see text
	87 100	Deep oceans, see text
Lithosphere	1.15×10^{10}	$757 \mu g\,g^{-1}$ P in continental crust
	2.40×10^{11}	$60 \mu g\,g^{-1}$ P in mantle

environments such as landfills and marshes and industrial waste gases are apparently PH_3 sources. Phosphine is probably destroyed by oxidation to phosphate, and lost by rainout and dry deposition.

Oceanic and terrestrial biomass is the next largest P reservoir. The P content of oceanic biota is computed taking a C/P molar ratio of 106, from the average C/N/P composition of 106 : 16 : 1 in oceanic biota (Redfield ratio). The P content of terrestrial biota is 0.052% (Table 6.5). The amount of P in seawater varies with depth due to uptake by biota in the surface mixed layer and release back into seawater from the sinking detritus. Chemical equilibria dependent on the pH, temperature and concentrations of more abundant elements control the speciation of P in seawater. At 20 °C, pH 8 and average salinity (34.8‰), the most abundant forms of dissolved P are $MgHPO_4^0$, HPO_4^{2-} and $NaHPO_4^-$.

Phosphorus in the crust occurs primarily in apatite, orthophosphate minerals and sedimentary deposits of phosphorite rock, which is a F-rich apatite solid solution, $Ca_5(PO_4)_3(F,OH,CO_3)$. Fluroapatite is also the major fluorine-bearing mineral in the terrestrial crust. In contrast chlorapatite is much less common, and is not a major Cl-bearing mineral. Less abundant phosphate minerals include whitlockite [$Ca_3(PO_4)_2$], found in granite pegmatites, monazite [$(Ce,La)PO_4$], which can also contain small amounts of U and Th, xenotime (YPO_4), which can also contain small amounts of Ce, La and Th, and berlinite ($AlPO_4$), which controls availability of phosphorus in acidic soils. Phosphorus is an important fertilizer (now obtained by mining phosphorite rock in Florida and other locales, formerly from mining guano deposits) and an essential nutrient.

As mentioned above, P is the sixth most abundant element (1.11%) in the human body. An apatite solid-solution $Ca_5(PO_4)_3X$ dominated by hydroxyapatite (X = OH, 75.0%) and carbonatoapatite (X = CO_3, 12.1%) with smaller amounts of chlorapatite (X = Cl, 4.4%) and fluorapatite (X = F, 0.67%) is a major constituent of tooth enamel. Calcium phosphates are key constituents of bones, and organo-phosphates are part of nucleic acids DNA and RNA, the molecules ADP and ATP, which supply energy in metabolic processes, and phospholipids in cellular membranes. The phosphate minerals in bones and teeth do not dissolve away because their solubility in water is very small. For example, $Ca_3(PO_4)_2$ solubility in pure water (pH 7) at 25 °C is about 27 µmol (0.008 g) per liter.

Likewise, the apatite and phosphate minerals in soil and the crust have very small solubilities in natural waters. Their solubilities depend upon the pH, with berlinite ($AlPO_4$) limiting the solubility of phosphate in acidic environments and apatite limiting the solubility in basic environments.

The major fluxes in the global phosphorus cycle are those involving productivity on land and in the oceans. The P flux from soils to terrestrial biota is $64\,\text{Tg-P}\,\text{yr}^{-1}$. The amount of P in soils is roughly $150\,000\,\text{Tg}$, so the phosphorus removed annually by plants and trees corresponds to about a 2300-year lifetime for P in soils. The death and decay of plants and trees returns P to soils at the same rate ($64\,\text{Tg-P}\,\text{yr}^{-1}$) as it is extracted by terrestrial biomass. This return flux corresponds to a P residence time of about 17 years in terrestrial biomass. The mining of phosphorite rock for production and application of phosphate fertilizers is an additional input to soils of about $13\,\text{Tg}\,\text{P}$ per year.

The uptake of P from the surface oceans by marine biota is about $1040\,\text{Tg-P}\,\text{yr}^{-1}$, and corresponds to a residence time of 2.6 years for P dissolved in surface seawater. Phosphorus uptake by marine biota occurs faster than that by plants and trees because the residence time of P in marine biomass is much shorter (26 days) than the residence time of 17 years in terrestrial biomass. However, not all the phosphorus taken up by marine biota is returned to the surface oceans because the continual sedimentation of dead organisms gives a flux of about 4% of that assimilated by biota ($42\,\text{Tg-P}\,\text{yr}^{-1}$) to the deep ocean. The downward circulation of surface water also carries $18\,\text{Tg-P}\,\text{yr}^{-1}$ into the deep ocean. The upward circulation of deep ocean water gives $58\,\text{Tg-P}\,\text{yr}^{-1}$ to the surface ocean. This upward flux corresponds to a residence time of 1500 years for P in deep ocean water. However, the upward flux does not balance the downward fluxes. The imbalance, which is about 5% of the downward flux of dead organisms, is the amount of phosphorus incorporated into marine sediments ($\sim 2\,\text{Tg-P}\,\text{yr}^{-1}$).

The oceans are essentially a closed system with respect to phosphorus, except for riverine transport from the continents ($1.7\,\text{Tg-P}\,\text{yr}^{-1}$, mainly as solid particulates), loss in sea spray ($0.3\,\text{Tg-P}\,\text{yr}^{-1}$) and deposition of atmospheric particulates ($0.5\,\text{Tg-P}\,\text{yr}^{-1}$).

6.3.5.2   Halogens (F, Cl, Br, I)

Fluorine.   Fluorine is a trace element in the solar system with an abundance of only $804\,\text{F}$ atoms per 10^6 Si atoms and is the 16th most abundant element in the crust ($525\,\mu\text{g}\,\text{g}^{-1}$). The major F-bearing minerals are fluorite (CaF_2), also known as fluorspar, and fluorapatite ($Ca_5(PO_4)_3F$), which is widely dispersed in rocks and contains most of the fluorine in the crust. There are several less common fluoride minerals, including cryolite (Na_3AlF_6), found mainly in western Greenland, topaz [$Al_2SiO_4(OH,F)_2$], sellaite (MgF_2), villiaumite (NaF) and bastnaesite (REE, Y)CO_3F, which is a source of yttrium and the lanthanide or rare earth elements (REE). Large bastnaesite deposits occur in the

Sierra Nevada Mountains in California, in China and in Inner Mongolia. Fluoride minerals are generally less soluble than chlorides, and seawater contains only $1.3\,\mu g\,g^{-1}$ F *versus* $19\,353\,\mu g\,g^{-1}$ Cl. The oceans contain a negligible amount of Earth's total fluorine (0.003%), which is mainly found in the mantle (89%) and crust (11%). Fluorine is also a biologically important element with an abundance of $37\,\mu g\,g^{-1}$ in the human body. It is primarily found in teeth (0.025% in tooth enamel) and bones. In contrast, phytomass contains only about $3.5\,\mu g\,g^{-1}$ fluorine.

As discussed in Chapter 5, the atmosphere contains several F-bearing gases, most of which are anthropogenic (Table 5.2). The total F atom mixing ratio is 2466 parts per trillion by volume (pptv, equivalent to $\sim 8.3\,Tg$-F), dominated by CF_2Cl_2 (533 pptv), $CFCl_3$ (268 pptv), $CHClF_2$ (132 pptv) and CF_4 (80 pptv). Hydrogen fluoride (27 pptv) is the only natural F-bearing gas in the atmosphere, but most of the observed HF comes from industrial emissions and downward mixing from the stratosphere where photolytic destruction of chlorofluorocarbon (CFC) gases forms HF. The residence time for HF in Earth's troposphere is about 5.5 days because it is removed by rainout. The global average HF volcanic flux is in the range of 0.06–$6\,Tg$-F yr^{-1}, but because of its short atmospheric lifetime none of the volcanic HF enters the stratosphere (unless directly injected there by very powerful volcanic eruptions).

The CFC and hydrochlorofluorocarbon (HCFC) gases are destroyed by UV photolysis in the stratosphere (Chapter 5). The HCFC gases are also susceptible to attack by OH radicals in the troposphere, *e.g.*, HCFC-123 ($CHCl_2CF_3$), which replaces CFC-11, is destroyed *via*:

$$OH + CHCl_2CF_3 \rightarrow CCl_2CF_3 + H_2O \ (k = 3.6 \times 10^{-14}\,cm^3\,s^{-1}) \qquad (6.56)$$

However, perfluorocarbon (PFC) gases such as CF_4 (perfluoromethane) and C_2F_6 (perfluoroethane) are destroyed by ion–molecule chemistry in the ionosphere. Both of these gases react rapidly with O^+ ions, for example:

$$CF_4 + O^+ \rightarrow CF_3^+ + FO \ (k = 1.4 \times 10^{-9}\,cm^3\,s^{-1}\,at\,300\,K) \qquad (6.57)$$

$$C_2F_6 + O^+ \rightarrow CF_3^+ + CF_3O \ (k = 1.1 \times 10^{-9}\,cm^3\,s^{-1}\,at\,300\,K) \qquad (6.58)$$

These reactions give minimum lifetimes of 330 000 years (for CF_4) and 420 000 years (for C_2F_6) in comparison to lifetimes greater than 10^6 years from neutral–neutral chemistry or photolysis.

Chlorine. Chlorine is a minor element in the solar system with an abundance of 5170 Cl atoms per 10^6 Si atoms. It is the 17th most abundant element in the crust (472 $\mu g\,g^{-1}$), and major Cl-bearing minerals include halite (NaCl), sylvite (KCl), carnallite (KMgCl$_3 \cdot$ 6H$_2$O) and apatite [Ca$_5$(PO$_4$)$_3$(F,Cl,OH)]. Chlorine is an incompatible element and is concentrated in seawater. It is the second most abundant dissolved element and most abundant anion (1.9353% by mass) in the oceans. In fact, about 75% of Earth's chlorine resides in seawater, with the remainder in the crust (20%) and mantle (5%). River water contains less chlorine (7.8 mg kg^{-1} = 7.8 $\mu g\,g^{-1}$), but it is still the third most abundant anion after bicarbonate and sulfate. If rivers contain much less Cl than seawater, why are the oceans salty?

The reason is that the rivers carry dissolved salts and particulates from continental weathering into the oceans and that these species remain there for much longer times than water, which is evaporated and rained out globally. As mentioned earlier, the residence times for Cl and other conservative elements in seawater are much longer than that of ocean water itself, *e.g.*, Cl (120 Ma), Br (100 Ma), F (0.5 Ma) *versus* 34 250 years for water. However, evaporation of river water gives alkaline water rich in dissolved NaHCO$_3$, and chemical reactions between seawater, hydrothermal vents, sediments and submarine volcanoes are required to adjust the concentrations of dissolved salts.

Phytomass contains about 500 $\mu g\,g^{-1}$ Cl (Table 6.5), and the human body contains 0.14% Cl. Chlorine is an essential nutrient found primarily as Cl$^-$ in fluids in humans, in tooth enamel (0.3%) and bones, and is concentrated in chloroplasts in plants.

Table 5.2 lists the Cl-bearing gases in Earth's atmosphere, which has a total Cl atom mixing ratio of 3665 pptv (equivalent to 23 Tg-Cl). Methyl chloride (CH$_3$Cl), emitted by oceanic biota and by biomass burning, is the major Cl-bearing gas (620 pptv). Most of the other important Cl-bearing gases are anthropogenic, *e.g.*, CF$_2$Cl$_2$ (533 pptv), CFCl$_3$ (268 pptv), CHClF$_2$ (132 pptv) and CCl$_4$ (102 pptv). Hydrogen chloride (100 pptv) is the second most abundant Cl-bearing gas with natural sources (sea salt, volcanoes). Estimates for the volcanic and sea salt HCl fluxes are 0.4–11 Tg-Cl yr^{-1} and 25 Tg-Cl yr^{-1}, respectively. Some HCl is also mixed downward from the stratosphere ($\sim$0.03 Tg-Cl yr^{-1}) where it is produced by reactions that scavenge Cl(g) and prevent it from destroying ozone:

$$Cl + CH_4 \rightarrow CH_3 + HCl \qquad\qquad (5.99)$$

$$Cl + HO_2 \rightarrow O_2 + HCl \qquad\qquad (5.100)$$

Hydrogen chloride is very soluble in water and is removed from the troposphere by rainout within 10 days.

Bromine. Bromine is a trace element in the solar system with an abundance of only 10.7 Br atoms per 10^6 Si atoms. Seawater contains $67 \,\mu g \, g^{-1}$ Br, while the crust contains about $1 \,\mu g \, g^{-1}$, and the mantle $4.6 \, ng \, g^{-1}$. Seawater is the major Br reservoir (75.1%), followed by the crust (12.5%) and mantle (12.4%). Bromine occurs as bromargyrite (AgBr), in evaporate deposits as bromocarnallite ($KMgBr_3 \cdot 6H_2O$), and dissolved in brines. The Br abundance in the human body is about $2.9 \,\mu g \, g^{-1}$ (where it is mainly in soft tissues), and is $1.7–20 \,\mu g \, g^{-1}$ in plants.

The total Br atom mixing ratio in the atmosphere is 30 pptv (0.42 Tg-Br), dominantly as methyl bromide (CH_3Br, Table 5.2). Methyl bromide has natural and anthropogenic sources. It is a fumigant for grain and soil and is an herbicide. Sources include biomass burning (anthropogenic and natural), marshes, oceanic biota and cruciferous plants. The biomass burning flux of CH_3Br is estimated as $10–50 \, Gg \, yr^{-1}$. The volcanic flux of Br is about $2.6–43 \, Gg \, yr^{-1}$. Methyl bromide is destroyed by reaction with OH radicals and has a lifetime of 1.5 yr. Soils are also a sink for CH_3Br.

Iodine. Iodine is a trace element in the solar system with an abundance of 1.10 I atoms per 10^6 Si atoms. Most of Earth's iodine occurs in organic-rich marine ($55 \,\mu g \, g^{-1}$, 87.8% of total I) and non-marine ($200 \, ng \, g^{-1}$, 0.8% of total I) sediments instead of in seawater ($59 \, ng \, g^{-1}$, 0.2% total I), crustal igneous rocks ($63 \, ng \, g^{-1}$, 3.2% total I) or mantle ($0.8 \, ng \, g^{-1}$, 8.0% total I). Iodine is also concentrated in seaweed and is extracted from it. Iodargyrite (AgI) and marshite (CuI) are the most important iodides, but commercial deposits such as the caliche beds in Chile contain 0.02–1% iodine as iodate minerals such as lautarite $[Ca(IO_3)_2]$ and dietzeite $[7Ca(IO_3)_2 \cdot CaCrO_4]$. Iodine has an abundance of $0.2 \,\mu g \, g^{-1}$ in the human body and is concentrated in the thyroid gland ($0.6 \, mg \, g^{-1}$). Seafood is rich in iodine, but about 2 billion people worldwide are at risk from iodine deficiency disorder (IDD), which reduces learning ability and causes cretinism and retardation. Adding KI or KIO_3 to table salt, giving an intake of about $100–300 \,\mu g$-I per day, prevents IDD.

Methyl iodide is the major I-bearing gas with a mixing ratio of about 0.8 pptv (18 Gg-I). Methyl iodide has natural (oceanic biota, rice paddies, biomass burning) and anthropogenic sources (release during use as a fumigant and herbicide, and from biomass burning). The oceans also emit other alkyl iodides, including CH_2I_2 [ethyl iodide (C_2H_5I)], CH_2ICl and CH_2IBr. Photolysis rapidly destroys CH_3I (5 days) and other alkyl iodides (minutes to days) and produces I atoms. These react with O_3 giving IO, which in turn photolyzes to I and O atoms that react further with ozone.

FURTHER READING

J. H. Bowes and M. M. Murray, *Biochem. J.*, 1935, **29**, 2721.

S. S. Butcher, R. J. Charlson, G. H. Orians and G. V. Wolfe, (eds.), *Global Biogeochemical Cycles*, Academic Press, London, 1992.

G. S. Callendar, *Q. J. Roy. Meteorol. Soc.*, 1938, **64**, 223.

G. Dreibus and H. Palme, *Geochim. Cosmochim. Acta*, 1996, **60**, 1125.

D. Ehhalt and M. Prather, lead authors, Chapter 4, Atmospheric chemistry and greenhouse gases, in IPCC Assessment Report, *Working Group 1: The Scientific Basis*, IPCC (available online: www.ipcc.ch/ipccreports//tar/wg1/)

T. E. Graedel and W. C. Keene, *Pure Appl. Chem.*, 1996, **68**, 1689.

J. T. Kiehl and K. E. Trenberth, *Bull. Am. Meteorol. Soc.*, 1997, **78**, 197.

G. P. Kuiper (Ed.), *The Earth as a Planet,* vol. II of *The Solar System*, University of Chicago Press, Chicago, IL, 1954.

S. W. Leavitt, *Environ. Geol.*, 1982, **4**, 15.

V. Smil, *Enriching the Earth*, The MIT Press, Cambridge, MA, 2001.

W. S. Snyder, M. J. Cook, E. S. Nasset, L. R. Karhausen, G. P. Howells and I. H. Tipton, *Report of the Task Group on Reference Man*, ICRP Publication 23, Pergamon Press, Oxford, 1975.

CHAPTER 7

The Outer Solar System

7.1 INTRODUCTION

Chapter 4 discussed the chemistry of objects in the inner solar system, and this chapter covers the chemistry of bodies in the outer solar system. We focus on the four giant planets (Jupiter, Saturn, Uranus and Neptune) and their large icy satellites (*e.g.*, the four Galilean satellites of Jupiter, Saturn's largest satellite, Titan, and Neptune's largest satellite, Triton). We talk in less detail – because less data are available – about the chemistry of other smaller bodies in the outer solar system, including the dwarf planets (*e.g.*, Ceres and Pluto).

In general bodies in the outer solar system are rich in gaseous and icy volatiles – especially H_2O, CO, CH_4, N_2, NH_3 and organic matter – unlike the inner planets. Water ice, organic matter and ices of methane, ammonia, N_2 and CO are significant constituents of various outer solar system bodies because the ices and organic matter were produced and/or preserved at the markedly lower temperatures in the outer regions of the solar nebula during planetary formation.

7.2 THE GIANT PLANETS

7.2.1 Overview

The four giant planets (Jupiter, Saturn, Uranus and Neptune) are qualitatively different from the four terrestrial planets (Mercury, Venus, Earth and Mars) in several respects. The terrestrial planets Venus, Earth

Chemistry of the Solar System
By Katharina Lodders and Bruce Fegley, Jr.
© K. Lodders and B. Fegley, Jr. 2011
Published by the Royal Society of Chemistry, www.rsc.org

and Mars have relatively oxidizing atmospheres that are $<0.01\%$ of their total mass and are terminated by sharp boundaries with their surfaces. In contrast, the reducing atmospheres of the four giant planets make significant fractions of the total planetary masses, there are no observable solid surfaces, and (apparently) bottomless atmospheres extend deep into the planets. Therefore, atmospheric properties that are referenced to sea level on Earth are referenced to the 1 bar level in the atmospheres of the giant planets. Although their atmospheres extend to great depth, liquid metallic H-He may occur deep inside Jupiter and Saturn, while ionic oceans of aqueous ammonia may occur deep inside Uranus and Neptune. All four planets have magnetic fields, which are probably generated by dynamo currents in electrically conductive fluids in their interiors (liquid metallic H-He inside Jupiter and Saturn and ionic oceans of aqueous ammonia inside Uranus and Neptune). At least three of the four giant planets emit more energy than they receive from the Sun (Uranus is the apparent exception for reasons discussed later.) The heat emitted by the giant planets arises from their continued gravitational contraction and cooling and also from phase separation and sedimentation of He from H in their deep interiors (on Saturn and Jupiter). In contrast, as discussed in Chapter 6, Earth and the other terrestrial planets are in radiative equilibrium and emit only as much energy as they absorb from the Sun.

Theoretical models suggest that the observed heat fluxes on Jupiter, Saturn and Neptune are transported by atmospheric convection. This has been demonstrated for Jupiter by *in situ* measurements down to 22 bar from the *Galileo* probe and for Saturn, Uranus and Neptune by determination of P, T profiles down to ~ 10 bar from *Voyager* IRIS data. The giant planets have low densities and volatile-rich bulk compositions, yet are much more massive than the terrestrial planets, which have higher densities characteristic of metal and silicate rock. Finally, all of the giant planets have extensive satellite and ring systems, *e.g.*, 63, 61, 27 and 13 satellites for Jupiter, Saturn, Uranus and Neptune, respectively. In contrast, Venus has no satellites, the Earth has one satellite (the Moon), Mars has two satellites (Phobos and Deimos) and none of the terrestrial planets have any rings.

Photochemistry and thermochemistry are the two major processes that control atmospheric chemistry on the giant planets. Ultraviolet sunlight (and other disequilibrating energy sources such as charged particles and cosmic rays) drives photochemical reactions in the upper atmospheres of all four planets. The heat released by gravitational contraction and cooling (and by He phase separation and sedimentation on Jupiter and Saturn) drives thermochemical reactions in the deep

Table 7.1 Some physical properties of the giant planets.

Property	Jupiter	Saturn	Uranus	Neptune
Mass (10^{24} kg)	1 898.6	568.46	86.832	102.43
Mass (M_E)[a]	318	95.2	14.5	17.2
Radius at 1 bar (km)[b]	71 492	60 268	25 559	24 764
Moment of inertia (C/MR^2)	0.26	0.22	0.23	0.24
Density (g cm^{-3})	1.326	0.687	1.318	1.638
T (K) at 1 bar	165	134	76	72
$-(dT/dZ)$ at 1 bar (K km^{-1})[c]	2.3	1.0	1.1	1.3
Atm. c_p at 1 bar (J g^{-1} K^{-1})[d]	11.1	10.7	8.2	8.5
Emitted/absorbed flux[e]	1.67	1.78	~1	2.61
Mean mol. wt. (g mol^{-1})[f]	2.28	2.25	2.64	2.53
Scale height at 1 bar (km)	24.4	51.5	27.0	21.1
Gravity at 1 bar (m s^{-2})	25.4	10.4	8.85	11.1

[a]$M_E = 1$ Earth mass $= 5.97 \times 10^{24}$ kg.
[b]Equatorial radius at the 1 bar level.
[c]Adiabatic dry lapse rate at the 1 bar level.
[d]Mean specific heat of atmospheric gas at the 1 bar level.
[e]Ratio of thermal emission to the absorbed solar flux.
[f]Mean molecular weight of atmospheric gas.

atmospheres. Thermochemical reactions also lead to formation of condensation clouds throughout the observable and deeper, unobservable regions of all four giant planets. Both photochemistry and thermochemistry affect chemistry in the intermediate regions of giant planet atmospheres, although the location of these intermediate regions varies from planet to planet.

The giant planets divide into two qualitatively different groups – the gas-rich giants Jupiter and Saturn and the gas-poor giants Uranus and Neptune. Jupiter and Saturn are larger, gas-rich and closer to solar composition (>50 mass% H and He), whereas Uranus and Neptune are smaller, gas-poor and farther from solar composition (<50 mass% H and He). Table 7.1 summarizes physical properties of the giant planets and their atmospheres.

7.2.2 Jupiter and Saturn: The Gas-rich Giant Planets

Jupiter is the largest planet in the solar system and has a mass (M_J) of about 10^{-3} that of the Sun, or approximately 318 times that of the Earth (M_E). Saturn is the second most massive gas giant planet with a mass of about 95 M_E. Although quite massive relative to the Earth and other planets in our solar system, Jupiter is much smaller than an object capable of "burning" deuterium by nuclear fusion, which requires at least 13 times Jupiter's mass. The 13 M_J mass limit is widely taken as the boundary between planets at lower masses and brown dwarfs at higher masses. (Brown dwarfs are sub-stellar objects with masses between 13 and 80 M_J that can "burn" deuterium but not hydrogen by nuclear fusion.) Jupiter is less massive than most extrasolar planets discovered to date, which have masses between that of Saturn (95 M_E) and 10 M_J. However, the currently known mass distribution of extrasolar planets is changing as observational methods become more sensitive to detect lower mass planets around other stars. For example, the Kepler Mission flown by NASA and the CoRoT Mission flown by ESA detected lower mass Earth-like planets.

Jupiter's low bulk density (1.33 g cm^{-3}) and Saturn's even lower density (0.69 g cm^{-3}) coupled with their atmospheric composition (Table 7.2), size, shape (*i.e.*, their oblateness) and interior structure models show that they are mainly H and He with smaller amounts of heavier elements. This has been suspected since the pioneering work of the British geophysicist Sir Harold Jeffreys (1891–1989) in the 1920s. Subsequent work by the German astronomer Rupert Wildt and his student the American physicist Wendell DeMarcus (1925–1982) in the 1950s demonstrated that Jupiter is mainly H and He.

Table 7.2   Chemical composition of the atmospheres of Jupiter and Saturn.[a]

Gas	Jupiter[b]	Saturn
H_2	86.4 ± 0.3%	88 ± 2%
^4He	13.6 ± 0.3%	12 ± 2%
CH_4	(1.81 ± 0.34) × 10^{-3}	(4.7 ± 0.2) × 10^{-3}
NH_3	(6.1 ± 2.8) × 10^{-4}	(1.6 ± 1.1) × 10^{-4}
H_2O	520^{+340}_{-240} ppmv	2–20 ppbv
H_2S	67 ± 4 ppmv	<0.4 ppmv
HD	45 ± 12 ppmv	110 ± 58 ppmv
$^{13}CH_4$	19 ± 1 ppmv	51 ± 2 ppmv
C_2H_6	5.8 ± 1.5 ppmv	7.0 ± 1.5 ppmv
PH_3	1.1 ± 0.4 ppmv	4.5 ± 1.4 ppmv
CH_3D	0.20 ± 0.04 ppmv	0.30 ± 0.02 ppmv
C_2H_2	0.11 ± 0.03 ppmv	0.30 ± 0.10 ppmv
HCN	60 ± 10 ppbv	<4 ppbv
C_2H_4	7 ± 3 ppbv	~0.2 ppbv[c]
CO_2	5 – 35 ppbv	0.3 ppbv
CH_3C_2H	2.5^{+2}_{-1} ppbv	0.6 ppbv
CO	1.6 ± 0.3 ppbv	1.4 ± 0.7 ppbv
GeH_4	$0.7^{+0.4}_{-0.2}$ ppbv	0.4 ± 0.4 ppbv
C_4H_2	0.3 ± 0.2 ppbv	0.09 ppbv
AsH_3	0.22 ± 0.11 ppbv	2.1 ± 1.3 ppbv

[a]The units ppmv and ppbv stand for parts per million by volume (1 ppmv = 10^{-4}%) and parts per billion by volume (1 ppbv = 10^{-3} ppmv = 10^{-7}%), respectively.
[b]^{3}He 22.6 ± 0.7 ppmv, Ne 21 ± 3 ppmv, Ar 16 ± 3 ppmv, Kr 8 ± 1 ppbv, Xe 0.8 ± 0.1 ppbv.
[c]Assumes a total stratospheric column density of 1.54 × 10^{25} cm^{-2}.

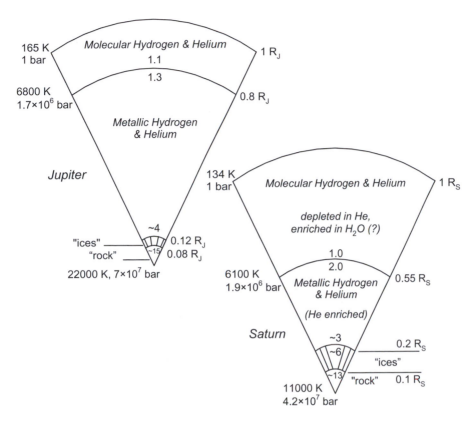

Figure 7.1 Contemporary ideas for the interior structure of Jupiter and Saturn.

Figure 7.1 shows current ideas for the interior structure of Jupiter and Saturn. Metallic H–He makes up the bulk of Jupiter and a significant fraction of Saturn. The moment of inertia (C/MR^2) for Jupiter (0.26) and Saturn (0.22) are significantly lower than the value of 0.40 for a uniform density sphere and require dense central regions. Thermal evolution and interior structure models suggest that distinct cores made up of "rocky" and "icy" elements exist inside Jupiter and Saturn. "Rocky" elements are those found in metal, silicate and sulfide rock (*e.g.*, Ca, Al, Fe, Mg, Si, S) and "icy" elements are those expected to condense as ices in the solar nebula (*e.g.*, Ar, C, N, O). However, whether distinct cores actually exist or all of the "rocky" and "icy" elements are dissolved in the molecular and/or metallic H–He phases is unclear at present. At least some of the "rocky" and "icy" elements are present in the molecular H–He layers of Jupiter and Saturn because CH_4, AsH_3 (arsine), GeH_4 (germane), H_2O, H_2S, NH_3 and PH_3 (phosphine) are observed in their atmospheres. The *Juno* polar orbiter, scheduled for launch in August 2011, will measure Jupiter's gravitational field with sufficient accuracy to decide whether a distinct core exists.

As shown in Figure 7.1, the molecular H–He layers of Jupiter and Saturn are only a small fraction of each planet. These layers are the only regions of the two planets that are observable from the Earth and spacecraft. In general remote sensing observations in the UV, visible and IR regions sample only the uppermost regions of Jupiter's and Saturn's atmospheres down to pressures of a few bars. Radio wave observations extend deeper to tens of bars.

In situ observations of Jupiter's atmosphere by the *Galileo* entry probe (7 December 1995) extended down to ~ 22 bar (at 429 K), about 126 km below the 1 bar level. Saturn's atmosphere has not yet been explored by entry probes. (As described later, the *Huygens* probe on the *Cassini* spacecraft descended to the surface of Titan, Saturn's largest satellite, on 14 January 2005.) Thus, our knowledge of atmospheric composition and structure on Jupiter and Saturn is based on a combination of Earth-based and spacecraft observations. The spacecraft that explored Jupiter and Saturn are *Pioneer* 10 (4 December 1973 Jupiter flyby), *Pioneer* 11 (3 December 1974 Jupiter flyby, 1 September 1979 Saturn flyby), *Voyager* 1 (5 March 1979 Jupiter flyby, 12 November 1980 Saturn flyby), *Voyager* 2 (9 July 1979 Jupiter flyby, 25 August 1981 Saturn flyby), *Galileo* (7 December 1995 arrival at Jupiter and entry probe launch – 21 Sept 2003 end of mission when orbiter was crashed into Jupiter), *Cassini* (1 July 2004 arrival at Saturn and continuing now), *New Horizons* (28 February 2007 Jupiter flyby *en route* to the dwarf planet Pluto and its three satellites Charon, Hydra and Nix).

Visually, Jupiter's atmosphere is composed of dark belts alternating with bright zones. These are the upward (bright zones) and downward (dark belts) arms of atmospheric convection cells. The Great Red Spot (GRS) in Jupiter's southern hemisphere has existed since at least 1664 when it (or a similar feature) was observed by the English scientist Robert Hooke (1635–1703). Good observations and records of the GRS extend over 160 years. It is about 32 000 km long and 13 000 km wide. The Great Red Spot is apparently a long-lived cyclonic storm and its red color is plausibly due to red phosphorus formed by UV sunlight photolysis of PH_3 gas. Other ovals and spots have also been observed on Jupiter for various periods of time, and its atmosphere is meteorologically active. Inorganic compounds of sulfur and other elements probably are the chromophores coloring the belts in Jupiter's atmosphere. The organic compounds produced by atmospheric photochemistry are colorless and do not provide the observed cloud colors.

The bright zones are high level NH_3 ice clouds at about 0.5 bar (138 K). The chemistry of the condensation cloud layers on Jupiter and Saturn is described in more detail below. Saturn's atmosphere is less colorful than Jupiter's but has a similar banded structure visible in spacecraft photographs.

Remote sensing observations by the *Voyager* spacecraft and the *in situ* measurements by the *Galileo* entry probe show that Jupiter's atmosphere is convective down to at least 22 bar, where the temperature is 429 K. The presence of CO, AsH_3 (arsine), GeH_4 (germane) and PH_3 (phosphine) at abundances orders of magnitude greater than their thermochemical equilibrium values in the upper tropospheres of Jupiter and Saturn shows that convective mixing extends down to at least kilobar levels in their atmospheres. The convective regions (*i.e.*, the tropospheres) of Jupiter's and Saturn's atmospheres extend up to their tropopauses, which are the radiative–convective boundaries and temperature minima in their atmospheres (~ 100 mb, 110 K on Jupiter and ~ 83 mb, 83 K on Saturn). Methane in the stratospheres of the giant planets plays the role that O_3 does in Earth's stratosphere and absorption of UV sunlight by CH_4 heats the stratospheres of the four gas giant planets.

Because the tropospheres of Jupiter and Saturn are convective, the temperature gradient (or lapse rate) is adiabatic. At the 1 bar level ($T = 165$ K on Jupiter and $T = 135$ K on Saturn), the dry adiabatic lapse rate (*i.e.*, the temperature gradient dT/dZ in cloud-free regions) given by Equation (5.24) is about 2.3 K km^{-1} on Jupiter and 1.0 K km^{-1} on Saturn. The dry adiabatic lapse rate on Jupiter and Saturn decreases slightly with increasing temperature because the specific heat of H_2 increases with increasing temperature (He has a constant specific heat). Thus, at 422 K, the lowest level reached by the Galileo entry probe, the dry adiabatic lapse is about 2.1 K km^{-1} on Jupiter.

The abundances of CH_4, H_2O, NH_3 and other atmospheric constituents are low enough that their specific heats do not appreciably affect the dry adiabatic lapse rates on Jupiter and Saturn. However, condensation of these species releases heat back into the surrounding atmospheric gas and makes the atmospheric warmer than it would be in the absence of cloud formation. Thus, the wet adiabatic lapse rates, *i.e.*, the temperature gradients in cloudy regions, on Jupiter and Saturn are lower than the dry adiabatic lapse rates because of cloud condensation. The exact value of the wet adiabatic lapse rate, given by Equation (5.25), depends upon the type of condensate [*e.g.*, Fe metal, $MgSiO_3$, liquid water, water ice, ammonium hydrosulfide (NH_4SH), NH_3 ice] and its abundance in the gas phase below its cloud base (see the discussion for water vapor in Chapter 5).

For example, Table 7.2 lists a NH_3 mixing ratio of 6.1×10^{-4} in Jupiter's atmosphere. As Jovian atmospheric gas rises and cools, it eventually becomes saturated with NH_3 and NH_3 ice starts to condense out of the gas forming a cloud. The location of the NH_3 ice cloud base is given by the equation:

$$\frac{10\,000}{T_{NH_3}} = 68.02 - 6.19 \log E_N - 6.31 \log P_T \qquad (7.1)$$

The E_N and P_T are the NH_3 enrichment relative to solar composition (explained below) and the total atmospheric pressure, respectively. Ammonia ice condenses instead of liquid NH_3 because the partial pressure curve intersects the vapor pressure curve at temperatures below the triple point of ammonia (195.5 K).

Formation of the NH_3 ice cloud releases heat, equal to the latent heat of sublimation of NH_3 ice ($\sim 30\,400 \, J \, mol^{-1}$) back into the surrounding gas. The average molecular weight of Jovian atmospheric gas is $2.28 \, g \, mol^{-1}$ and its specific heat is $11.1 \, J \, g^{-1} \, K^{-1}$ (Table 7.1). Using these values and Equation (5.25), the calculated wet adiabatic lapse rate at the NH_3 ice cloud base on Jupiter is about $2.0 \, K \, km^{-1}$, about 0.3-degree smaller than the dry lapse rate. The difference between the wet and dry lapse rates is small (about 13% of the dry lapse rate) because only a small amount of NH_3 in present in Jupiter's atmosphere. Larger wet lapse rates occur if larger amounts of condensable gas are present and if the latent heats of sublimation or vaporization are large. Uranus and Neptune should have large wet lapse rates in the water cloud condensation region due to the large abundance of water vapor expected in their lower atmospheres.

Table 7.2 gives the abundances of gases observed in the atmospheres of Jupiter and Saturn. Percentages by volume are given for H_2 and He and mixing ratios by volume are listed for the other gases. The uncertainties in the nominal abundances are also given. We now use these data to discuss atmospheric chemistry on Jupiter and Saturn.

Many of the gases observed in their atmospheres are hydrides, e.g., CH_4, NH_3, H_2O, H_2S, PH_3, GeH_4 and AsH_3. All of these gases (except H_2O and H_2S) are photochemically destroyed by UV sunlight in the stratospheres of Jupiter and Saturn. Ammonia is removed by formation of NH_3 ice clouds and photolysis above these clouds in the atmospheres of Jupiter and Saturn. Water vapor and H_2S condense to form clouds (liquid water, NH_4SH) before they reach the stratospheres where they can be photolyzed by UV sunlight. As shown in Table 7.2, NH_3 is more abundant than H_2S on Jupiter and NH_4SH condensation removes all H_2S, but not all NH_3 from Jupiter's atmosphere above the NH_4SH cloud. As long as nitrogen and sulfur are equally enriched (or depleted) relative to solar composition this is probably also true on the other giant planets because the atomic N/S ratio is 5.0 in solar composition material. The locations of the water and NH_4SH cloud bases are given by the equations:

$$\frac{10\,000}{T_{water}} = 38.84 - 3.93 \log E_O - 3.83 \log P_T - 0.20 \log E_O \log P_T \qquad (7.2)$$

$$\frac{10\,000}{T_{NH_4SH}} = 48.94 - 4.27 \log E_S - 4.15 \log P_T \qquad (7.3)$$

The E_O and E_S are the enrichments of H_2O and H_2S relative to solar abundances (described below) and P_T is the total atmospheric pressure. Equation (7.3) implicitly assumes equal enrichments of NH_3 and H_2S, which is true within the large uncertainties on the NH_3 enrichment. Equation (7.3) gives a temperature of about 220 K for the NH_4SH cloud base on Jupiter.

Hydride gases are important in the atmospheres of all four giant planets because H_2 is the major gas and high atmospheric pressures exist in their hot, deep atmospheres where thermochemical reactions proceed rapidly. Thus, any gases or solids produced by photochemistry in the upper atmospheres of Jupiter and Saturn are rapidly converted back into hydrides by thermochemical reactions deep inside these planets. For example, as discussed later, CH_4 photochemistry forms ethane (C_2H_6), which has an abundance of 5.8–7 ppmv on Jupiter and Saturn. At ~ 1000 K, deep inside these two planets, it reacts with H_2 to reform CH_4 *via* the net reaction:

$$C_2H_6 + H_2 = 2CH_4 \qquad (7.4)$$

Other photochemically produced species (*e.g.*, C_2H_2, C_2H_4, CH_3C_2H, C_4H_2, N_2, elemental As, Ge and P) that are transported downward into the hot, high-pressure regions of the atmospheres of Jupiter and Saturn also react with H_2 to reform hydrides. (We discuss atmospheric photochemistry in more detail later.)

The presence of hydride gases is important for several other reasons. Hydrogen sulfide was first observed and measured in Jupiter's atmosphere by the *Galileo* probe mass spectrometer (GPMS). The observed H_2S/H_2 ratio of $\sim 7.8 \times 10^{-5}$ is about 2.7 times higher than the S/H_2 ratio of $\sim 2.9 \times 10^{-5}$ based on solar elemental abundances. (The S/H_2

ratio and other element/H_2 ratios based on solar abundances are calculated using $\frac{1}{2}$ the H elemental abundance, which gives the equivalent H_2 abundance in solar material.) The presence of H_2S in Jupiter's atmosphere requires depletion of iron by Fe metal cloud condensation at high temperatures deep in Jupiter's atmosphere. If Fe cloud formation did not occur, H_2S would be completely absent from Jupiter's atmosphere because FeS (troilite) formation at $\sim 700\,K$ would consume all H_2S gas (because the atomic ratio of Fe/S≈ 2 in solar composition material) *via* the net reaction:

$$H_2S(\text{gas}) + Fe(\text{metal}) = FeS\ (\text{troilite}) + H_2(\text{gas}) \tag{7.5}$$

Hence, Fe metal condensation clouds form at great depth inside Jupiter according to thermochemical equilibrium models.

Similar thermochemical equilibrium models predict H_2S is the dominant sulfur gas in the atmospheres of Saturn, Uranus and Neptune. Hydrogen sulfide is not observed on these planets because formation of NH_4SH clouds (*via* reaction of H_2S with NH_3) depletes H_2S from the observable regions of their atmospheres. The NH_4SH cloud base temperature on each planet is computed with Equation (7.3) by assuming that H_2S and CH_4 are equally enriched with respect to solar composition and using the observed CH_4 enrichment. These estimates show that planetary entry probes will be needed to detect H_2S on Saturn, Uranus and Neptune because astronomical observations cannot penetrate to atmospheric levels below the NH_4SH cloud bases.

The presence of CH_4 and GeH_4 (germane), but not of SiH_4 (silane), on Jupiter and Saturn is due to condensation of magnesium silicates such as enstatite ($MgSiO_3$) and forsterite (Mg_2SiO_4) deep in their atmospheres. Silicon is much more abundant than germanium in solar composition material (the atomic Si/Ge ratio is ~ 8700), but it is more refractory than Ge and forms silicate clouds deep in the atmospheres of the giant planets *via* net thermochemical reactions such as:

$$SiH_4(g) + Mg(OH)_2(g) + H_2O(g) = MgSiO_3(\text{enstatite}) + 3H_2(g) \tag{7.6}$$

$$SiH_4(g) + 2Mg(OH)_2(g) + H_2O(g) = Mg_2SiO_4(\text{forsterite}) + 4H_2(g) \tag{7.7}$$

Silane is not observed on either Jupiter or Saturn and the observational upper limits for it are about 1 ppbv (SiH_4/H_2 ratio $< 10^{-9}$) on both planets. This is $68\,000$ times smaller than the Si/H_2 ratio of 6.8×10^{-5} from solar elemental abundances. In contrast, the observed CH_4/H_2 ratios of 2.1×10^{-3} on Jupiter and 5.3×10^{-3} on Saturn are 4–11 times higher than the C/H_2 ratio of 4.9×10^{-4} from solar elemental abundances. The GeH_4/H_2 ratios of 8.1×10^{-10} (Jupiter) and 4.6×10^{-10} (Saturn) are 0.1 to 0.06 times the Ge/H_2 ratio of 78×10^{-10} from solar abundances. The Ge/H_2 ratios in the atmospheres of Jupiter and Saturn are lower than the solar value because not all Ge in their atmospheres is present as germane. Most Ge condenses out of the atmospheres of Jupiter and Saturn as elemental Ge and Ge chalcogenides at high temperatures.

Leaving aside photochemically produced gases, the hydride/H_2 ratios in the atmospheres of Jupiter and Saturn give elemental enrichments (or depletions) relative to solar composition. We have already seen that carbon and sulfur are enriched and germanium is depleted relative to solar composition in Jupiter's observable atmosphere. Carbon is also enriched on Saturn (~ 11 times solar) but H_2S is not observed, as discussed earlier. Nitrogen is enriched 4.9 times solar (with 46% uncertainties) on Jupiter, but only 1.2 times solar on Saturn where condensation of NH_3 into NH_3 ice clouds depletes ammonia in the observable region of its atmosphere. Water is depleted and is only 0.6 times solar on Jupiter. This depletion is observed below the level where water clouds are predicted to form and is probably a depletion of water (and oxygen) throughout Jupiter's atmosphere and interior. Alternatively, the depletion of water may be a meteorological effect because the Galileo probe entered Jupiter's atmosphere in the North Equatorial Belt (NEB), which is one of the downward arms of atmospheric convection cells in Jupiter's atmosphere. However, if this were the case one would also expect depletions of other cloud-forming gases such as NH_3 and H_2S, which is not observed (these gases are enriched relative to solar).

On Saturn, H_2O is removed by condensation of water clouds below the observable region of Saturn's troposphere. The small amounts of H_2O observed in Saturn's stratosphere are due to oxygen coming into Saturn's upper atmosphere from its icy rings and provide no information on the water content of the planet.

Phosphorus is enriched 2.2 times solar on Jupiter and 9 times solar on Saturn. Arsenic is depleted on Jupiter (~ 0.6 times solar) and enriched on Saturn (~ 5.8 times solar). Arsine is the major As-bearing gas on Jupiter and Saturn, but condensation of elemental arsenic at $400 \, K$ depletes the AsH_3 abundance in the cooler, observable region of Jupiter's atmosphere. Arsenic and phosphorus behave similarly in meteorites and in the solar nebula, so their enrichment factors on Jupiter and Saturn are plausibly the same. This is the case within the uncertainties on the enrichment factors for As and P on Saturn.

These data on hydride/H_2 ratios show that the atmospheres of Jupiter and Saturn are close to solar composition and that Saturn is more enriched in heavy elements relative to solar composition than Jupiter. This trend is continued at least qualitatively by Uranus and Neptune where CH_4/H_2 ratios are about 57 and 38 times larger relative to solar composition, respectively. Given their mean densities, similar sizes and uncertainties in the CH_4 observations, it is plausible that Neptune is more enriched than Uranus. Except for photochemically produced gases and isotopomers (*e.g.*, CH_3D), no other hydride gases are observed on Uranus or Neptune. Thus the heavy element enrichment trend on the four giant planets depends on observations of methane.

Deuterium (heavy hydrogen) is also enriched on Uranus and Neptune in comparison to Jupiter and Saturn. Both HD and CH_3D are observed in the atmospheres of Jupiter, Saturn, Uranus and Neptune. Hydrogen deuteride gas (HD) is the major reservoir of deuterium in the atmospheres of these planets because hydrogen is much more abundant than any other gas. Other less abundant D-bearing gases are formed by thermochemical reactions such as:

$$HD + CH_4 = CH_3D + H_2 \qquad (7.8)$$

$$HD + NH_3 = NH_2D + H_2 \qquad (7.9)$$

$$HD + H_2O = HDO + H_2 \qquad (7.10)$$

$$HD + H_2S = HDS + H_2 \qquad (7.11)$$

To date only CH_3D has been observed; the other monodeuterated hydrides have not been observed because their abundances are too small. For example, $^{15}NH_3$ is observed in Jupiter's atmosphere with a $^{15}NH_3/^{14}NH_3$ ratio of about 2.23×10^{-3}. In contrast, the expected NH_2D/NH_3 ratio is $\sim 2.3 \times 10^{-5}$, about 100 times smaller and unobservable until higher resolution infrared spectra are obtained from Earth-based observatories or spacecraft.

The atomic D/H ratio on the four giant planets is calculated from the observed HD/H_2 ratio using the equation:

$$\frac{D}{H} = \frac{1}{2} \frac{X_{HD}}{X_{H_2}} \qquad (7.12)$$

The HD/H_2 ratio is divided by two because H_2 has two H atoms while HD has only one H atom. Equation (7.12) and the H_2 and HD abundances in Table 7.2 give D/H ratios of 2.6×10^{-5} on Jupiter and 6.2×10^{-5}

on Saturn. The D/H value for Saturn is more uncertain because neither the HD nor the H_2 abundances are as well known as on Jupiter. The corresponding D/H ratios on Uranus and Neptune are about 9×10^{-5} and 1.2×10^{-4}, respectively, and have larger uncertainties. However, there is a clear trend of increasing D/H ratio from Jupiter outward to Neptune.

It is also possible to compute the D/H ratio from the observed CH_3D/CH_4 ratios, using an equation that takes into account the isotopic fractionation factor for exchanging hydrogen and deuterium between hydrogen and methane (α value) and its temperature dependence. [In other words the equilibrium constants for Reactions (7.8)–(7.11) are different from unity.] This equation is:

$$\frac{D}{H} = \frac{1}{4\alpha} \frac{X_{CH_3D}}{X_{CH_4}} \tag{7.13}$$

In general, α values such as that in Equation (7.13), are defined as the D/H atomic ratio in a hydride relative to that in molecular hydrogen:

$$\alpha = \frac{(D/H)_{hydride}}{(D/H)_{H_2}} \tag{7.14}$$

An α of unity means that the D/H ratio is the same in H_2 and the hydride, while an α greater than unity means that the hydride has a higher D/H ratio than H_2, *i.e.*, deuterium is concentrated in the hydride. This is usually the case, and D/H ratios in hydrides and α values increase with decreasing temperature.

The lowest temperature at which deuterium exchange between HD and CH_4 occurs in the atmospheres of the giant planets is called the quench temperature and it determines the α value. The quench temperature occurs at the atmospheric level where the time required for isotopic exchange is equal to the time for vertical mixing from hotter to cooler regions in the planetary atmospheres. Below this level, isotopic exchange occurs faster than vertical mixing and isotopic equilibrium is maintained. Above this level isotopic exchange occurs slower than vertical mixing and isotopic equilibrium is not reached.

The time required for Reaction (7.8) to alter the D/H ratio in methane is the chemical lifetime (t_{chem}) for isotopic exchange. It is defined by the CH_3D concentration divided by the rate of change of that concentration:

$$t_{chem} = \frac{[CH_3D]}{\left(\frac{d}{dt}[CH_3D]\right)} \tag{7.15}$$

The square brackets indicate molecular number density, the number of CH_3D molecules per cubic centimeter, and is calculated using Equation (5.2). The t_{chem} value is calculated from reaction rate data in the literature for D/H exchange between HD and CH_4. The reaction rate varies exponentially with temperature and so does the chemical lifetime.

The vertical mixing time (t_{mix}) is computed from the pressure scale height (H) and the vertical eddy diffusion coefficient (K_{eddy}) using Equation (5.21). The scale height (H) is computed using Equation (5.15) and the vertical eddy diffusion coefficient is estimated from the observed heat flow and atmospheric properties. A value of about $10^4 \, m^2 \, s^{-1}$ is typical for K_{eddy} in the tropospheres of Jupiter and Saturn. The vertical mixing time varies as H^2 and is less sensitive to temperature than the chemical lifetimes. For example, a comparison of t_{chem} and t_{mix} shows that CH_3D formation is quenched at 790 K on Jupiter where the fractionation factor $\alpha = 1.205$. The corresponding D/H ratio given by Equation (7.13) and the abundances in Table 7.2 is about 2.3×10^{-5}, which is the same within the uncertainties, as that from the HD/H_2 ratio.

As discussed earlier, the D/H ratio increases from Jupiter to Saturn to Uranus to Neptune. This increase is consistent with the increased CH_4/H_2 ratios from Jupiter to Neptune because the D/H ratio in methane is greater than that in H_2 ($\alpha > 1$) and increases with decreasing temperature. (Even though the CH_4/H_2 ratio on Neptune is apparently lower than that on Uranus, the higher bulk density of Neptune shows it contains more heavy elements than Uranus.) In Chapter 2 we showed that the temperature in the solar nebula decreased with increasing radial distance so Uranus and Neptune accreted lower temperature icy material than Jupiter and Saturn. The lower temperature hydride ices will contain more deuterium because of the larger α values at lower temperatures. Furthermore, Uranus and Neptune have higher bulk densities and contain more icy and rocky material than Jupiter and Saturn. Thus, these ices (containing CH_4, H_2O, NH_3, *etc.*) provided more deuterium to Uranus and Neptune than to Jupiter and Saturn and give higher D/H ratios on Uranus and Neptune.

Chemical reaction rates and vertical mixing rates also explain the abundances of CO, PH_3, GeH_4 and AsH_3 in the atmospheres of Jupiter and Saturn. The abundances of these gases are significantly higher than those predicted at chemical equilibrium in the cool, observable regions of the atmospheres of Jupiter and Saturn. Carbon monoxide and PH_3 are the most dramatic examples because their observed abundances are about 32–36 orders of magnitude greater than their chemical equilibrium values. The reason for this is rapid vertical transport of CO and

PH_3 from hot, high pressure atmospheric regions where they are more abundant due to favorable chemical equilibria.

For example, consider CO in Jupiter's atmosphere. Carbon monoxide is produced by oxidation of methane *via*:

$$H_2O + CH_4 = 3H_2 + CO \qquad (7.16)$$

The equilibrium constant for Reaction (7.16) in terms of partial pressures is:

$$K = \frac{P_{CO} P_{H_2}^3}{P_{CH_4} P_{H_2O}} \qquad (7.17)$$

If we rewrite this to solve for the CO/CH_4 ratio at equilibrium we obtain:

$$\frac{X_{CO}}{X_{CH_4}} = K \left(\frac{X_{H_2O}}{X_{H_2}^3} \right) \frac{1}{P_T^2} \qquad (7.18)$$

Thermodynamic data show that the equilibrium constant for Reaction (7.16) increases with increasing temperature. Furthermore, the adiabatic gradient (dP/dT) in the convective lower atmosphere of Jupiter is easily calculated using the temperature dependent heat capacity of H_2–He gas mixtures. Thus, the CO/CH_4 ratio as a function of temperature (and thus of depth and pressure) can be computed from the observed H_2O and H_2 abundances. These calculations show that the observed CO/CH_4 ratio of $\sim 8.8 \times 10^{-7}$ occurs at about the 1100 K level in Jupiter's atmosphere.

The *Galileo* probe mass spectrometer also observed Ne, Ar, Kr and Xe in Jupiter's atmosphere (Table 7.2). The observed noble gas/H_2 mixing ratios are 0.1 (Ne), 2.9 (Ar), 2.4 (Kr) and 2.5 (Xe) times relative to solar composition. Neon and Ge are depleted by the same factor (0.1) but for different reasons – Ne is thought to partition preferentially into helium in Jupiter's interior while most Ge condenses out of Jupiter's atmosphere as elemental Ge and Ge chalcogenides at high temperatures. In contrast, Ar, Kr and Xe are about as enriched as S (2.7 times) relative to solar composition.

7.2.3 Uranus and Neptune: The Gas-poor Giant Planets

Uranus and Neptune are less massive than Jupiter and Saturn because they contain significantly less H and He, but they are still significantly

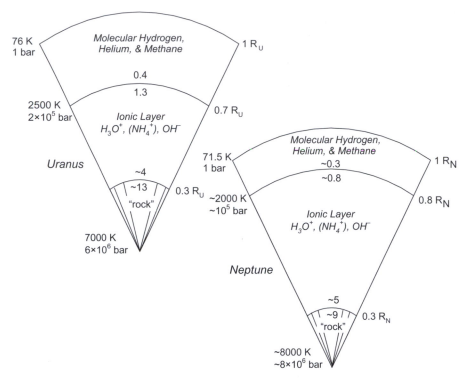

Figure 7.2   Contemporary ideas for the interior structure of Uranus and Neptune.

more massive than the Earth – about 14.5 and 17.2 M_E, respectively. Their greater mass relative to Earth is plausibly due to the incorporation of large amounts of water, CH_4, NH_3, organic matter and rocky material during their formation. Figure 7.2 schematically illustrates current ideas about their internal structure.

Neither Uranus nor Neptune was known to the ancients. The discovery of Uranus, which is the first planet discovered with a telescope, subsequently led to the discovery of Neptune, which is the first planet discovered by mathematical predictions. The German–English astronomer and composer Sir William Herschel (1738–1822) discovered Uranus on March 13, 1781. Herschel originally named Uranus the Georgian Planet after his patron King George III, but this name was unpopular outside of England and was eventually discarded in favor of the name Uranus. During the 60 years following Herschel's discovery of Uranus, it became clear that there were large discrepancies between the observed and calculated positions of the planet. Analyses of the discrepancies were carried out by the British astronomer and mathematician John Couch Adams (1819–1892) in England and the French mathematician Urbain Jean Joseph Leverrier (1811–1877) in France. Their work led to predictions for the mass and orbit of a trans-Uranian planet and culminated in the discovery of Neptune by the German astronomers Johann Gottfried Galle (1812–1910) and Heinrich Louis d'Arrest (1822–1875) on September 23, 1846. Shortly thereafter, the British brewer and amateur astronomer William Lassell (1799–1880) discovered Triton, Neptune's largest satellite on October 10, 1846. However, the story of Neptune's discovery does not end there because a 160 year controversy has arisen over whether Adams and Leverrier should share priority or if one of them should have sole priority for the calculations predicting Neptune's position. Several books describe this story and the one by Grosser (1962)[1] is recommended.

Uranus has an obliquity of 97.9° and its rotational and orbital axes are nearly aligned – it is tilted on its side. This unique orientation was originally inferred from Earth-based telescopic observations of the orientation of its five major satellites (Miranda, Ariel, Umbriel, Titania and Oberon). Uranus rotates around the Sun in about 84 years. As a result of its tilted orientation, one pole faces the Sun for 42 years and the other pole has 42 years of darkness. Uranus also rotates in a retrograde (*i.e.*, east to west) direction with a magnetic rotation period of 17.24 h.

Unlike the other three giant planets, Uranus may not have an internal heat source. The upper limit on the internal heat flux (from *Voyager* 2) is 14% of the absorbed solar flux. This is consistent with heating from decay of radionuclides in the rocky material inside Uranus. However, at the time of the *Voyager* 2 flyby in 1986, the South Pole had recently come into sunlight. Thus, the apparent lack of an internal heat flux is possibly connected to Uranus' unique orbital orientation.

Table 7.1 shows that Uranus has a bulk density of $1.318\,\mathrm{g\,cm^{-3}}$. This bulk density combined with its size shows that Uranus is enriched in elements heavier than He and does not have solar composition. Spectroscopic observations of CH_4 lead to the same conclusion as do observations of the D/H ratio on Uranus. However, within error, the He/H_2 ratio of ~ 0.18 is the same as the solar value and suggests that the Uranian atmosphere is not depleted in He (as on Saturn) (Table 7.3).

In contrast to the solar He/H ratio and enrichment of methane, Earth-based microwave observations indicate an apparent NH_3 depletion of 0.005–0.01 times the solar N/H_2 ratio in the 150–200 K region of Uranus' atmosphere. Atmospheric circulation patterns may cause temporal and latitudinal variations of the NH_3/H_2 ratio in this region of the atmosphere. The apparent NH_3 depletion may be due to the lack of nitrogen on Uranus, the loss of NH_3 in water- and NH_4SH-cloud layers

Table 7.3   Chemical composition of the atmospheres of Uranus and Neptune.

Gas	Uranus	Neptune
H_2	$\sim 82.5 \pm 3.3\%$	$\sim 80 \pm 3.2\%$
He	$15.2 \pm 3.3\%$	$19.0 \pm 3.2\%$
CH_4	$\sim 2.3\%$	$\sim 1\text{--}2\%$
HD	~ 148 ppmv	~ 192 ppmv
CH_3D	~ 8.3 ppmv	~ 12 ppmv
H_2S^a	<0.8 ppmv	<3 ppmv
$NH_3{}^a$	<100 ppb	<600 ppb
CO	<27 ppb	$0.6\text{--}22$ ppmv[b]
C_2H_6	10 ± 1 ppb	$1.5^{+2.5}_{-0.5}$ ppmv
C_2H_2	~ 10 ppb	60^{+140}_{-40} ppb
CH_3CN	–	<5 ppb
HCN	<15 ppb	0.3 ± 0.15 ppb
HC_3N	<0.8 ppb	<0.4 ppb
CH_3C_2H	0.25 ± 0.03 ppb	0.12 ± 0.01 ppb
C_4H_2	0.16 ± 0.02 ppb	3 ± 1 ppt
CO_2	40 ± 5 ppt	–

[a]Converted into a mixing ratio using an H_2 column abundance of 400 km-amagat. (1 km amagat $= 2.69 \times 10^{24}$ cm^{-2}).
[b]0.6 ppmv in the troposphere and lower stratosphere, increasing to 22 ppmv in the upper stratosphere.

deeper in Uranus' atmosphere or a combination of these (and other) factors.

The upper atmosphere of Uranus is cold enough ($T = 76$ K at 1 bar) that the dry adiabatic lapse rate is influenced by the temperature-dependent distribution of hydrogen between its two nuclear spin isomers – ortho (with parallel nuclear spins) and para (with antiparallel nuclear spins). This is also true for the upper atmosphere of Neptune ($T = 72$ K at 1 bar), but not for the warmer atmospheres of Jupiter and Saturn.

The *Voyager 2* flyby provided conclusive evidence of Uranus' magnetic field, which can be modeled as a dipole offset $0.3 R_U$ from the center of Uranus and tilted $\sim 60°$ from the spin axis. The surface field strength varies from 0.1 to 1 gauss. The nature of the magnetic field suggests that it is generated by dynamo activity in a region of Uranus that extends out to $\sim 70\%$ of the planet's radius. In turn, this implies a convective, partially fluid interior, possibly an ionic ocean of ammonia in water. Uranus is too small to contain metallic hydrogen, which requires much higher pressures than exist inside Uranus or Neptune.

Neptune is smaller than Uranus, but is more massive (17.2 M_E) and has a mean density of 1.638 g cm^{-3}. Thus, Neptune is even more enriched than Uranus in elements heavier than He. A large enrichment of heavy elements is also suggested by the observed CH_4/H_2 molar ratio of $\sim 1\text{--}2\%$, which is 14 to 28 times larger than the solar C/H_2 ratio.

However, with the exception of CH_3D, no other hydrides are observed in Neptune's atmosphere and it is generally assumed that all heavy elements are enriched as carbon. Whether or not this assumption is correct remains to be seen, and it may be questioned because of the apparent depletions of water (and oxygen) on Jupiter and of ammonia (and nitrogen) on Uranus.

In fact, the H_2O abundance deduced for Neptune's atmosphere suggests that carbon and oxygen are not equally enriched. Water vapor is not observed at Neptune's cloud tops because the atmosphere is too cold. However, the CO/H_2 molar ratio of $\sim 0.6 \times 10^{-6}$ requires a water abundance several hundred times the solar O/H_2 ratio to produce the observed CO *via* the net reaction:

$$H_2O + CH_4 = 3H_2 + CO \tag{7.16}$$

in Neptune's deep atmosphere. The observed HCN/H_2 molar ratio of 0.3×10^{-9} requires an ammonia abundance greater than the solar N/H_2 ratio to produce sufficient amounts of N_2 (the precursor to HCN) *via* the net reaction (7.19) in Neptune's deep atmosphere:

$$2NH_3 = 3H_2 + N_2 \tag{7.19}$$

Vertical mixing transports the N_2 into Neptune's upper atmosphere where it reacts to form the observed HCN.

Interior structure models suggest that Neptune is water-rich. Dynamo action in a conductive, water-rich interior is also apparently required to produce the observed magnetic field, which is offset $0.55R_N$ from Neptune's center and tilted $47°$ to the rotational axis. In addition, as mentioned earlier, the high D/H ratio of Neptune suggests a large abundance of water and other hydrides.

7.2.4 Photochemistry on the Giant Planets

Ultraviolet sunlight drives photochemistry in the upper atmospheres of the four giant planets Jupiter, Saturn, Uranus and Neptune. In turn, photochemistry moves the upper atmospheres of these planets away from equilibrium and produces disequilibrium species [*e.g.*, ethane (C_2H_6), acetylene (C_2H_2) and ethylene (C_2H_4)] from methane and hydrazine (N_2H_4) from ammonia, the thermodynamically stable forms of carbon and nitrogen in the H_2-rich atmospheres of the giant planets. In addition, photochemistry of sulfur and phosphorus compounds is plausibly responsible for the colored belts and bands on Jupiter.

Just over 100 years ago in 1909 the American astronomer V. M. Slipher (1875–1969) photographed bands with an unknown origin in the atmosphere of Jupiter. Slipher speculated about the nature of these bands but was unable to identify their origin. Nearly 30 years later, the German astronomer Rupert Wildt (1905–1976) finally identified these features as CH_4 and NH_3 bands in the spectrum of Jupiter. At about the same time Sydney Chapman proposed his photochemical model for O_3 production and loss on Earth (discussed earlier in Chapter 5). Chapman's model and Wildt's identification of methane and ammonia on Jupiter led Wildt to his pioneering study (in 1937) of photochemistry of CH_4 and NH_3 on the outer planets. Wildt's concise statement of the problem is worth repetition:

"All polyatomic molecules detected in planetary atmospheres are highly sensitive to ultraviolet solar radiation. Their photochemical decomposition must be followed by secondary chemical reactions, reuniting the products of dissociation, in order to maintain the observed stationary composition of the atmospheres."

However, very little progress followed Wildt's qualitative suggestions until the 1960s. The reasons for this are that only after World War II were the necessary advances made in several areas:

- Rocket-borne spectroscopic observations of the UV solar flux,
- earth-based spectroscopic observations of gas abundances on Jupiter,
- laboratory measurement of absorption coefficients for CH_4, NH_3, *etc.*,
- laboratory measurement of rate constants for reactions of interest,
- application of digital computers to chemical kinetic modeling,
- development of algorithms for chemical kinetic networks.

With the advantage of these developments, the first modern models of hydrocarbon and ammonia photochemistry on Jupiter emerged in the late 1960s. The photochemical models have improved and grown in complexity over time. This is a natural consequence of the continuing improvements in spectroscopic observations of the solar flux and the giant planets, and in the laboratory studies of absorption coefficients and chemical reaction rates.

7.2.4.1 Methane. Methane is the most abundant hydrocarbon in the atmospheres of Jupiter, Saturn, Uranus and Neptune, and it is photolyzed by UV sunlight ($\lambda < 160$ nm) to initiate the formation of more complex hydrocarbons. The Lyman α line at 121.6 nm provides most of the photons in this wavelength region. Methane photochemistry occurs high in Jupiter's atmosphere at pressures of 10^{-5} to 10^{-7} bar, in the middle and upper stratomesosphere. This is just below the homopause (or turbopause) level at about 10^{-7} bar where diffusive separation begins. Methane absorbs essentially all UV sunlight ($\lambda < 145$ nm) by 10^{-5} bar, and direct photolysis of CH_4 stops below this level (*i.e.*, at higher pressures). However, energetic H atoms produced by photolysis of acetylene (C_2H_2) and other hydrocarbons can still destroy CH_4 in Jupiter's middle and lower stratomesosphere.

Ethane is the major hydrocarbon produced by methane photochemistry. The elementary reactions forming C_2H_6 and the net photochemical reaction are:

$$CH_4 + h\nu \rightarrow {}^1CH_2 + H_2 \quad (q \sim 0.41) \tag{7.20}$$

$$CH_4 + h\nu \rightarrow {}^1CH_2 + H_2 \quad (q \sim 0.41) \tag{7.20}$$

$$^1CH_2 + H_2 \rightarrow H + CH_3 \tag{7.21}$$

$$^1CH_2 + H_2 \rightarrow H + CH_3 \tag{7.21}$$

$$CH_3 + CH_3 + M \rightarrow C_2H_6 + M \tag{7.22}$$

$$\textit{Net reaction} : 2CH_4 \rightarrow C_2H_6 + 2H \tag{7.23}$$

The $^1CH_2$, CH_3 and M in the reactions above are the methylene radical in its singlet excited energy state, the methyl CH_3 radical and a third body (or collision partner). Molecular H_2 and He are the most probable third bodies because they are the most abundant gases on Jupiter and the other giant planets. The q in the first reaction is the quantum yield, or branching ratio, and it shows that $^1CH_2$ plus H_2 form 41% of the time when CH_4 is photolyzed.

The termolecular recombination of two methyl CH_3 radicals is responsible for most ethane production, and this reaction occurs throughout Jupiter's stratomesosphere. The methyl radical is observed in the stratomesospheres of Jupiter, Saturn and Neptune with abundances of 1.5–7.5×10^{13} molecules cm^{-2} (Saturn) and 0.7–2.8×10^{13} molecules cm^{-2} (Neptune).

Ethane absorbs UV sunlight ($\lambda < 160$ nm), but most C_2H_6 is not destroyed photochemically because it is shielded by methane, which is more abundant and also absorbs light with wavelengths < 160 nm. Instead, downward transport into Jupiter's troposphere is the major loss process for ethane. Deep in the tropospheres of the giant planets at ~ 1000 K, thermally driven reactions are fast enough to destroy ethane and methane reforms *via* the net reaction:

$$C_2H_6 + H_2 = 2CH_4 \qquad (7.4)$$

Acetylene (C_2H_2) is the next most abundant product of CH_4 photochemistry. The reaction scheme leading to acetylene is:

$$CH_4 + h\nu \rightarrow CH_2 + 2H \quad (q \sim 0.51) \qquad (7.24)$$

$$CH_4 + h\nu \rightarrow CH_2 + 2H \quad (q \sim 0.51) \qquad (7.24)$$

$$CH_2 + CH_2 \rightarrow C_2H_2 + H_2 \qquad (7.25)$$

$$Net\, reaction: 2CH_4 \rightarrow C_2H_2 + H_2 + 4H \qquad (7.26)$$

The q in Reaction (7.24) shows that two H atoms plus methylene radicals in their ground energy state forms about 51% of the time when CH_4 is photolyzed. Photolysis or H atom addition destroys most of the acetylene. The latter reaction converts the triple bond in acetylene into the C=C double bond in ethylene:

$$C_2H_2 + H + M \rightarrow C_2H_3 + M \qquad (7.27)$$

$$C_2H_3 + H_2 \rightarrow C_2H_4 + H \qquad (7.28)$$

$$Net\, reaction: C_2H_2 + H_2 \rightarrow C_2H_4 \qquad (7.29)$$

The C_2H_3 in the reactions above is the vinyl radical.

Acetylene photolysis initiates reaction chains that produce either alkanes such as propane (C_3H_8) and butane (C_4H_{10}) or polyynes (polyacetylenes) such as diacetylene (C_4H_2). The reactions leading to propane are:

$$C_2H_2 + h\nu \rightarrow C_2H + H \qquad (7.30)$$

$$C_2H_2 + h\nu \rightarrow C_2H + H \qquad (7.30)$$

$$C_2H + CH_4 \rightarrow C_2H_2 + CH_3 \qquad (7.31)$$

$$C_2H + C_2H_6 \rightarrow C_2H_2 + C_2H_5 \qquad (7.32)$$

$$CH_3 + C_2H_5 + M \rightarrow C_3H_8 + M \qquad (7.33)$$

$$Net\, reaction: CH_4 + C_2H_6 \rightarrow C_3H_8 + 2H \qquad (7.34)$$

The C_2H and C_2H_5 in the reactions above are the ethynyl (C_2H) and ethyl (C_2H_5) radicals. Propane occurs in Saturn's stratomesosphere with an abundance of about 26 ppbv. Some propane is lost by photolysis ($\lambda < 170$ nm) and the rest by transport downward into Saturn's troposphere where it eventually reacts with H_2 reforming methane. The net thermochemical reaction for this is:

$$C_3H_8 + 2H_2 = 3CH_4 \tag{7.35}$$

The reactions leading to diacetylene start with acetylene photolysis:

$$C_2H_2 + h\nu \rightarrow C_2H + H \tag{7.30}$$

$$C_2H + C_2H_2 \rightarrow C_4H_2 + H \tag{7.36}$$

$$Net\,reaction : 2C_2H_2 \rightarrow C_4H_2 + 2H \tag{7.37}$$

The IRIS spectrometer on the Voyager spacecraft observed diacetylene with an abundance of 1–40 ppbv in the atmosphere of Titan, Saturn's largest satellite. However, C_4H_2 has not yet been detected on the giant planets.

Ethylene (C_2H_4) is the third most abundant product of CH_4 photochemistry. It is produced by several sets of reactions, for example:

$$CH_4 + h\nu \rightarrow CH_2 + 2H \quad (q \sim 0.51) \tag{7.24}$$

$$CH_2 + H \rightarrow H_2 + CH \tag{7.38}$$

$$CH + CH_4 \rightarrow C_2H_4 + H \tag{7.39}$$

$$Net\,reaction : 2CH_4 \rightarrow C_2H_4 + H_2 + 2H \tag{7.40}$$

Ethylene is short lived and is destroyed by hydrogenation back to methane:

$$C_2H_4 + H + M \rightarrow C_2H_5 + M \tag{7.41}$$

$$C_2H_5 + H \rightarrow CH_3 + CH_3 \tag{7.42}$$

$$CH_3 + H + M \rightarrow CH_4 + M \tag{7.43}$$

$$CH_3 + H + M \rightarrow CH_4 + M \tag{7.43}$$

$$Net\,reaction : C_2H_4 + 4H \rightarrow 2CH_4 \tag{7.44}$$

The reactions listed above are only a small subset of those involved in hydrocarbon photochemistry on the giant planets. Methane photolysis by UV sunlight initiates the production of many higher hydrocarbons. Several of these are observed in the upper atmospheres of the giant planets, some are seen in the atmosphere of Titan, Saturn's largest satellite, while others remain to be discovered.

7.2.4.2 Ammonia. Ammonia is the major nitrogen-bearing gas observed in the atmospheres of the giant planets Jupiter, Saturn, Uranus and Neptune. On the one hand, thermodynamic equilibrium at the high temperatures and pressures in the deep atmospheres of the giant planets shifts the equilibrium between NH_3 and its constituent elements strongly to the ammonia side of the equation:

$$N_2 + 3H_2 = 2NH_3 \qquad (7.19)$$

However, on the other hand, the presence of NH_3 is somewhat surprising because UV sunlight ($\lambda < 230$ nm) readily destroys ammonia. The photodissociation rate constant for NH_3 photolysis at 1 AU is $J_1 = 1.80 \times 10^{-4} \, s^{-1}$, corresponding to a lifetime of only 1.6 h (96 min). Although the photodissociation rate constant at Jupiter's distance from the Sun (5.2 AU) will be ~ 27-times smaller, the lifetime of NH_3 is still relatively short (~ 42 h). These values are appropriate at the top of Jupiter's atmosphere and neglect absorption and/or scattering of UV sunlight by other gases in the atmosphere. When these effects are taken into account the photochemical lifetime of NH_3 increases to about 38 days. Nevertheless, the destruction and reformation of NH_3 on the giant planets must be a dynamic process that is going on at all times.

The major features of NH_3 photochemistry on the giant planets are simple. Ammonia has a higher melting point, higher boiling point and higher vapor pressure than CH_4 and condenses out of the atmospheres of Jupiter and Saturn as NH_3 ice while CH_4 does not. At even greater depths in these atmospheres, some NH_3 is lost by dissolution in aqueous water clouds and by formation of ammonium hydrosulfide (NH_4SH). Although CH_4 may condense in the colder atmospheres of Uranus and Neptune, it does so at a much higher altitude than NH_3 is removed from the gas by dissolution in water clouds, formation of NH_4SH or condensation of NH_3 ice.

Ammonia photolysis mainly occurs in the 200–700 mbar region on Jupiter in the vicinity of the NH_3 ice cloud ($P \sim 500$ mbar, $T \sim 138$ K) where the abundance of NH_3 gas is limited by its saturation vapor pressure. As discussed in the last section, CH_4 photolysis occurs much higher in Jupiter's atmosphere, in the 10^{-5}–10^{-7} bar region. Thus, CH_4 absorbs the UV sunlight below 160 nm and only UV sunlight in the 160–230 nm range is responsible for NH_3 photolysis on the giant planets. The initial step is dominated by the formation of the NH_2 (amidogen) radical plus H:

$$NH_3 + h\nu \rightarrow NH_2 + H \tag{7.45}$$

Another pathway forms the NH (imidogen) radical plus H_2:

$$NH_3 + h\nu \rightarrow NH(a^1\Delta) + H_2 \tag{7.46}$$

However, Reaction (7.46), forming excited NH radicals, is much less important than Reaction (7.45). Any NH formed is rapidly converted into NH_2 *via* reaction with H_2, which makes up most of the atmospheres of the giant planets:

$$NH(a^1\Delta) + H_2 \rightarrow NH_2 + H \tag{7.47}$$

Some of the NH_2 radicals are recycled to ammonia *via* reaction with H atoms:

$$NH_2 + H + M \rightarrow NH_3 + M \tag{7.48}$$

The rest of the NH_2 reacts with itself to form hydrazine (N_2H_4):

$$NH_2 + NH_2 + M \rightarrow N_2H_4 + M \tag{7.49}$$

The hydrazine formed in this reaction has two fates. The more likely fate is condensation of hydrazine vapor into hydrazine ice crystals in cool regions of Jupiter's upper atmosphere where NH_3 photolysis occurs:

$$N_2H_4 \text{ (gas)} = N_2H_4 \text{ (ice)} \tag{7.50}$$

Otherwise, gaseous hydrazine is photolyzed by UV sunlight to generate N_2 *via* a series of reactions:

$$N_2H_4 + h\nu \rightarrow N_2H_3 + H \tag{7.51}$$

$$N_2H_4 + H \rightarrow N_2H_3 + H_2 \tag{7.52}$$

$$N_2H_3 + H \rightarrow N_2H_2 + H_2 \tag{7.53}$$

$$N_2H_3 + H \rightarrow 2NH_2 \tag{7.54}$$

$$N_2H_3 + N_2H_3 \rightarrow N_2H_4 + N_2H_2 \tag{7.55}$$

$$N_2H_3 + N_2H_3 \rightarrow 2NH_3 + N_2 \tag{7.56}$$

$$N_2H_2 + M \rightarrow N_2 + H_2 + M \tag{7.57}$$

Ultimately, the N_2H_4 and N_2 produced by ammonia photolysis are reconverted back into NH_3 *via* thermally driven reactions in the hot, high-pressure deep tropospheres of the giant planets. The net thermochemical reactions are:

$$H_2 + N_2H_4 = 2NH_3 \tag{7.58}$$

$$3H_2 + N_2 = 2NH_3 \tag{7.19}$$

As mentioned earlier this recycling must occur continually or else all NH_3 in Jupiter's atmosphere would be depleted on a short time scale relative to the age of the solar system.

7.2.4.3 Phosphine. Phosphine (PH_3) is the major phosphorus-bearing gas observed in the atmospheres of the giant planets. Phosphine, unlike ammonia, is not thermodynamically stable in the upper tropospheres of the giant planets. Instead, as discussed earlier, it is thermodynamically stable in the deep atmospheres of the giant planets but becomes unstable at higher altitudes. Phosphine persists because its oxidation to P_4O_6 gas by water vapor is much slower than vertical mixing to cooler regions where the oxidation is too slow to occur. Thus, vertical mixing quenches PH_3 destruction before it can proceed.

Phosphine absorbs UV light in about the same wavelength region as NH_3, and its photochemistry is similar to that of ammonia. Initially, PH_3 absorbs UV sunlight and dissociates into a phosphino (PH_2) radical and an H atom:

$$PH_3 + h\nu \rightarrow PH_2 + H \tag{7.59}$$

The H atoms formed in this step, or by photolysis of NH_3, have two fates. The more likely fate is that they attack PH_3, giving another

PH_2 plus H_2

$$PH_3 + H \rightarrow PH_2 + H_2 \tag{7.60}$$

The less likely fate is recombination of two H atoms giving H_2:

$$H + H + M \rightarrow H_2 + M \tag{7.61}$$

Statistically, M is probably either H_2 or He because these are the two most abundant gases on the giant planets.

Subsequent chemistry leading to the formation of other phosphorus-bearing compounds depends upon the PH_2 radicals formed in Reactions (7.59) and (7.60). Their recombination gives diphosphine (P_2H_4):

$$PH_2 + PH_2 + M \rightarrow P_2H_4 + M \tag{7.62}$$

In turn, H atoms and PH_2 radicals attack diphosphine, which leads to a series of reactions ending with the formation of elemental phosphorus vapor:

$$P_2H_4 + H \rightarrow P_2H_3 + H_2 \tag{7.63}$$

$$P_2H_4 + PH_2 \rightarrow P_2H_3 + PH_3 \tag{7.64}$$

$$P_2H_3 + PH_2 \rightarrow P_2H_2 + PH_3 \tag{7.65}$$

$$P_2H_3 + P_2H_3 \rightarrow P_2H_2 + P_2H_4 \tag{7.66}$$

$$P_2H_2 + M \rightarrow P_2 + H_2 \tag{7.67}$$

$$P_2 + P_2 + M \rightarrow P_4 + M \tag{7.68}$$

$$P_4(gas) = P_4(solid) \tag{7.69}$$

This reaction sequence produces solid red phosphorus, which may be responsible for the color of the Great Red Spot on Jupiter.

Coupled photochemistry of PH_3 and NH_3 also occurs because both gases are present in the same altitude regions of the giant planet atmospheres. Aminophosphine (NH_2PH_2) is a predicted, as yet, unobserved product of these coupled reactions. The production of thermally excited, or hot, hydrogen atoms (H^*) by photolysis of NH_3 leads to photochemical reaction schemes involving methane, which yield methylamine (CH_3NH_2):

$$NH_3 + h\nu \rightarrow NH_2 + H^* \tag{7.70}$$

$$H^* + CH_4 \rightarrow H_2 + CH_3 \tag{7.71}$$

$$CH_3 + NH_2 + M \rightarrow CH_3NH_2 + M \tag{7.72}$$

$$\textit{Net reaction} : NH_3 + CH_4 \rightarrow CH_3NH_2 + H_2 \tag{7.73}$$

Likewise, the production of hot hydrogen atoms by photolysis of PH_3 initiates a reaction scheme producing methylphosphine (CH_3PH_2):

$$PH_3 + h\nu \rightarrow PH_2 + H^* \tag{7.74}$$

$$H^* + CH_4 \rightarrow H_2 + CH_3 \tag{7.75}$$

$$CH_3 + PH_2 + M \rightarrow CH_3PH_2 + M \tag{7.76}$$

$$\textit{Net reaction} : CH_4 + PH_3 \rightarrow CH_3PH_2 + H_2 \tag{7.77}$$

The chemical kinetic data for coupled photochemistry of NH_3, PH_3 and CH_4 are uncertain and these possible species are yet unobserved. Furthermore, several of the products probably condense out of the giant planet atmospheres as aerosol droplets or ice crystals.

7.2.4.4 Hydrogen Sulfide. As discussed earlier, at thermochemical equilibrium H_2S is the expected dominant sulfur-bearing gas on the giant planets because the large H_2 partial pressure strongly shifts the equilibrium toward the right-hand side:

$$H_2 + S_2 = 2H_2S \tag{7.78}$$

The mass spectrometer on the Galileo entry probe into Jupiter's atmosphere found that H_2S was the most abundant sulfur-bearing gas, verifying the predictions of the chemical equilibrium calculations.

Hydrogen sulfide absorbs UV light with wavelengths $\lambda < 270$ nm:

$$H_2S + h\nu \rightarrow SH + H \ (\lambda < 270 \text{ nm}) \tag{7.79}$$

In the reaction above, SH is the mercapto radical. The H atoms formed by H_2S photolysis can attack H_2S to produce more

SH radicals:

$$H_2S + H \rightarrow SH + H_2 \tag{7.80}$$

Some of the H_2S destroyed by photolysis and H atom attack reforms *via*:

$$H + SH + M \rightarrow H_2S + M \tag{7.81}$$

$$SH + SH \rightarrow H_2S + S \tag{7.82}$$

However, most of the H atoms and mercapto SH radicals formed by photolysis react to form elemental sulfur vapor:

$$H + SH \rightarrow H_2 + S \tag{7.83}$$

$$SH + SH \rightarrow H_2 + S_2 \tag{7.84}$$

Reactions between S and S_2 eventually lead to S_8 vapor, which subsequently condenses as elemental sulfur (S_8), *e.g.*, the sequence:

$$S + S + M \rightarrow S_2 + M \tag{7.85}$$

$$S + S_2 + M \rightarrow S_3 + M \tag{7.86}$$

$$S_3 + S_3 + M \rightarrow S_6 + M \tag{7.87}$$

$$S_2 + S_6 + M \rightarrow S_8 + M \tag{7.88}$$

$$S_8(gas) = S_8(solid) \tag{7.89}$$

Elemental sulfur is yellow at room temperature and may account for the yellow coloration of clouds near this temperature in Jupiter's atmosphere.

7.3 SATELLITES AND RINGS OF THE GIANT PLANETS

All of the giant planets have rings and extensive satellite systems – 63 (Jupiter), 61 (Saturn), 27 (Uranus) and 13 (Neptune) as of the time of writing. Many of these satellites are fascinating worlds with diverse properties that were revealed only in the past three decades with the *Voyager*, *Galileo* and *Cassini* missions. Tables 7.4–7.7 list some physical properties for some satellites of Jupiter, Saturn, Uranus and Neptune,

Table 7.4   Some physical properties of the Galilean satellites.

Property	Io	Europa	Ganymede	Callisto
Semimajor axis (R_{Jup})	5.897	8.969	14.98	26.34
Mean radius (km)	1821.6	1565.0	2631.2	2410.3
Mass (10^{23} kg)	0.893	0.480	1.4815	1.076
Mean density (g cm^{-3})	3.528	2.989	1.942	1.834
GM (km^3 s^{-2})	5959.91	3202.72	9887.83	7179.29
$g = GM/R^2$ (m s^{-2})	1.797	1.31	1.425	1.24
C/MR^2	0.37824	0.346	0.3115	0.3549

respectively. The sizes, masses and thus the densities of most of the satellites of the giant planets are unknown. We refer the reader to the web pages at the National Space Science Data Center (http://nssdc.gsfc.nasa.gov/planetary/) for the available information on the rest of the 164 satellites of the giant planets.

The satellites of the giant planets are diverse and span a wide size range. For example, Tables 7.4–7.7 show that six satellites have radii larger than 1000 km. These are Ganymede (2631 km), Titan (2575 km), Callisto (2410 km), Io (1822 km), Europa (1565 km) and Triton (1353 km). Ganymede and Titan are larger than and Callisto is about the same size as Mercury (2438 km radius). Io is comparable in size to Earth's Moon (1737 km). Several of the large satellites are larger than or comparable in size to dwarf planets, *e.g.*, Eris ($\sim$1250 km), Pluto (1153 km), Makemake ($\sim$750 km), Haumea ($\sim$575 km) and Ceres (475 km).

However, other satellites of the giant planets are only a few km across and may be irregularly shaped. Some of these small satellites may be objects captured by the strong gravitational fields of the giant planets. This is probably the case for the outermost satellites of Jupiter. Jupiter has four small irregularly shaped satellites, which are Metis, Adrastea, Amalthea and Thebe. They orbit at 1.8–3.1 Jovian radii and are sources of dust for the rings around Jupiter. The three smaller ones (Metis, Adrastea and Thebe) plausibly formed in the Jovian subnebula along with the four Galilean satellites, but Amalthea may be a captured object.

Jupiter's rings were only discovered in March 1979 (by Voyager 1) and are much fainter than Saturn's rings, which were discovered by Galileo in 1610. The rings of Jupiter are probably made of organic and rocky material and are much darker than Saturn's rings. The latter are mainly water ice, but organic and rocky material is probably also present.

Most of Saturn's satellites are smaller than 200 km radius and irregularly shaped. The five largest ones are Titan (2575 km), Rhea

Table 7.5 Physical properties of Saturnian satellites (sources: refs 2–5).

Satellite	Radius (km)	Mass (10^{19} kg)	Density ($g\,cm^{-3}$)	Semimajor axis (R_{Sat})	Shape[a]
Pan	14.1	0.000 495	0.420	2.217	I
Daphnis	3.8	0.000 0077	0.340	2.265	I
Atlas	15.1	0.000 66	0.460	2.285	I
Prometheus	43.1	0.015 95	0.480	2.313	I
Pandora	40.7	0.013 71	0.490	2.351	I
Epimetheus	58.1	0.052 66	0.640	2.512	I
Janus	89.5	0.190	0.630	2.514	I
Mimas	198.2	3.749	1.149	3.078	E
Enceladus	252.1	10.80	1.609	3.949	E
Tethys	531.0	61.74	0.985	4.890	E
Dione	561.4	109.54	1.478	6.262	E
Rhea	763.5	230.65	1.237	8.744	E
Titan	2574.73	13 452.	1.880	20.273	T
Hyperion	135.	0.562	0.544	24.575	I
Iapetus	734.3	180.56	1.088	59.091	E
Phoebe	106.5	0.829	1.638	214.9	I

[a]E – ellipsoidal, I – irregular, T – triaxial ellipsoid.

Table 7.6 Physical properties of Uranian satellites (source: ref. 6).

Satellite	Radius (km)	Mass (10^{20} kg)	Density ($g\,cm^{-3}$)	Semimajor axis (R_{Ura})
Miranda	236	0.659	1.20	5.063
Ariel	579	13.53	1.66	7.473
Umbriel	586	11.72	1.39	10.42
Titania	790	35.27	1.71	17.06
Oberon	762	30.14	1.63	22.83

Table 7.7 Some physical properties of Triton.

Property	Value
Semimajor axis (R_{Nep})	14.33
Mean radius (km)	1352.6
Mass (10^{22} kg)	2.147
Mean density ($g\,cm^{-3}$)	2.061
$g = GM/R^2$ ($m\,s^{-2}$)	0.78
Mass% ices	30–45
Surface T (K)	~38
Surface P (μbar)	~16
Atmospheric gases	N_2, CO, CH_4

(764 km), Iapetus (734 km), Dione (561 km) and Tethys (531 km). Enceladus is the sixth largest satellite and emits gas plumes (cryovolcanism) discovered by the *Cassini* orbiter. The plumes are mainly water vapor with smaller amounts of N_2, CH_4, CO_2 and NH_3. Saturn's

small inner satellites (Pan, Daphnis, Atlas, Prometheus and Pandora) have low densities ($<1\,\mathrm{g\,cm^{-3}}$) and may be objects formed by accretion of smaller "moonlets" from the rings onto a pre-existing core. These small inner satellites orbit within gaps in Saturn's rings, *e.g.*, Pan is in the Encke gap and Daphnis is in the Keeler gap, and interact with the rings to maintain the gaps. Several other satellites also have densities less than $1\,\mathrm{g\,cm^{-3}}$ (*e.g.*, Epimetheus, Hyperion, Janus) and presumably have significant porosity.

Astronomical observations (*e.g.*, of size, shape, mass, albedo and reflection spectra), solar elemental abundances, cosmochemical considerations and interior structure models show that to first approximation the satellites of the giant planets are composed of varying amounts of four components – ices, rocks, organics and porosity. This is probably also true for other small bodies in the outer solar system such as Centaur objects, comets and dwarf planets (*e.g.*, Ceres, Eris, Haumea, Makemake and Pluto) and their satellites (*e.g.*, Charon, Hydra, Nix, Haumea I Hi'aka and Haumea II Namaka).

The rocky material in the satellites of the giant planets may be hydrous rock similar to that in CI chondrites or anhydrous rock similar to that in ordinary (H, L, LL) chondrites. The hydrous rock in CI chondrites does not contain metal and is mainly serpentine, talc and magnetite with smaller amounts of carbonates and sulfates. The anhydrous rock in ordinary chondrites is a mixture of Fe-Ni metal alloy, silicates (mainly olivine, orthopyroxene, feldspar, clinopyroxene) and troilite with smaller amounts of oxide and phosphate minerals. (Chapter 2 gives a detailed discussion of the mineralogy of chondritic meteorites.)

The icy material is generally dominated by water ice without (or with) more volatile C- and N-bearing ices such as CO, CO_2, CH_4, N_2, NH_3, NH_4OH and clathrate hydrates. Quantitatively, water ice is the most important ice because oxygen is the third most abundant element in the solar system. Water ice has the lowest vapor pressure and highest condensation temperature of any of the ices present on satellites (or other bodies) in the outer solar system. Thus, it is the first ice to condense with decreasing temperature and the last ice to evaporate with increasing temperature in the solar nebula. The water ice condensation temperature (T_C) as a function of the total pressure (P_T) in the solar nebula is given by the equation:

$$\frac{10\,000}{T_C} = 38.84 - 3.83 \log_{10} P_T$$

This gives a temperature of 185 K at 10^{-4} bar total pressure (*e.g.*, see Figure 3.8). Fifty percent of all water ice condenses by 180 K. As discussed earlier in Chapter 3, liquid water can form if the total pressure is at least 3.8 bar. However, it is unlikely that pressures were this high in the solar nebula.

The amount of water ice formed depends on the amount of oxygen consumed by other compounds. Rock consumes about 20–25% of all oxygen in solar composition material, leaving 80–75% for formation of O-bearing gases such as H_2O and CO. (The amount of oxygen consumed by rock varies from $\sim 20\%$ for anhydrous rock plus Fe metal, like that found in ordinary chondrites, to $\sim 25\%$ for hydrous rock plus magnetite, like that found in CI chondrites.) Carbon has a solar elemental abundance that is 46% that of oxygen. Thus, the amount of CO formed in the solar nebula has a large effect on the amount of oxygen present in H_2O. Figure 3.11 shows that carbon occurs as CO at low pressures and high temperatures while carbon is in CH_4 at high pressures and low temperatures. This thermodynamic partitioning and the slow reduction of CO to CH_4 at low pressures mean that most carbon is in CO in the solar nebula. Consequently, water ice is about 35% and rock is about 65% by mass of the solid material in the outer solar nebula (until other more volatile ices form). The water ice/rock mass ratio is predicted to be about 0.55 for icy bodies formed in the solar nebula.

In contrast, Figure 3.11 and the kinetics of CO reduction to CH_4 mean that most carbon is in CH_4 in the circumplanetary subnebulae around Jupiter, Saturn, Uranus and Neptune during their formation. In this case water ice is about 56% and rock is about 44% by mass of solid material (until more volatile ices form). The water ice/rock mass ratio is predicted to be ~ 1.3 for icy bodies formed in circumplanetary subnebulae around the giant planets. This is over twice as large as the water ice/rock ratio for icy bodies formed in the solar nebula. Thus, in principle, the water ice/rock ratio of an icy body is diagnostic of its formation location (solar nebula *versus* circumplanetary subnebula).

Condensation of C- and N-bearing ices at lower temperatures increases the mass fraction of ice and the ice/rock ratio. As discussed in Chapter 3 and illustrated in Figure 7.3, different C- and N-bearing ices form in the solar nebula and in the circumplanetary subnebulae around the giant planets. Carbon monoxide is the major C-bearing gas in the solar nebula and condenses as CO clathrate hydrate $CO \cdot 6H_2O$ *via* reaction of CO with water ice. This occurs at ~ 60 K at 10^{-4} bar total pressure. However, there is not enough water ice to condense all CO as a clathrate hydrate because the CO to H_2O ratio is about 1 : 6 in the clathrate hydrate while the solar C/O atomic ratio is about 1 : 2. Most CO condenses as CO ice at ~ 20 K (at 10^{-4} bar total pressure). The condensation temperatures for CO clathrate hydrate and CO ice

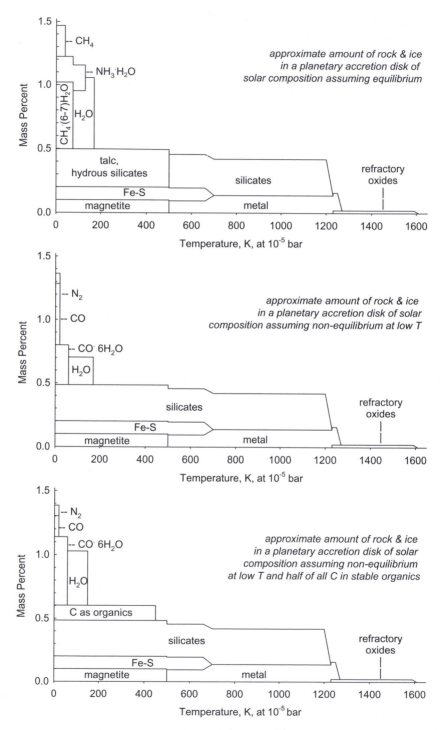

Figure 7.3   Carbon- and nitrogen-bearing ices that form in the solar nebula and in the circumplanetary subnebulae around the giant planets.

increase with increasing total pressure in the solar nebula with a slope similar to that for water ice.

Molecular nitrogen is the major N-bearing gas in the solar nebula and it forms N_2 clathrate hydrate at about the same temperature as CO clathrate hydrate. A mixed clathrate hydrate containing both CO and N_2 may form because the two gases have similar properties. The remaining N_2 condenses as N_2 ice at $\sim 20\,K$ (at 10^{-4} bar total pressure) and at higher temperatures at higher total pressures. The complete condensation of CO and N_2 ices in the solar nebula increases the ice/rock mass ratio from 0.55 to almost 3 (Figure 7.3).

Condensation of C- and N-bearing solids occurs at higher temperatures in the circumplanetary subnebulae around the giant planets during their formation and the condensates are dominated by reduced C- and N-bearing compounds such as methane and ammonia.

With decreasing temperature, ammonia first condenses as ammonium bicarbonate NH_4HCO_3 (at 168–190 K) or as ammonium carbamate $NH_4CO_2NH_2$ (at 173–188 K) in the Jovian circumplanetary subnebula. The formation of either one consumes all available CO_2 (which is a small fraction of the total gaseous carbon) and prevents formation of the other compound, but their stabilities are so similar that which forms first is unclear. Ammonium bicarbonate (or ammonium carbamate) make up a tiny fraction (10^{-9} to 10^{-6}) of the total condensate mass, but are important because they are one of two carbon sources in the condensed material (HCN is the other one). Most NH_3 condenses as ammonia monohydrate ($NH_3 \cdot H_2O$), also known as ammonium hydroxide (NH_4OH), at 160 K. Hydrogen cyanide (HCN) ice condenses at slightly lower temperatures (110–130 K) and makes up 10^{-6} to 10^{-4} of the total condensate mass. This tiny amount of HCN is important because it is a precursor for production of adenine, one of the four nucleotide bases in DNA and RNA.

Small amounts of carbon first condense in ammonium bicarbonate or ammonium carbamate as discussed above. However, most carbon condenses either as methane clathrate hydrate (94 K) or methane ice (41 K). This plausibly took place in circumplanetary subnebulae around Saturn, Uranus or Neptune but not around Jupiter because there is no evidence that the Jovian circumplanetary subnebula was this cold. Thus, ammonium bicarbonate or ammonium carbamate and HCN are the only carbon sources for satellites formed in the Jovian circumplanetary subnebula.

The organic material may be similar to the insoluble aromatic polymer found in CI and CM carbonaceous chondrites. The elemental composition of the polymer in the Orgueil CI chondrite is

$C_{100}H_{72}N_2O_{10}S_4$, corresponding to 75% carbon, 10% oxygen, 8% sulfur, 5% hydrogen and 2% nitrogen by mass. The structure of the insoluble aromatic polymer in CI and CM carbonaceous chondrites is complex and is an active research subject. Some researchers think the polymer is similar to coal, others think it is similar to humic acids in soil and still others think it is similar to bitumins. Low albedos and reflection spectra indicate that spectrally dark carbonaceous material (possibly similar to the insoluble aromatic polymer) is present on several satellites of the giant planets, the rings of Uranus and other small bodies in the outer solar system.

7.3.1 Galilean Satellites

The Galilean satellites of Jupiter (Io, Europa, Ganymede and Callisto) are the classic example for illustrating variable contents of icy and rocky material. As Table 7.4 shows, their densities decrease outward from Jupiter. Io has a high density of $3.528 \, \text{g cm}^{-3}$ and it is composed of metal, silicate and sulfide. Europa's density of $2.989 \, \text{g cm}^{-3}$ indicates a water-bearing, rock-rich composition. Models indicate that Ganymede (with a density of $1.942 \, \text{g cm}^{-3}$) is 27–57% water ice and Callisto (with a density of $1.834 \, \text{g cm}^{-3}$) is 34–58% water ice. All of the Galilean satellites are differentiated and liquid water layers ("oceans") are thought to exist inside Europa, Ganymede and possibly Callisto. All four satellites probably have metal–sulfide cores.

Water ice is present on the surfaces of Europa, Ganymede and Callisto and is the source of the tenuous oxygen (O, O_2) atmospheres on these satellites. (Europa's atmosphere is mainly CO_2 with smaller amounts of O and O_2.) Charged particle irradiation of the water ice produces O_2 and O_3, which are seen in water ice on the surfaces of Ganymede and Callisto. Sulfuric acid hydrates are also thought to be produced by charged particle reactions with water ice on Europa's surface, which is bombarded with SO_2 from Io.

Io is innermost Galilean satellite of Jupiter and it is the most volcanically active object in the solar system. Io is about the same size as Earth's Moon (1.05 times larger), has about the same mass (1.22 times larger), the same density (within 0.5%), but is geologically active with an enormous volcanic eruption rate $> 500 \, \text{km}^3$ of magma per year. The Moon is not geologically active and has no active volcanism. Earth is volcanically active and has a global average volcanism rate of about $20 \, \text{km}^3$ magma per year, which is dominated by plate creation at mid ocean ridges. Subaerial volcanism on Earth is only $\sim 1 \, \text{km}^3$ magma per year. The high rate of volcanism on Io is equivalent to an average global resurfacing rate of ~ 1.2 cm per year. Io's volcanic activity is due to intense tidal heating by Jupiter, Europa and Ganymede. The heat flux on Io is over 25 times that on Earth (2.5 *versus* $0.09\,\mathrm{W\,m^{-2}}$) and over 80 times that on the Moon.

JPL engineer Linda Morabito discovered Io's volcanism on images taken by the *Voyager 1* spacecraft in March 1979. Infrared observations made by the Voyager spacecraft showed eruption of SO_2 and sulfur vapor. The sulfur vapor condenses out on Io's surface as colorful patches of different allotropes (*e.g.*, S_6, S_8). Voyager observed volcanic vent temperatures up to 650 K. Subsequent Earth-based and Galileo observations showed significantly higher volcanic vent temperatures (up to ~ 1500 K) that indicate eruption of silicate magmas.

Geochemists have demonstrated that terrestrial volcanic gases erupted at similar temperatures (>900 K) are complex gas mixtures in chemical equilibrium. For example, in January 1983 the Kilauea basaltic volcano in Hawaii erupted volcanic gas composed mainly of H_2O (80.4%) with smaller amounts of SO_2 (14.0%), CO_2 (3.52%), H_2 (0.93%), H_2S (0.51%), S_2 (0.20%), HF (0.19%), HCl (0.17%) plus trace amounts ($<0.1\%$) of other species. Thus it is likely that volcanic gases erupted on Io are not pure SO_2 but are also chemically equilibrated gas mixtures. Chemical equilibrium calculations and astronomical observations have shown that this is the case. For example, volcanic gas erupted at the Pele vent has a temperature of ~ 1440 K and a pressure of $\sim 10\,\mu$bar. The gas is mainly SO_2 (82%), with smaller amounts of S_2 (12%) and SO (6%). Sodium chloride vapor, K gas and probably KCl vapor are also erupted by Pele with abundances up to a few percent. The possible identification of SO_2Cl_2 around the Marduk vent is consistent with calculations that sulfur chlorides may be emitted from low temperature vents too cold to vaporize Na and K from the rock.

Volcanism on Io is the ultimate source of its thin, patchy SO_2-rich atmosphere, which has a maximum pressure of about 10 nbar and is controlled by vapor pressure equilibrium with SO_2 ice on Io's surface. Sulfur monoxide is the second most abundant gas observed in Io's atmosphere and is produced volcanically and by photolysis of SO_2.

Io orbits within Jupiter's magnetosphere and loses ~ 1000 kg mass per second into the magnetosphere. This material is mainly S, O, Na, Cl and K and forms the Io plasma torus within which Io orbits and the neutral Na and K clouds that surround Io. Other elements that are plausibly present on Io's surface (*e.g.*, Mg, Si, Fe, Ca, Al, P) and in its volcanic gases (*e.g.*, F, Br, Se, As) are probably also lost into the plasma torus.

Table 7.8 Chemical composition of the atmosphere of Titan.

Gas	Abundance	Notes	Reference
N_2	95.1–98.4%	By difference	–
CH_4	1.41×10^{-2}(strat.)	Huygens GC-MS	7
	4.90×10^{-2}(surf.)		
H_2	$9.6 \pm 2.4 \times 10^{-4}$	Cassini CIRS, troposphere	8
CO	$4.7 \pm 0.8 \times 10^{-5}$	Cassini CIRS, stratosphere	9
^{40}Ar	$4.32 \pm 0.1 \times 10^{-5}$	<18 km, Huygens GC-MS	7
C_2H_6	$1.3 \pm 0.3 \times 10^{-5}$	Cassini CIRS, stratosphere, equatorial abundance, spatially variable	10
CH_3D	$7.5^{+1.5}_{-1.8} \times 10^{-6}$	Cassini CIRS, stratosphere	10
C_2H_2	$3.7 \pm 0.8 \times 10^{-6}$	Cassini CIRS, stratosphere, equatorial abundance, spatially variable	10
^{36}Ar	$2.8 \pm 0.3 \times 10^{-7}$	Huygens GC-MS, stratosphere	7
C_3H_8	$6.0 \pm 1.8 \times 10^{-7}$	Cassini CIRS, stratosphere, equatorial abundance, spatially variable	10
C_2H_4	$1.1–2.2 \times 10^{-7}$	Cassini CIRS, stratosphere, spatially variable	10
HCN	$1.5 \pm 0.5 \times 10^{-7}$	Cassini CIRS, stratosphere, equatorial abundance, spatially variable	10
CH_3CN	$\sim 4 \times 10^{-8}$	Earth-based mm-observations, stratosphere, spatially variable	11
CO_2	$1.6 \pm 0.2 \times 10^{-8}$	Cassini CIRS, stratosphere, vertically variable	9
HC_3N	$0.013–2.5 \times 10^{-8}$	Cassini CIRS, stratosphere, spatially variable	10
C_3H_4	$1.2–8.0 \times 10^{-9}$	Cassini CIRS, stratosphere, spatially variable	10
C_4H_2	$1.18–4.2 \times 10^{-9}$	Cassini CIRS, stratosphere, spatially variable	10
C_6H_6	$0.1–3.5 \times 10^{-9}$	Cassini CIRS, stratosphere, spatially variable	10
C_2N_2	9×10^{-10}	Cassini CIRS, stratosphere, spatially variable	12
H_2O	4×10^{-10}	SWIS/ISO, stratosphere	13

7.3.2 Titan

Titan is the largest satellite of Saturn and was the first one to be discovered – by the Dutch scientist Christiaan Huygens (1629–1695) in 1655. Nearly 300 years later, in 1944 the Dutch–American astronomer Gerard P. Kuiper (1905–1973) discovered Titan's methane atmosphere. However, as Table 7.8 shows, Titan's atmosphere is mainly N_2 with a few percent of CH_4 and smaller abundances of many other species.

Earth-based observations of Titan (at visual wavelengths) are difficult because of its thick, hazy atmosphere, which shield its surface from our view. Our knowledge of Titan expanded greatly with the *Voyager* flybys in 1979 and 1980 and more recently with the *Cassini–Huygens* mission. The *Huygens* probe descended through the atmosphere of Titan on 14 January 2005 and the *Cassini* orbiter has been making observations (*e.g.*, UV, visible, IR, radar, mass spectrometer) since that time.

The pressure and temperature at the surface of Titan are 94 K and 1.5 bar. The total atmospheric mass m_T is:

$$m_T = \frac{P_T}{g} \cong \frac{1.5 \times 10^6 \, \text{dyn cm}^{-2}}{135.4 \, \text{cm s}^{-2}} \cong 11\,080 \, \text{g cm}^{-2} \qquad (5.6)$$

This is about ten times larger than on Earth ($1033 \, \text{g cm}^{-2}$). Titan's atmosphere, like that of Earth, is N_2-rich (95–98% on Titan *versus* ~78% on Earth) but the similarity ends there because the other abundant gases in Earth's atmosphere are either absent or present at much lower abundances in Titan's atmosphere. For example, there is no observable O_2 in Titan's atmosphere (*versus* 20.946% on Earth), virtually no water vapor (0.4 ppbv *versus* 1–4% in the terrestrial troposphere), very little Ar (~40 ppmv *versus* 9340 ppmv on Earth) and very little CO_2 (0.16 ppmv *versus* ~390 ppmv on Earth).

In contrast, Titan's atmosphere is similar to the reducing atmosphere (H_2, CH_4, NH_3 and H_2O without any O_2) thought to have existed on the early Earth when abiotic synthesis of organic compounds led to the origin of life. However, even this analogy is not exact because N_2 and not NH_3 is observed in Titan's atmosphere. Ammonia is absent because the NH_3 gas vapor pressure over NH_3 ice is ~10^{-10} bar at Titan's surface temperature. There are two reasons why water vapor is only a trace species. First, the water vapor pressure over water ice is insignificant (~10^{-21} bar) at 94 K. Second, the water vapor that is observed in Titan's stratosphere is produced by vaporization of infalling icy particles and it is photolyzed by UV light. The OH radicals formed are responsible for production of the observed CO_2 *via* the reaction:

$$OH + CO \rightarrow H + CO_2 \; (k = 1.5 \times 10^{-13}\,cm^3\,s^{-1}) \tag{5.29}$$

If NH_3 were also released from infalling icy particles, it would be photolyzed to N_2 and H_2 on a fairly short timescale.

Methane plays the role of ozone in Titan's atmosphere (as it does in the atmospheres of the giant planets) and is responsible for the presence of a warm stratosphere and a temperature inversion at the tropopause. Methane also acts as a greenhouse gas (playing the role that H_2O and CO_2 do in Earth's atmosphere) and keeps the surface warmer than it otherwise would be. The meteorological cycle on Titan revolves around CH_4 instead of around H_2O as on Earth. Methane rain falls and CH_4 clouds form in Titan's troposphere. However, these are not the clouds that block Titan's surface from our view.

The photochemical haze that envelops Titan is produced by CH_4 photolysis. One problem is to understand the photochemical reactions that convert CH_4, N_2, H_2 and H_2O (from infalling icy grains) into the photochemical haze and the complex suite of organic compounds observed in Titan's atmosphere. However, a second, equally difficult problem is to explain the persistence of CH_4 in Titan's atmosphere. As discussed earlier, CH_4 in the atmospheres of the gas giant planets is photochemically converted into hydrogen and C_2 hydrocarbons, *e.g.*:

$$\textit{Net reaction} : 2CH_4 \rightarrow C_2H_6 + 2H \tag{7.23}$$

$$\textit{Net reaction} : 2CH_4 \rightarrow C_2H_2 + H_2 + 4H \tag{7.26}$$

$$\textit{Net reaction} : 2CH_4 \rightarrow C_2H_4 + H_2 + 2H \tag{7.40}$$

Most of the C_2H_2 and C_2H_4 formed by CH_4 photolysis are photochemically converted back into CH_4 in the stratospheres of the giant planets. However, formation of C_2H_6 *via* Reaction (7.23) is a net sink for methane. Ethane is transported downward into Jupiter's atmosphere where thermally driven reactions at $\sim 1000\,K$ reconvert it back into CH_4 *via* the net reaction:

$$C_2H_6 + H_2 = 2CH_4 \tag{7.4}$$

Reaction (7.4) cannot occur on Titan because the base of the atmosphere is cold – 94 K – and C_2H_6 condenses as liquid ethane on the surface. Thus, ethane is not recycled to CH_4 in Titan's atmosphere the way it is in the atmospheres of the gas giant planets. Consequently, the photochemical lifetime of CH_4 in Titan's atmosphere is geologically short and all CH_4 would be destroyed in about 10 million years unless it is replenished in another manner.

Originally it was suggested that a global ethane "ocean" was the methane source that was needed. The global ethane ocean was a ternary liquid solution of C_2H_6, N_2 and CH_4, which would have a sufficiently high CH_4 vapor pressure to replenish CH_4 lost by atmospheric photochemistry. However, neither Earth-based radar, the *Huygens* probe, nor radar on the *Cassini* orbiter found evidence for a global ethane ocean. The *Huygens* probe apparently landed in "swamp land" and the *Cassini* radar has detected hydrocarbon lakes in Titan's northern hemisphere. These may be refilled by CH_4 and/or intersect the liquid methane table on Titan.

7.4 DWARF PLANETS

The dwarf planets include Ceres, Pluto and several Kuiper Belt Objects (KBOs, also known as trans-Neptunian objects or TNOs). Ceres was the first asteroid discovered, on New Year's Day January 1801 by the Italian astronomer Giuseppe Piazzi (1746–1826). It is the largest object in the Asteroid Belt with a mean radius of 476.2 km, a mass of 9.40×10^{20} kg, and a mean density of $2.077 \, \mathrm{g\,cm^{-3}}$. Ceres (formerly known as the first asteroid) and Pluto (formally known as the ninth planet) were reclassified as dwarf planets in 2006. HST observations of Ceres' shape and topography indicate that it is gravitationally relaxed, and flatter than expected for a homogeneous, undifferentiated body. However, geophysical and thermochemical equilibrium models indicate that Ceres' shape may be consistent with an undifferentiated carbonaceous chondritic object. Observations to be made with the *Darwin* spacecraft in 2015 are required to resolve this issue.

Pluto was discovered by the American astronomer Clyde W. Tombaugh (1906–1997) at Lowell Observatory on 18 February 1930. Pluto's discovery was the result of searches for a ninth planet, which was predicted by the American astronomer Percival Lowell (1855–1916). Lowell calculated that apparent discrepancies between the observed and predicted motions of Uranus and Neptune required the presence of a trans-Neptunian planet. However, with the discovery of Charon, Pluto's largest satellite, in 1978 it became clear that Pluto's mass was too small to cause the apparent discrepancies in the motions of Uranus and Neptune. We now know that the apparent discrepancies arose because of errors in the calculations and measurements of the predicted and observed motions of Uranus and Neptune. Pluto's discovery – at about the same position as that predicted by Lowell – was thus a coincidence. After Pluto's discovery, photographic plates taken in 1919 by the American astronomer M. L. Humason (1891–1972) at Wilson Observatory were re-examined and also showed Pluto.

James W. Christy of the U.S. Naval Observatory discovered Charon on 22 June 1978. In retrospect his discovery was serendipitous because it allowed astronomers to observe mutual eclipses of Pluto and Charon to obtain detailed information on both objects. The mutual eclipses, which started shortly after Charon's discovery, occur in a five year series about once every 124 years, with the next series starting in 2109. The eclipses are transits of Pluto by Charon and occultations of Charon by Pluto about once every 3.2 days, which is half of Charon's orbital period (~ 6.4 days). Two smaller and more distant satellites, Nix and Hydra, were discovered in 2005, and orbit the center of mass of the Pluto–Charon system (which is between Pluto and Charon with a Pluto–Charon distance of about 19 540 km).

Pluto and Charon form a double planet system because Charon is nearly as large and massive as Pluto. For example, the derived mass of the Pluto–Charon system is about 1.48×10^{22} kg, and the Pluto/Charon mass ratio is ~ 8.6. Pluto's mass is $\sim 1.31 \times 10^{22}$ kg, and Charon's mass is $\sim 1.7 \times 10^{21}$ kg. Pluto's radius is ~ 1175 km, Charon's radius is ~ 617 km, and their radius ratio is about 1.90. For comparison, the Earth/Moon mass ratio is ~ 81 and their radius ratio is ~ 3.67.

The mean densities of Pluto (~ 1.9 g cm^{-3}) and Charon (~ 1.7 g cm^{-3}) indicate that they are icy bodies. Spectroscopic observations confirm this deduction. Pluto's surface is dominantly N_2 ice with smaller amounts of CH_4, CO and C_2H_6. The surface temperature of 40 K and the vapor pressure of N_2 gas over N_2 ice imply a surface pressure of about 58 μbar. However, the atmospheric pressure and structure changes dramatically with seasons due to Pluto's large orbital eccentricity. Small amounts of CH_4 are also observed in Pluto's atmosphere. In contrast, Charon's surface is dominantly water ice without detectable CH_4, N_2, CO or ethane. No atmosphere has been detected, probably because of the very low vapor pressure of water vapor over water ice at Charon's surface temperature of ~ 40 K. Interestingly, Triton also has a N_2-rich atmosphere with minor amounts of methane and a surface pressure of about 16 μbar. Nitrogen, CH_4, CO and C_2H_6 ices are observed on Triton's surface, but water ice is apparently absent. Triton's mean density is also similar to that of Pluto. These similarities have led scientists to argue that Triton is a Kuiper Belt Object that was captured by Neptune. At the time of writing, the other three known dwarf planets (Eris, Haumea and

Makemake) are also Kuiper Belt Objects like Pluto. Eris is larger than Pluto while Haumea and Makemake are smaller than Pluto. We do not discuss these objects further here. Hopefully the *New Horizons* spacecraft will provide more information about the Pluto–Charon system, Triton and other dwarf planets in the Kuiper Belt.

FURTHER READING

T. Gehrels (ed.), *Jupiter*, University of Arizona Press, Tucson, AZ, 1976.

T. Gehrels and M. S. Matthews (eds), *Saturn*, University of Arizona Press, Tucson, AZ, 1984.

J. S. Levine (ed.), *The Photochemistry of Atmospheres*, Academic Press, Orlando, FL, 1985.

J. S. Lewis and R. G. Prinn, *Planets and their Atmospheres Origin and Evolution*, Academic Press, Orlando, FL, 1984.

K. Lodders and B. Fegley, Jr., *The Planetary Scientist's Companion*, Oxford University Press, New York, 1998.

E. Stofan and 37 others, *Nature*, 2007, **445**, 61.

P. C. Thomas, J. W. Parker, L. A. McFadden, C. T. Russell, S. A. Stern, M. V. Sykes and E. F. Young, *Nature*, 2005, **437**, 224.

M. Yu. Zolotov, *Icarus*, 2009, **204**, 183.

REFERENCES

1. M. Grosser, *The Discovery of Neptune*, Harvard University Press, 1962.
2. R. A. Jacobson, J. Spitale, C. C. Porco, K. Beurle, N. J. Cooper, M. W. Evans and C. D. Murray, *Astron. J.*, 2008, **135**, 261.
3. C. C. Porco, P. C. Thomas, J. W. Weiss and D. C. Richardson, *Science*, 2007, **318**, 1602.
4. P. C. Thomas, *Icarus*, 2010, **208**, 395.
5. H. Zebker, B. Stiles. S. Hensley, R. Lorenz, R. L. Kirk and J. Lunine, *Science*, 2009, **324**, 921.
6. R. A. Jacobson, J. K. Campbell, A. H. Taylor and S. P. Synnott, *Astron. J.*, 1992, **103**, 2068.
7. H. B. Niemann and 17 others, *Nature*, 2005, **438**, 779.
8. R. D. Courtin, C. K. Sim, S. J. Kim and D. Gautier, *Bull. Am. Astron. Soc.*, 2007, **39**, abstract 56.05.
9. R. De Kok and 12 others, *Icarus*, 2007, **186**, 354.
10. A. Coustenis and 24 others, *Icarus*, 2007, **189**, 35.

11. A. Marten, T. Hidayat, Y. Biraud and R. Moreno, *Icarus*, 2002, **158**, 532.
12. N. A. Teanby and 11 others, *Icarus*, 2007, **181**, 243.
13. A. Coustenis, A. Salama, E. Lellouch, Th. Encrenaz, G. L. Bjoraker, R. E. Samuelson, Th. de Graauw, H. Feuchtgruber and M. F. Kessler, *Astron. Astrophys.*, 1998, **336**, L85.

Table of Abundances of Nuclides in the Solar System

Table A.1   Abundances of the nuclides in the solar system.

Z		A	Atom %	Abundance
1	H	1	99.998 1	2.59×10^{10}
1	H	2	0.001 94	5.03×10^{5}
			100	2.59×10^{10}
2	He	3	0.016 6	1.03×10^{6}
2	He	4	99.983 4	2.51×10^{9}
			100	2.51×10^{9}
3	Li	6	7.589	4.2
3	Li	7	92.411	51.4
			100	55.6
4	Be	9	100	0.612
5	B	10	19.820	3.7
5	B	11	80.180	15.1
			100	18.8
6	C	12	98.889	7.11×10^{6}
6	C	13	1.111	7.99×10^{4}
			100	7.19×10^{6}
7	N	14	99.634	2.12×10^{6}
7	N	15	0.366	7.78×10^{3}
			100	2.12×10^{6}
8	O	16	99.763	1.57×10^{7}
8	O	17	0.037	5.90×10^{3}
8	O	18	0.200	3.15×10^{4}

Chemistry of the Solar System
By Katharina Lodders and Bruce Fegley, Jr.
© K. Lodders and B. Fegley, Jr. 2011
Published by the Royal Society of Chemistry, www.rsc.org

Table A.1 (*Continued*).

Z		A	Atom %	Abundance
			100	1.57×10^7
9	F	19	100	804
10	Ne	20	92.943 1	3.06×10^6
10	Ne	21	0.222 8	7.33×10^3
10	Ne	22	6.834 1	2.25×10^5
			100	3.29×10^6
11	Na	23	100	57 700
12	Mg	24	78.992	8.10×10^5
12	Mg	25	10.003	1.03×10^5
12	Mg	26	11.005	1.13×10^5
			100	1.03×10^6
13	Al	27	100	8.46×10^4
14	Si	28	92.230	9.22×10^5
14	Si	29	4.683	4.68×10^4
14	Si	30	3.087	3.09×10^4
			100	1.00×10^6
15	P	31	100	8300
16	S	32	95.018	400 300
16	S	33	0.75	3 160
16	S	34	4.215	17 800
16	S	36	0.017	72
			100	421 200
17	Cl	35	75.771	3 920
17	Cl	37	24.229	1 250
			100	5 170
18	Ar	36	84.595	78 400
18	Ar	38	15.381	14 300
18	Ar	40	0.024	22
			100	92 700
19	K	39	93.132	3 500
19	K*	40	0.147	6
	K*	40		(0.4)
19	K	41	6.721	253
			100	3 760
20	Ca	40	96.941	58 500
20	Ca	42	0.647	391
20	Ca	43	0.135	82
20	Ca	44	2.086	1 260
20	Ca	46	0.004	2
20	Ca	48	0.187	113
			100	60 400
21	Sc	45	100	34.4
22	Ti	46	8.249	204

Table A.1 (*Continued*).

Z		A	Atom %	Abundance
22	Ti	47	7.437	184
22	Ti	48	73.72	1 820
22	Ti	49	5.409	134
22	Ti	50	5.185	128
			100	2 470
23	V	50	0.249 7	0.7
23	V	51	99.750 3	285.7
			100	286.4
24	Cr	50	4.345 2	569
24	Cr	52	83.789 5	11 000
24	Cr	53	9.500 6	1 240
24	Cr	54	2.364 7	309
			100	13 100
25	Mn	55	100	9 220
26	Fe	54	5.845	49 600
26	Fe	56	91.754	7.78×10^5
26	Fe	57	2.119 1	18 000
26	Fe	58	0.281 9	2 390
			100	8.48×10^5
27	Co	59	100	2 350
28	Ni	58	68.076 9	33 400
28	Ni	60	26.223 1	12 900
28	Ni	61	1.139 9	559
28	Ni	62	3.634 5	1 780
28	Ni	64	0.925 6	454
			100	49 000
29	Cu	63	69.174	374
29	Cu	65	30.826	167
			100	541
30	Zn	64	48.63	630
30	Zn	66	27.9	362
30	Zn	67	4.1	53
30	Zn	68	18.75	243
30	Zn	70	0.62	8
			100	1 300
31	Ga	69	60.108	22.0
31	Ga	71	39.892	14.6
			100	36.6
32	Ge	70	21.234	24.3
32	Ge	72	27.662	31.7
32	Ge	73	7.717	8.8
32	Ge	74	35.943	41.2
32	Ge	76	7.444	8.5

Table A.1   (*Continued*).

Z		A	Atom %	Abundance
			100	115
33	As	75	100	6.10
34	Se	74	0.89	0.60
34	Se	76	9.37	6.32
34	Se	77	7.64	5.15
34	Se	78	23.77	16.04
34	Se	80	49.61	33.48
34	Se	82	8.73	5.89
			100	67.5
35	Br	79	50.686	5.43
35	Br	81	49.314	5.28
			100	10.7
36	Kr	78	0.362	0.20
36	Kr	80	2.326	1.30
36	Kr	82	11.655	6.51
36	Kr	83	11.546	6.45
36	Kr	84	56.903	31.78
36	Kr	86	17.208	9.61
			100	55.8
37	Rb	85	70.844	5.12
37	Rb*	87	29.156	2.11
	Rb*	87		(*1.98*)
			100	7.23
38	Sr	84	0.558 0	0.13
38	Sr	86	9.867 8	2.30
38	Sr	87	6.896 1	1.60
	Sr	87		(*1.74*)
38	Sr	88	82.678 1	19.2
			100	23.3
39	Y	89	100	4.63
40	Zr	90	51.452	5.546
40	Zr	91	11.223	1.210
40	Zr	92	17.146	1.848
40	Zr	94	17.38	1.873
40	Zr	96	2.799	0.302
			100	10.78
41	Nb	93	100	0.780
42	Mo	92	14.525	0.370
42	Mo	94	9.151	0.233
42	Mo	95	15.838	0.404
42	Mo	96	16.672	0.425
42	Mo	97	9.599	0.245
42	Mo	98	24.391	0.622

Table A.1 (*Continued*).

Z		A	Atom %	Abundance
42	Mo	100	9.824	0.250
		100		2.55
44	Ru	96	5.542	0.099
44	Ru	98	1.869	0.033
44	Ru	99	12.758	0.227
44	Ru	100	12.599	0.224
44	Ru	101	17.060	0.304
44	Ru	102	31.552	0.562
44	Ru	104	18.621	0.332
		100		1.78
45	Rh	103	100	0.370
46	Pd	102	1.02	0.0139
46	Pd	104	11.14	0.1513
46	Pd	105	22.33	0.3032
46	Pd	106	27.33	0.371
46	Pd	108	26.46	0.359
46	Pd	110	11.72	0.159
		100		1.36
47	Ag	107	51.839	0.254
47	Ag	109	48.161	0.236
		100		0.489
48	Cd	106	1.25	0.020
48	Cd	108	0.89	0.014
48	Cd	110	12.49	0.197
48	Cd	111	12.8	0.201
48	Cd	112	24.13	0.380
48	Cd	113	12.22	0.192
48	Cd	114	28.73	0.452
48	Cd	116	7.49	0.118
		100		1.57
49	In	113	4.288	0.008
49	In	115	95.712	0.170
		100		0.178
50	Sn	112	0.971	0.035
50	Sn	114	0.659	0.024
50	Sn	115	0.339	0.012
50	Sn	116	14.536	0.524
50	Sn	117	7.676	0.277
50	Sn	118	24.223	0.873
50	Sn	119	8.585	0.309
50	Sn	120	32.593	1.175
50	Sn	122	4.629	0.167
50	Sn	124	5.789	0.209

Table A.1 (*Continued*).

Z		A	Atom %	Abundance
			100	3.60
51	Sb	121	57.213	0.179
51	Sb	123	42.787	0.134
			100	0.313
52	Te	120	0.096	0.005
52	Te	122	2.603	0.122
52	Te	123	0.908	0.043
52	Te	124	4.816	0.226
52	Te	125	7.139	0.335
52	Te	126	18.952	0.889
52	Te	128	31.687	1.486
52	Te	130	33.799	1.585
			100	4.69
53	I	127	100	1.10
54	Xe	124	0.129	0.007
54	Xe	126	0.112	0.006
54	Xe	128	2.23	0.122
54	Xe	129	27.463	1.499
54	Xe	130	4.378	0.239
54	Xe	131	21.802	1.190
54	Xe	132	26.355	1.438
54	Xe	134	9.661	0.527
54	Xe	136	7.868	0.429
			100	5.457
55	Cs	133	100	0.371
56	Ba	130	0.106	0.005
56	Ba	132	0.101	0.005
56	Ba	134	2.417	0.108
56	Ba	135	6.592	0.295
56	Ba	136	7.853	0.351
56	Ba	137	11.232	0.502
56	Ba	138	71.699	3.205
			100	4.471
57	La*	138	0.091	0.000 4
57	La	139	99.909	0.457
			100	0.457
58	Ce	136	0.186	0.002
58	Ce	138	0.250	0.003
58	Ce	140	88.450	1.043
58	Ce	142	11.114	0.131
			100	1.180
59	Pr	141	100	0.172
60	Nd	142	27.044	0.231

Table A.1 (*Continued*).

Z		A	Atom %	Abundance
60	Nd	143	12.023	0.103
	Nd	143		(*0.104*)
60	Nd	144	23.729	0.203
60	Nd	145	8.763	0.075
60	Nd	146	17.130	0.147
60	Nd	148	5.716	0.049
60	Nd	150	5.596	0.048
			100	0.856
62	Sm	144	3.073	0.008
62	Sm*	147	14.993	0.041
	Sm*	147		(*0.040*)
62	Sm*	148	11.241	0.030
62	Sm	149	13.819	0.037
62	Sm	150	7.380	0.020
62	Sm	152	26.742	0.071
62	Sm	154	22.752	0.060
			100	0.267
63	Eu	151	47.81	0.0471
63	Eu	153	52.19	0.0514
			100	0.098 4
64	Gd	152	0.203	0.0007
64	Gd	154	2.181	0.0078
64	Gd	155	14.800	0.0533
64	Gd	156	20.466	0.0736
64	Gd	157	15.652	0.0563
64	Gd	158	24.835	0.0894
64	Gd	160	21.864	0.0787
			100	0.360
65	Tb	159	100	0.0634
66	Dy	156	0.056	0.0002
66	Dy	158	0.095	0.0004
66	Dy	160	2.329	0.0094
66	Dy	161	18.889	0.0762
66	Dy	162	25.475	0.1028
66	Dy	163	24.896	0.1005
66	Dy	164	28.260	0.1141
			100	0.404
67	Ho	165	100	0.0910
68	Er	162	0.139	0.0004
68	Er	164	1.601	0.0042
68	Er	166	33.503	0.088
68	Er	167	22.869	0.060
68	Er	168	26.978	0.071

Table A.1 (*Continued*).

Z		A	Atom %	Abundance
68	Er	170	14.910	0.039
			100	0.262
69	Tm	169	100	0.040 6
70	Yb	168	0.12	0.000 3
70	Yb	170	2.98	0.007 6
70	Yb	171	14.09	0.036 1
70	Yb	172	21.69	0.055 6
70	Yb	173	16.10	0.0413
70	Yb	174	32.03	0.082 1
70	Yb	176	13.00	0.033 3
			100	0.256
71	Lu	175	97.179 5	0.037 0
71	Lu*	176	2.820 5	0.001 1
	Lu*	176		(0.001 0)
			100	0.038 0
72	Hf	174	0.162	0.000 3
72	Hf	176	5.206	0.008 1
	Hf	176		(*0.008 2*)
72	Hf	177	18.606	0.029 0
72	Hf	178	27.297	0.0425
72	Hf	179	13.629	0.021 2
72	Hf	180	35.100	0.054 7
			100	0.156
73	Ta	180	0.0123	2.6×10^{-6}
73	Ta	181	99.9877	0.021 0
			100	0.021 0
74	W	180	0.120	0.000 2
74	W	182	26.499	0.036 3
74	W	183	14.314	0.019 6
74	W	184	30.642	0.042 0
74	W	186	28.426	0.039 0
			100	0.137
75	Re	185	35.662	0.020 7
75	Re*	187	64.338	0.037 4
	Re*	187		(*0.034 7*)
			100	0.058 1
76	Os	184	0.020	0.000 13
76	Os	186	1.598	0.010 8
76	Os	187	1.271	0.008 6
	Os	187		(*0.011 3*)
76	Os	188	13.337	0.090 4
76	Os	189	16.261	0.110
76	Os	190	26.444	0.179

Table A.1 (*Continued*).

Z		A	Atom %	Abundance
76	Os	192	41.070	0.278
			100	0.678
77	Ir	191	37.272	0.250
77	Ir	193	62.728	0.421
			100	0.672
78	Pt*	190	0.014	0.000 2
78	Pt	192	0.783	0.010
78	Pt	194	32.967	0.420
78	Pt	195	33.832	0.431
78	Pt	196	25.242	0.322
78	Pt	198	7.163	0.091
			100	1.27
79	Au	197	100	0.195
80	Hg	196	0.15	0.001
80	Hg	198	9.97	0.046
80	Hg	199	16.87	0.077
80	Hg	200	23.10	0.106
80	Hg	201	13.18	0.060
80	Hg	202	29.86	0.137
80	Hg	204	6.87	0.031
			100	0.458
81	Tl	203	29.524	0.054
81	Tl	205	70.476	0.129
			100	0.182
82	Pb	204	1.997	0.066
82	Pb	206	18.582	0.614
	Pb	206		(*0.623*)
82	Pb	207	20.563	0.680
	Pb	207		(*0.686*)
82	Pb	208	58.858	1.946
	Pb	208		(*1.955*)
			100	3.306
83	Bi	209	100	0.138 2
90	Th*	232	100	0.0440
	Th*	232		(*0.351*)
92	U*	234	0.002	4.9×10^{-7}
92	U*	235	24.286	0.005 78
	U*	235		(*6.4×10^{-5}*)
92	U*	238	75.712	0.018 0
	U*	238		(*0.008 9*)
			100	0.023 8

[a]Abundances on a scale where Si $\equiv 10^6$. An * marks radioactive nuclides. Values in italics refer to *present* day abundances if significant decay occurred since 4.57 Ga ago. The 100% totals refer to protosolar abundances.

Table of Average Element Concentrations in Major Chondritic Meteorite Groups

Chemistry of the Solar System
By Katharina Lodders and Bruce Fegley, Jr.
© K. Lodders and B. Fegley, Jr. 2011
Published by the Royal Society of Chemistry, www.rsc.org

Table B.1 Average element concentrations in major chondritic meteorite groups [part per million (ppm) by mass, $\mu g\,g^{-1}$].

	CI	CM	CV	CO	CK	CR	H	L	LL	EH	EL
H	19 700	14 000	2800	70	3400	3700	400	900	1300	–	–
Li	1.47	1.5	1.7	1.8	1.4	–	1.7	1.85	1.8	1.9	0.7
Be	0.021	0.031	0.05	–	–	–	0.03	0.04	0.045	0.021	–
B	0.775	0.48	0.3	–	–	–	0.4	0.4	0.7	1	–
C	34 800	22 000	5300	4400	2200	14 400	1200	1600	2350	3900	4300
N	2950	1520	80	90	–	620	34	34	50	–	240
O	458 500	432 000	370 000	370 000	–	–	338 600	369 300	383 700	310 000	310 000
F	58.2	38	24	30	22	–	27	28	58	180	150
Na	4990	3900	3400	4200	3130	2900	6110	6900	6840	6880	5770
Mg	95 800	115 000	143 000	145 000	147 000	138 000	141 000	149 000	153 000	107 300	137 500
Al	8500	11 300	16 800	14 000	14 700	12 300	10 600	11 600	11 800	8200	10 000
Si	107 000	127 000	157 000	158 000	158 000	151 000	171 000	186 000	189 000	165 900	188 000
P	967	1030	1120	1210	1100	1330	1200	1030	910	2130	1250
S	53 500	27 000	22 000	22 000	17 000	15 750	20 000	22 000	21 000	56 000	31 000
Cl	698	430	250	280	260	–	77	76	126	570	230
K	544	370	360	360	290	315	780	920	880	840	700
Ca	9220	12 900	18 400	15 800	17 000	13 400	12 200	13 300	13 200	8500	10 200
Sc	5.9	8.2	10.2	9.5	11	7.8	7.8	8.1	8	6	7.7
Ti	451	550	870	730	940	750	630	670	680	460	550
V	54.3	75	97	95	96	74	73	75	76	56	64
Cr	2650	3050	3480	3520	3530	3590	3500	3690	3680	3320	3030
Mn	1930	1650	1520	1620	1440	1830	2340	2590	2600	2170	1580
Fe	185 000	213 000	235 000	250 000	230 000	238 000	272 000	217 500	198 000	304 000	248 000
Co	506	560	640	680	620	640	830	580	480	870	720
Ni	10 800	12 300	13 200	14 200	13 100	13 200	17 100	12 400	10 600	18 400	14 700
Cu	131	130	104	130	90	100	97	95	86	215	120
Zn	323	180	110	110	80	90	47	52	55	305	18
Ga	9.71	7.6	6.1	7.1	5.2	6	6	5.4	5.1	16.9	11
Ge	32.6	26	16	20	14	18	10	11	10	39	30
As	1.74	1.8	1.5	2	1.4	1.5	2.2	1.4	1.3	3.55	2.2
Se	20.3	12	8.7	8	8	8.2	8.2	9.2	8.6	25.4	15

Table B.1 (*Continued*).

	CI	CM	CV	CO	CK	CR	H	L	LL	EH	EL
Br	3.26	2.6	1.6	1.4	0.6	0.97	0.55	0.9	0.8	2.7	0.8
Rb	2.31	1.6	1.2	1.3	–	1.1	2.2	2.3	2	3.1	2.3
Sr	7.81	10	14.8	13	15	10	8.8	11	13	7	9.4
Y	1.53	2	2.6	2.4	2.7	–	2	1.8	2	1.2	–
Zr	3.62	5.24	6.87	9	8	5.4	7.3	6.4	7.4	6.6	7.2
Nb	0.279	0.4	0.5	–	0.43	0.5	0.4	0.4	–	–	–
Mo	0.973	1.4	1.8	1.7	0.38	1.4	1.4	1.2	1.1	–	–
Ru	0.686	0.87	1.2	1.08	1.1	0.97	1.1	0.75	–	0.93	0.77
Rh	0.139	0.16	0.17	–	0.18	–	0.21	0.155	–	–	–
Pd	0.558	0.63	0.71	0.71	0.58	0.69	0.845	0.62	0.56	0.84	0.73
Ag	0.201	0.16	0.1	0.1	–	0.095	0.036	0.066	0.065	0.27	0.027
Cd	0.674	0.42	0.35	0.008	–	0.3	0.04	0.03	0.02	0.72	0.035
In	0.0778	0.05	0.032	0.025	–	0.03	0.001	0.0005	0.004	0.083	0.004
Sn	1.63	0.79	0.68	0.89	0.49	0.73	0.39	0.6	0.33	0.88	–
Sb	0.145	0.13	0.085	0.11	0.06	0.08	0.071	0.076	0.07	0.2	0.09
Te	2.28	1.3	1	0.95	0.88	1	0.35	0.38	0.45	0.24	0.93
I	0.53	0.27	0.16	0.2	0.22	–	0.06	0.07	–	0.165	0.08
Cs	0.188	0.11	0.09	0.08	–	0.084	0.06	0.013	0.12	0.21	0.125
Ba	2.41	3.1	4.55	4.3	4.7	3.4	4.4	4.1	4	4.7	5.1
La	0.242	0.32	0.469	0.38	0.49	0.31	0.301	0.318	0.33	0.24	0.196
Ce	0.622	0.94	1.19	1.14	1.27	0.75	0.763	0.97	0.88	0.65	0.58
Pr	0.0946	0.137	0.174	0.14	–	–	0.12	0.14	0.13	0.1	0.07
Nd	0.471	0.626	0.919	0.85	0.99	0.79	0.581	0.7	0.65	0.44	0.37

Sm	0.152	0.204	0.294	0.25	0.29	0.23	0.194	0.203	0.205	0.14	0.149
Eu	0.05778	0.078	0.105	0.096	0.11	0.08	0.074	0.08	0.078	0.052	0.054
Gd	0.205	0.29	0.405	0.39	0.44	0.32	0.275	0.317	0.29	0.21	0.196
Tb	0.0384	0.051	0.071	0.06	—	0.05	0.049	0.059	0.054	0.03355	0.032
Dy	0.255	0.332	0.454	0.42	0.49	0.28	0.305	0.372	0.36	0.23	0.245
Ho	0.0572	0.077	0.097	0.096	0.1	0.1	0.074	0.089	0.082	0.05	0.051
Er	0.163	0.221	0.277	0.305	0.35	—	0.213	0.252	0.24	0.16	0.16
Tm	0.0261	0.035	0.048	0.04	—	—	0.033	0.038	0.035	0.02355	0.023
Yb	0.169	0.215	0.312	0.27	0.32	0.22	0.203	0.226	0.23	0.154	0.157
Lu	0.02533	0.033	0.046	0.039	0.046	0.032	0.033	0.034	0.034	0.025	0.025
Hf	0.106	0.15	0.23	0.22	0.25	0.15	0.15	0.17	0.17	0.14	0.21
Ta	0.01455	0.019	—	—	—	—	0.021	0.021	—	—	—
W	0.096	0.15	0.16	0.15	0.19	0.11	0.164	0.138	0.115	0.14	0.14
Re	0.03933	0.05	0.057	0.058	0.06	0.05	0.078	0.047	0.032	0.055	0.057
Os	0.493	0.67	0.8	0.805	0.815	0.71	0.835	0.53	0.41	0.667	0.67
Ir	0.469	0.58	0.73	0.74	0.76	0.67	0.77	0.49	0.38	0.54	0.56
Pt	0.947	1.1	1.25	1.24	1.3	0.98	1.58	1.09	0.88	1.23	1.25
Au	0.146	0.15	0.153	0.19	0.12	0.16	0.22	0.16	0.142	0.33	0.24
Hg	0.350	0.2	0.05	—	—	—	—	0.03	0.022	0.06	—
Tl	0.142	0.092	0.058	0.04	—	0.06	0.004	0.002	0.002	0.101	0.007
Pb	2.63	1.6	1.1	2.15	0.84	—	0.14	0.06	0.06	1.5	0.24
Bi	0.110	0.071	0.054	0.035	0.013	0.04	0.009	0.004	0.012	0.09	0.013
Th	0.031	0.041	0.058	0.08	0.058	0.042	0.038	0.042	0.047	0.03	0.038
U	0.00810	0.012	0.017	0.018	0.015	0.013	0.013	0.015	0.015	0.0092	0.007

APPENDIX C

Review of Chemical Kinetics

In this appendix we give a brief review of chemical kinetics with a focus on important reactions in atmospheric chemistry of the Earth and other planets. The examples used are from the chapters in the book and the equation numbers correspond to the equation numbers used in those chapters. New equations are denoted as (C.1), (C.2) and so on.

The reactions that actually take place between atoms, radicals and molecules are elementary reactions. A net photochemical reaction in Earth's atmosphere or a net thermochemical reaction in Jupiter's deep atmosphere is the result of a sequence of elementary reactions. This sequence of reactions is the reaction mechanism. For example, each of the five reactions in the Chapman cycle for O_3 formation and loss is an elementary reaction and the net effect of these reactions is the photochemical conversion of ozone into molecular oxygen:

O_2 photolysis : $O_2 + h\nu \rightarrow O + O \, (\lambda = 180 - 240 \, nm)$ (5.59)

Ozone production : $O + O_2 + M \rightarrow O_3 + M$ (5.37)

Ozone photolysis : $O_3 + h\nu \rightarrow O(^1D) + O_2 \, (\lambda = 200 - 300 \, nm)$ (5.35)

Oxygen production : $O + O + M \rightarrow O_2 + M$ (5.60)

Ozone destruction : $O + O_3 \rightarrow O_2 + O_2$ (5.61)

Net reaction : $O + O_3 \rightarrow O_2 + O_2$ (5.61)

Chemistry of the Solar System
By Katharina Lodders and Bruce Fegley, Jr.
© K. Lodders and B. Fegley, Jr. 2011
Published by the Royal Society of Chemistry, www.rsc.org

The M that appears in some reactions is a third body, which can be any other gas. The third body is a collision partner that removes excess energy but does not take part in the chemical reaction. Statistically M is one of the most abundant gases in a planet's atmosphere. Thus, M is likely to be N_2, O_2, Ar or CO_2 in Earth's stratosphere and M is likely to be H_2 or He in Jupiter's atmosphere. The $h\nu$ in some reactions represents ultraviolet, visible or infrared sunlight that initiates a photochemical reaction.

Elementary reactions are classified as unimolecular, bimolecular or termolecular reactions according to the number of reactant molecules. The rates of elementary reactions are given by the decrease in reactant concentration per unit time or the increase in product concentration per unit time. Typically, atmospheric chemists express concentration in terms of number density, which is the number of molecules (or atoms) per cm^3. The units for the reaction rate are then molecules $cm^{-3} s^{-1}$. Reaction rates are proportional to the reactant concentrations and the proportionality constant is the reaction rate constant, or simply the rate constant. A few examples of the different types of elementary reactions, their reaction rates and rate constants follow.

C.1 UNIMOLECULAR REACTIONS

The photolysis of O_2 by UV sunlight is an exemplary unimolecular reaction:

$$O_2 \text{ photolysis} : O_2 + h\nu \rightarrow O + O \ (\lambda = 180 - 240 \text{ nm}) \tag{5.59}$$

The reaction rate (R) is:

$$R = -\frac{d[O_2]}{dt} = \frac{1}{2}\frac{d[O]}{dt} = J[O_2] \text{ molecule cm}^{-3} \text{ s}^{-1} \tag{C.1}$$

The photodissociation rate constant is J with units of s^{-1}. The J values take into account the wavelength-dependent absorption cross sections, photon fluxes (as a function of heliocentric distance, zenith angle and altitude) and quantum yields for the reaction. Larger J values denote faster reactions. The reciprocal of the J value gives the chemical lifetime of a molecule for the particular photochemical reaction written. For example, the J values for O_2 photolysis in Earth's atmosphere as a function of altitude are $3.0 \times 10^{-9} s^{-1}$ (80 km in the mesosphere), $1.2 \times 10^{-12} s^{-1}$ (25 km in the O_3 layer in the stratosphere) and 0 (1 km in the troposphere). The corresponding chemical lifetimes for O_2 loss *via* photolysis are ~ 10.5 years (80 km) and $\sim 26\,410$ years (25 km).

However, a unimolecular reaction does not have to be a photochemical reaction. Many unimolecular reactions are thermochemical reactions that are driven by heat. For example, the dissociation of nitrogen pentoxide (N_2O_5) is a unimolecular reaction that occurs in Earth's troposphere:

$$N_2O_5 \rightarrow NO_2 + NO_3 \tag{C.2}$$

The rate of this reaction is:

$$R = -\frac{d[N_2O_5]}{dt} = \frac{d[NO_2]}{dt} = \frac{d[NO_3]}{dt} = k[N_2O_5] \text{ molecules cm}^{-3} \text{ s}^{-1} \tag{C.3}$$

The unimolecular rate constant k has units of inverse time (s^{-1}) and has a temperature dependent value given by:

$$k = 5.49 \times 10^{14} T^{0.10} \exp(-11\,080/T) \, s^{-1} \tag{C.4}$$

At 288 K, the average tropospheric temperature at sea level on Earth, $k = 0.0189 \, s^{-1}$.

The rate constant for nitrogen pentoxide thermal decomposition has three parts. The first number (5.49×10^{14}) is the pre-exponential or frequency factor (A). The $T^{0.10}$ term arises from the temperature dependence of the enthalpy of reaction. The exponent equals $\Delta C_P/R$, where ΔC_P is the difference in heat capacities of the products and reactants and R is the ideal gas constant in the same units. The third term, which is the exponential term, is $\exp[E_a/(RT)]$, where E_a is the activation energy. In Equation (C.4) the numerator of the fraction is $11\,080 = E_a/R$ with units of temperature.

The chemical lifetime of N_2O_5 is:

$$t_{chem}(N_2O_5) = k^{-1} \, s \tag{C.5}$$

At 288 K, the average tropospheric temperature at sea level, the N_2O_5 chemical lifetime is thus about 53 s. In general, the chemical lifetime of a species "i" is an e-folding time, *i.e.*, the time to reduce its concentration by a factor of e (~ 2.718), and is given by its concentration divided by the rate of change of its concentration:

$$t_{chem}(i) = \frac{[i]}{d[i]/dt} \text{ s} \tag{C.6}$$

C.2 BIMOLECULAR OR TWO-BODY REACTIONS

The reaction of O atoms with O_3 molecules is an exemplary bimolecular reaction:

$$O + O_3 \rightarrow O_2 + O_2 \tag{5.61}$$

The reaction rate for this bimolecular reaction is:

$$R = -\frac{d[O]}{dt} = -\frac{d[O_3]}{dt} = \frac{1}{2}\frac{d[O_2]}{dt} = k[O][O_3] \text{ molecule cm}^{-3}\text{ s}^{-1} \tag{C.7}$$

The rate constant k is a bimolecular rate constant with units of cm^3 molecule^{-1} s^{-1} and is:

$$k = 8.0 \times 10^{-12} \exp(-2060/T) \text{ cm}^3 \text{ molecule}^{-1}\text{ s}^{-1} \tag{C.8}$$

The rate constant for the $O + O_3$ reaction has the general form:

$$k = A \exp\left(-\frac{E_a}{RT}\right) \tag{C.9}$$

The pre-exponential factor $A = 8.0 \times 10^{-12}$, the T term equals unity and the exponential term has an activation energy $E_a = 17\,127\,\text{J mol}^{-1}$ ($E_a = 2060 \times R$). Equation (C.9) is the Arrhenius equation, named after the great Swedish physical chemist Svante Arrhenius (1859–1927). Thus, the rate constant in equation (C.8) follows Arrhenius behavior.

The chemical lifetimes of monatomic oxygen and ozone are given by:

$$t_{chem}(O) = \frac{1}{k[O_3]} \text{ s} \tag{C.10}$$

$$t_{chem}(O_3) = \frac{1}{k[O]} \text{ s} \tag{C.11}$$

C.3 TERMOLECULAR OR THREE-BODY REACTIONS

The recombination of two O atoms to form O_2 is an exemplary termolecular reaction:

$$\text{Oxygen production}: O + O + M \rightarrow O_2 + M \tag{5.60}$$

The rate of this reaction is:

$$R = -\frac{d[O]}{dt} = 2\frac{d[O_2]}{dt} = k[O][O][M]\,\text{cm}^3\,\text{s}^{-1} \tag{C.12}$$

The rate constant k is a termolecular rate constant with units of cm^6 $\text{molecule}^{-2}\,\text{s}^{-1}$ and is:

$$k = 4.7 \times 10^{-33}(300/T)^2\,\text{cm}^6\,\text{molecule}^{-2}\,\text{s}^{-1} \tag{C.13}$$

The chemical lifetime of monatomic oxygen is:

$$t_{\text{chem}}(O) = \frac{1}{k[O][M]}\,\text{s} \tag{C.14}$$

The reaction of an O atom and O_2 molecule to form O_3 is also a termolecular reaction:

$$\text{Ozone production}: O + O_2 + M \rightarrow O_3 + M \tag{5.37}$$

The rate of this reaction is:

$$R = -\frac{d[O]}{dt} = -\frac{d[O_2]}{dt} = \frac{d[O_3]}{dt} = k[O][O_2][M]\,\text{cm}^3\,\text{s}^{-1} \tag{C.15}$$

The rate constant k is a termolecular rate constant with units of cm^6 $\text{molecule}^{-2}\,\text{s}^{-1}$ and is:

$$k = 6.0 \times 10^{-34}(T/300)^{-2.3}\,\text{cm}^6\,\text{molecule}^{-2}\,\text{s}^{-1} \tag{C.16}$$

The chemical lifetimes of monatomic and molecular oxygen for Reaction (5.37) are:

$$t_{\text{chem}}(O) = \frac{1}{k[O_2][M]}\,\text{s} \tag{C.17}$$

$$t_{\text{chem}}(O_2) = \frac{1}{k[\text{O}][\text{M}]} \text{ s} \qquad \text{(C.18)}$$

There are no four-body reactions.

FURTHER READING

S. W. Benson, *The Foundations of Chemical Kinetics*, McGraw-Hill, New York, 1960 (reprinted by Krieger Publishing, Malabar, FL, 1982).

K. J. Laidler, *Chemical Kinetics*, 2nd edn, McGraw-Hill, New York, 1965.

Subject Index